Using Mineral Chemistry to Characterize Ore-Forming Processes

Using Mineral Chemistry to Characterize Ore-Forming Processes

Guest Editors

Xinlu Hu
Xinbiao Lv
Yafei Wu
Zhenjie Zhang

MDPI

Basel • Beijing • Wuhan • Barcelona • Belgrade • Novi Sad • Cluj • Manchester

Guest Editors
Xinlu Hu
School of Earth Resources
China University of Geosciences
Wuhan
China

Xinbiao Lv
School of Earth Resources
China University of Geosciences
Wuhan
China

Yafei Wu
School of Earth Resources
China University of Geosciences
Wuhan
China

Zhenjie Zhang
School of Earth Sciences and Resources
China University of Geosciences
Beijing
China

Editorial Office
MDPI AG
Grosspeteranlage 5
4052 Basel, Switzerland

This is a reprint of the Special Issue, published open access by the journal *Minerals* (ISSN 2075-163X), freely accessible at: https://www.mdpi.com/journal/minerals/special_issues/Y2W24UBFP5.

For citation purposes, cite each article independently as indicated on the article page online and as indicated below:

Lastname, A.A.; Lastname, B.B. Article Title. *Journal Name* **Year**, *Volume Number*, Page Range.

ISBN 978-3-7258-6628-1 (Hbk)
ISBN 978-3-7258-6629-8 (PDF)
https://doi.org/10.3390/books978-3-7258-6629-8

Contents

About the Editors

Xinlu Hu

Xinlu Hu is an Associate Professor at the School of Earth Resources, China University of Geosciences, Wuhan, China. He received his Ph.D. in Mineralogy, Petrology, and Ore Deposits from China University of Geosciences, Wuhan, in June 2015. His research focuses on metallogenic regularities and mineralization prediction, structural controls on ore systems, genetic mechanisms of strategic mineral resources, and regional metallogeny in key tectonic settings of China—including Northeast China, the Qinling Orogenic Belt, and the Sichuan–Yunnan–Guizhou region. He has led and participated in more than ten national- and provincial-level research projects. He has co-authored four academic monographs and one textbook, and published over 50 peer-reviewed papers.

Xinbiao Lv

Xinbiao Lv is a Professor at the School of Earth Resources, China University of Geosciences, Wuhan, where he has been teaching and conducting research in ore deposit geology and mineral exploration since 1988. His research focuses on metallogenic mechanisms, regional metallogeny, mineralization prediction, geological fluids in ore formation, and mineral resource economics. He has led over 30 national and provincial projects, including key tasks of the National Science & Technology Support Program and grants from the National Natural Science Foundation of China. He has published more than 120 papers and authored five academic monographs. His work has been recognized with national and provincial awards in both teaching and scientific research, including the Geological Society of China's Silver Hammer Award in 2002.

Yafei Wu

Yafei Wu is a Professor specializing in the genesis of orogenic gold deposits at the School of Earth Resources, China University of Geosciences, Wuhan. He earned his Ph.D. in Mineral Resource Prospecting and Exploration from China University of Geosciences, Wuhan in 2019 and a second Ph.D. in Applied Geology from Curtin University, Australia, in 2020. He completed postdoctoral fellowships at the University of Hong Kong and the University of Saint Joseph (Macau) from 2019 to 2021. His research integrates multiple microanalytical techniques to unravel the processes and mechanisms of gold mineralization. He also investigates Mo–Ni–platinum-group element mineralization in black shales and seafloor massive sulfide systems. He has led or participated in numerous research projects and published over 20 SCI papers, including 11 as first or corresponding author in top journals such as *Geology*, *Economic Geology*, *Geochimica et Cosmochimica Acta*, *Mineralium Deposita*, and *Chemical Geology*, with more than 850 total citations.

Zhenjie Zhang

Zhenjie Zhang is an Associate Professor at the School of Earth Resources, China University of Geosciences, Beijing. He received his Ph.D. in Mineral Resource Prospecting and Exploration in 2015. His research focuses on Earth system co-evolution and intelligent mineral prospecting using geoscience big data, multivariate statistics, and machine learning. He serves as Associate Editor for *Geoscience Data* Journal and *Geological* Journal, and is an active member of several editorial boards and professional committees in AI and geoscience data. He has participated in multiple projects funded by the National Natural Science Foundation of China and the National Key R&D Program. He was recognized as an Outstanding Young Scientist in Geological Prospecting by the Ministry of Natural Resources (2025).

Preface

This Reprint brings together nine original research articles from the *Minerals* Special Issue "Using Mineral Chemistry to Characterize Ore-Forming Processes." It showcases how in situ chemical and textural analyses of ore-related minerals—including sulfides, oxides, silicates, and accessory phases—can provide key insights into the sources, pathways, and evolution of mineralizing fluids and magmas. Through advanced microanalytical techniques such as electron probe microanalysis (EPMA), laser ablation–inductively coupled plasma–mass spectrometry (LA-ICP-MS), and Raman spectroscopy, the studies illustrate how mineral chemistry preserves a record of changing physical and chemical conditions—such as temperature, pressure, redox state, and fluid composition—during ore formation.

The Reprint offers a concise and accessible collection that highlights the power of mineral-scale data for unraveling genetic processes across a range of deposit types, including porphyry, skarn, epithermal, and magmatic systems. It seeks to bridge fundamental research and practical exploration by demonstrating how detailed mineralogical investigations can refine genetic models and enhance vectoring strategies toward undiscovered mineralization.

This volume is intended for economic geologists, geochemists, mineralogists, graduate students, and industry professionals engaged in ore deposit research and exploration. We hope it serves as a valuable reference for advancing the integration of mineral chemistry into the study of ore-forming processes.

Xinlu Hu, Xinbiao Lv, Yafei Wu, and Zhenjie Zhang
Guest Editors

minerals

MDPI

Editorial

Using Mineral Chemistry to Characterize Ore-Forming Processes: An Introduction

Xinlu Hu [1,*], Zhenjie Zhang [2] and Xinbiao Lv [1]

[1] School of Earth Resources, China University of Geosciences, Wuhan 430074, China
[2] School of Earth Sciences and Resources, China University of Geosciences, Beijing 100083, China
* Correspondence: huxinlu00@foxmail.com

This Special Issue brings together nine studies that employ mineral-based investigations to advance the understanding of ore-forming processes. The contributions demonstrate how integrated approaches combining microtextural analysis, in situ geochemistry, and isotopic systems can unravel complex mineralization processes across diverse geological settings. These studies successfully trace fluid evolution pathways, constrain metal sources, determine precise mineralization ages, and reveal petrogenetic processes controlling metal endowment. The findings demonstrate how mineral-scale signatures can reveal large-scale geological processes, from crust–mantle interactions in magmatic systems to fluid–rock reactions in hydrothermal environments. The methodological advances in micro-analytical techniques presented here enable the resolution of previously undetectable chemical signatures and temporal relationships in complex ore systems.

Li et al. (Contribution 1) conducted a comprehensive petrographic and geochemical study of magnetite from the Wuan orbicular diorite in Hebei Province, China. Their results indicate that the core magnetite crystallized from iron-rich melts formed through liquid immiscibility within the magma chamber. This immiscibility was triggered by a significant increase in oxygen fugacity, caused by CO_2 released during the reaction between ascending silicate magma and carbonate xenoliths. The study proposes a stepwise metallogenic model where Fe^{2+} in the original magma is oxidized under high oxygen fugacity, leading to the separation of an Fe-rich oxide melt and the early crystallization of iron oxides like magnetite, alongside iron-poor silicate minerals such as tremolite and actinolite. The compositional variations in magnetite suggest that the formation of Hanxing-type Fe deposits is closely linked to magmatic–carbonate rock interactions.

Filimon et al. (Contribution 2) employed a suite of microtextural and geochemical techniques—including SEM-EDS, EPMA, and LA-ICP-MS—to investigate sulfide minerals from the Matra As–Sb deposit in Alpine Corsica, France. Their results indicate that hydrothermal mineralization occurred within a system of reactivated faults. Sulfide precipitation is interpreted to have been primarily controlled by variations in sulfur fugacity (fS_2), driven by the influx of oxidizing meteoric fluids. This process led to a progressive evolution in arsenic speciation, from reduced forms (As^{1-} in pyrite) to oxidized species (As^{2+} in realgar, As^{3+} in orpiment, and As^{5+} in hörnesite). The authors identified a diagnostic trace-element signature of As–Sb–Tl–Ni in pyrite and As–Fe–Tl in stibnite, and compared the mineralization with Tuscan Carlin-type deposits, highlighting shared tectonic settings.

Sun et al. (Contribution 3) carried out an integrated mineralogical study of pyrite and quartz from the Qiubudong Ag deposit in the central North China Craton. Their analytical approach included crystal morphology, trace-element geochemistry, thermoelectric and thermoluminescence properties, and fluid inclusion microthermometry. The

https://doi.org/10.3390/min16010084

study concludes that the deposit formed from medium–low temperature, F-rich, low-salinity hydrothermal fluids of volcanic–magmatic origin, with limited meteoric water input. Key mineralogical indicators for mineralization comprise the predominance of pentagonal dodecahedral pyrite, elevated Au content, low Co/Ni ratios, and complex thermoluminescence patterns in quartz.

Ouyang et al. (Contribution 4) performed zircon U–Pb geochronology and whole-rock geochemical analyses on a newly identified biotite granite porphyry from the Zhuxi W–Cu polymetallic deposit in Jiangxi Province, South China, and compared it with previously documented diorite porphyrite and biotite quartz monzonite porphyry in the region. They conclude that these three intrusive rocks were emplaced contemporaneously around 160 Ma but are not comagmatic. It is inferred that the biotite granite porphyry originates from partial melting of the thickened lower crust, the diorite porphyrite from melting of the delaminated lower crust, and the biotite quartz monzonite porphyry through mixing of mantle and crustal derived melts, all formed in a post-orogenic extensional tectonic setting.

Xiang et al. (Contribution 5) present new geochemical data from Early Cretaceous Kongco granitic porphyry dykes in the northern Central Lhasa Microblock, Tibet. These dykes are classified as A2-type granites, formed by partial melting of the lower crust induced by subduction-related slab break-off and asthenospheric upwelling during the collision between the Qiangtang and Lhasa terranes. The close spatial relationship between the dykes and copper mineralization suggests a genetic link between this magmatic event and regional metallogenesis.

Khedr et al. (Contribution 6) report mineralogical and geochemical analyses of titaniferous iron ores and their host gabbro, along with interstitial clinopyroxene, from the El-Baroud Layered Gabbros. The iron ores are mainly composed of titanomagnetite, ilmenite, and magnetite. Geochemical data suggest that the gabbro formed in an arc rifting environment, and crystallized under high oxygen fugacity conditions. The Fe-Ti oxide ores are interpreted to have originated from in situ crystallization and liquid immiscibility from ferropicritic parent melts. The distribution of the ores is inferred to be structurally controlled.

Li et al. (Contribution 7) presented in situ LA-ICP-MS U–Pb ages of carbonate minerals from the Lannigou Carlin-type Au deposit, Lanmuchang Hg-(Tl) deposit, and Sixiangchang Hg deposit in South China. The results constrain gold mineralization to ca. 137 ± 9 Ma, Hg-Tl mineralization to ~97 Ma, and Hg mineralization to 454 ± 21 Ma. Integrating these data with previous geochronology, the authors identify two main episodes of low-temperature Au–As–Sb–Hg–Tl mineralization in the Youjiang metallogenic province. The study confirms the utility of carbonate U–Pb dating to constrain the timing of low-temperature mineralization.

Fan et al. (Contribution 8) conducted a petrographic and mineral chemical study of amphiboles from the Dalaku'an mafic–ultramafic intrusion in the western East Kunlun orogen. The amphiboles are classified into three types: tschermakitic hornblende and magnesio-hornblende of mantle affinity, magnesio-hornblende with lower Al–Ti contents, and actinolitic hornblende indicative of crustal input. Thermobarometric estimates reveal declining temperatures and pressures, consistent with multi-stage magma chamber evolution at different depths. High oxygen fugacity and abundant water suggest fluid metasomatism related to subducted oceanic crust. These findings collectively indicate that the Dalaku'an intrusion formed in a post-collisional extensional setting with contributions from both mantle and crustal sources.

Zhang et al. (Contribution 9) present a comprehensive study incorporating geological mapping, geochronology, and isotope geochemistry of the Tietangdong breccia pipe in the Yixingzhai gold deposit, Central Taihangshan District (CTD) of the North

https://doi.org/10.3390/min16010084

China Craton. Three breccia facies are identified: massive skarn breccia, polymictic skarn matrix-supported breccia, and polymictic intrusive rock cement chaotic breccia. $^{40}Ar/^{39}Ar$ dating of adularia yields an age of 136 $\pm$ 1.5 Ma, synchronous with post-breccia felsite dike emplacement. Pyrite $\delta^{34}S$ values (~2.7‰) support a magmatic–hydrothermal origin. The results establish a genetic relationship between the breccia pipe and adjacent lode gold mineralization.

We hope this Special Issue serves as a valuable reference and inspiration for researchers in economic geology and related fields. Looking ahead, several exciting research directions stand out. First, continued improvements in micro-analytical tools—like high-resolution imaging and in situ chemical analysis—will allow us to study mineral formation in greater detail than ever before. Second, combining mineral chemistry data with advanced computational methods, such as thermodynamic modeling and machine learning, opens new ways to understand ore-forming processes and uncover hidden patterns in complex datasets. Finally, linking fine-scale mineral observations to large-scale tectonic and geodynamic settings will be key to building reliable models of how different types of ore deposits form. By bringing together these interdisciplinary approaches, mineral chemistry will continue to play a central role in unraveling the origins of ore deposits and guiding more effective mineral exploration in the future.

Funding: This study was financially supported by the Hubei Provincial Natural Science Foundation of China (No. 2023AFD210), and Key Research and Development Program of Jiangxi Province (No. 20252BCF320015).

Conflicts of Interest: The authors declare no conflict of interest.

List of Contributions:

1. Li, R.; Su, S.; Wang, P. Chemistry and Fe Isotopes of Magnetites in the Orbicular Bodies in the Tanling Diorite and Implications for the Skarn Iron Mineralization in the North China Craton. *Minerals* **2025**, *15*, 1061. https://doi.org/10.3390/min15101061.
2. Filimon, D.I.; Groff, J.A.; Saccani, E.; Di Rosa, M. Ore Genesis Based on Microtextural and Geochemical Evidence from the Hydrothermal As–Sb Mineralization of the Matra Deposit (Alpine Corsica, France). *Minerals* **2025**, *15*, 814. https://doi.org/10.3390/min15080814.
3. Sun, W.; Xue, J.; Tong, Z.; Zhang, X.; Wang, J.; Li, S.; Wang, M. Genetic Mineralogical Characteristics of Pyrite and Quartz from the Qiubudong Silver Deposit, Central North China Craton: Implications for Ore Genesis and Exploration. *Minerals* **2025**, *15*, 769. https://doi.org/10.3390/min15080769.
4. Ouyang, Y.; Chen, Q.; Zeng, R.; Li, T. Chronological and Geochemical Characteristics of a Newly Discovered Biotite Granite Porphyry in the Zhuxi W-Cu Polymetallic Deposit, Jiangxi Province, South China: Implications for Cu Mineralization. *Minerals* **2025**, *15*, 624. https://doi.org/10.3390/min15060624.
5. Xiang, A.; Liu, H.; Fan, W.; Zhou, Q.; Wang, H.; Li, K. Petrogenesis, Geochemistry, and Geological Significance of the Kongco Granitic Porphyry Dykes in the Northern Part of the Central Lhasa Microblock, Tibet. *Minerals* **2025**, *15*, 283. https://doi.org/10.3390/min15030283.
6. Khedr, M.Z.; Moftah, A.; El-Shibiny, N.H.; Tamura, A.; Tan, W.; Ichiyama, Y.; Takazawa, E.; Kahal, A.Y.; Abdelrahman, K. Mineralogy and Geochemistry of Titaniferous Iron Ores in El-Baroud Layered Gabbros: Fe-Ti Ore Genesis and Tectono-Metallogenetic Setting. *Minerals* **2024**, *14*, 679. https://doi.org/10.3390/min14070679.
7. Li, J.; Zhuo, Y.; Guo, Y.; Lu, X.; Hu, X. In Situ Carbonate U-Pb Dating of Gold and Mercury Deposits in the Youjiang Metallogenic Province, SW China, and Implications for Multistage Mineralization. *Minerals* **2024**, *14*, 669. https://doi.org/10.3390/min14070669.

https://doi.org/10.3390/min16010084

8. Fan, Y.; Deng, Y.; Xia, Z.; Ren, M.; Huang, J. Petrogenesis of the Dalaku'an Mafic–Ultramafic Intrusion in the East Kunlun, Xinjiang: Constraints from the Mineralogy of Amphiboles. *Minerals* **2024**, *14*, 651. https://doi.org/10.3390/min14070651.
9. Zhang, L.; Gao, W.; Deng, X. Geology and Geochronology of Magmatic–Hydrothermal Breccia Pipes in the Yixingzhai Gold Deposit: Implications for Ore Genesis and Regional Exploration. *Minerals* **2024**, *14*, 496. https://doi.org/10.3390/min14050496.

minerals

MDPI

Article

Chemistry and Fe Isotopes of Magnetites in the Orbicular Bodies in the Tanling Diorite and Implications for the Skarn Iron Mineralization in the North China Craton

Ruipeng Li, Shangguo Su * and Peng Wang

School of Earth Sciences and Resources, China University of Geosciences (Beijing), Beijing 100083, China; liruipengcugb@126.com (R.L.); wangpeng9937@163.com (P.W.)
* Correspondence: susg@cugb.edu.cn

Abstract: Skarn-type iron ore is economically significant, and numerous skarn ore deposits have been identified in the North China Craton. The newly discovered orbicular diorite in this region is distinguished from other analogous rocks due to the accumulation of large magnetite particles, which may shed new light on the genesis of this ore type. The magnetite in different parts of the orbicular structure exhibits distinct compositional differences. For example, magnetite at the edge has a small particle size (200 μm) and is associated with the minerals plagioclase and hornblende, indicating that it crystallized from normal diorite magma. By contrast, magnetite in the core has a relatively large particle size (>1000 μm), is associated with apatite and actinolite, and contains apatite inclusions as well as numerous pores. The size of magnetite in the mantle falls between that of the edge and the core. The syngenetic minerals of magnetite in the mantle include epidote and plagioclase. The magnetites in the cores of orbicules have a higher content of Ti, Al, Ni, Cr, Sc, Zn, Co, Ga, and Nb than those in the rim. The δ^{56}Fe value of the core magnetite (0.46‰–0.78‰) is much higher than that of the mantle and rim magnetite in orbicules. Moreover, the δ^{56}Fe value of magnetite increases as the V content of magnetite gradually decreases. This large iron isotope fractionation is likely driven by liquid immiscibility that forms iron-rich melts under high oxygen fugacity. The reaction between magma and carbonate xenoliths (Ca, Mg)CO_3 during magma migration generates abundant CO_2, which significantly increases the oxygen fugacity of the magmatic system. Under the action of CO_2 and other volatile components, liquid immiscibility occurs in the magma chamber, and Fe-rich oxide melts are formed by the melting of carbonate xenoliths. Iron oxides (Fe_3O_4/Fe_2O_3) will crystallize close to the liquidus due to high oxygen fugacity. These characteristics of magnetite in the Tanling orbicular diorite (Wuan, China) indicate that diorite magma reacts with carbonate xenoliths to form "Fe-rich melts", and skarn iron deposits are probably formed by the reaction of intermediate-basic magma with carbonate rocks that generate such "Fe-rich melts". A possible reaction is as follows: diorite magma + carbonate → (magnetite-actinolite-apatite) + garnet + epidote + feldspar + hornblende + CO_2↑.

Keywords: orbicular diorite; Fe-rich melts; magnetite; skarn iron deposits; Wuan; North China Craton

1. Introduction

Skarn-type iron deposits are an important type of iron ore deposit in China. This type of deposit is commonly characterized by large reserves, high grades, and facilitated mining, thus holding important economic value. At present, the genesis of skarn-type iron deposits remains controversial, with two primary genetic models proposed: (1) the contact hydrothermal metasomatism model [1,2]. According to this model, after the upper part of the intermediate-to-acidic intrusions is consolidated, the residual magma exsolves a fluid containing volatiles; this fluid rises along structurally weak zones and reacts with surrounding country rocks to form the iron-bearing hydrothermal solution, which crystallizes to form large amounts of magnetite under reduced pressure and eventually accumulates to form a deposit [3]. However, the iron in the ore and that in the magma in this model cannot achieve mass balance [4], and the phenomena of vesicular structure and flat contact between the ore body and the surrounding rocks cannot be explained; (2) the genetic model of iron-rich magma: Iron-rich magma forms via differentiation of parent magma at depth, and this iron-rich magma then intrudes and crystallizes along weak structural zones to form magnetite [3,5–7]. This model can explain the occurrence of some iron veins and the porous structure of ores, but it is difficult to explain the rapid ascent of high-density iron-rich magma.

It can be seen that there are still issues with the two proposed genetic mechanisms of skarn-type iron deposits; thus, their genesis remains to be further clarified. In recent years, we have discovered an orbicular diorite in the Tanling area of the North China Craton. This orbicular diorite is distinguished from other orbicular rocks by the accumulation of large magnetite particles and the presence of dark actinolite aggregates in its core. Because the study area is located in the famous "Hanxing-type iron ore belt" and the Tanling orbicular diorite occurs near the contact zone between diorite intrusions and carbonate rocks, it is inferred that iron enrichment in this orbicular diorite is related to regional skarn iron mineralization.

Previous studies have shown that the major and trace element compositions of magnetite, as well as its iron isotopic composition, are helpful for understanding the processes of magnetite formation [8–16] and associated metallogeny. Magnetite in the Tanling orbicular diorite exhibits compositional variations from core to edge; therefore, this study analyzes the petrographic and geochemical characteristics of the magnetite in the Tanling area to clarify the origin of the magnetite and further explores the mechanism of iron enrichment in the orbicular diorite as well as its implications for regional skarn iron mineralization.

2. Regional Geology

The Wuan area is located in the central part of the North China Craton and the southern segment of the Taihang Mountains [17], as shown in Figure 1. Magmatic rocks are widely distributed in the region. Based on their distribution characteristics, these magmatic rocks can be divided into three representative magmatic belts from east to west: the Hongshan, Wuan, and Fushan belts. Previous zircon U-Pb dating results show that the intrusive complex in the Wuan area has an age between 120 Ma and 136 Ma [18–21], and that it formed during the Early Cretaceous magmatic event in eastern China. This indicates that the Wuan complex is located within the tectonic setting of strong lithospheric extension and thinning in the eastern North China Craton. The North China Craton, a major Archean craton in China, was destroyed during the Mesozoic—possibly due to the collision between the North China and South China blocks and the collapse of associated collisional orogens [22]. The Early Cretaceous magmatism—including the Tanling diorite in this study—marks the culmination of the cratonic destruction of the North China Craton [23].

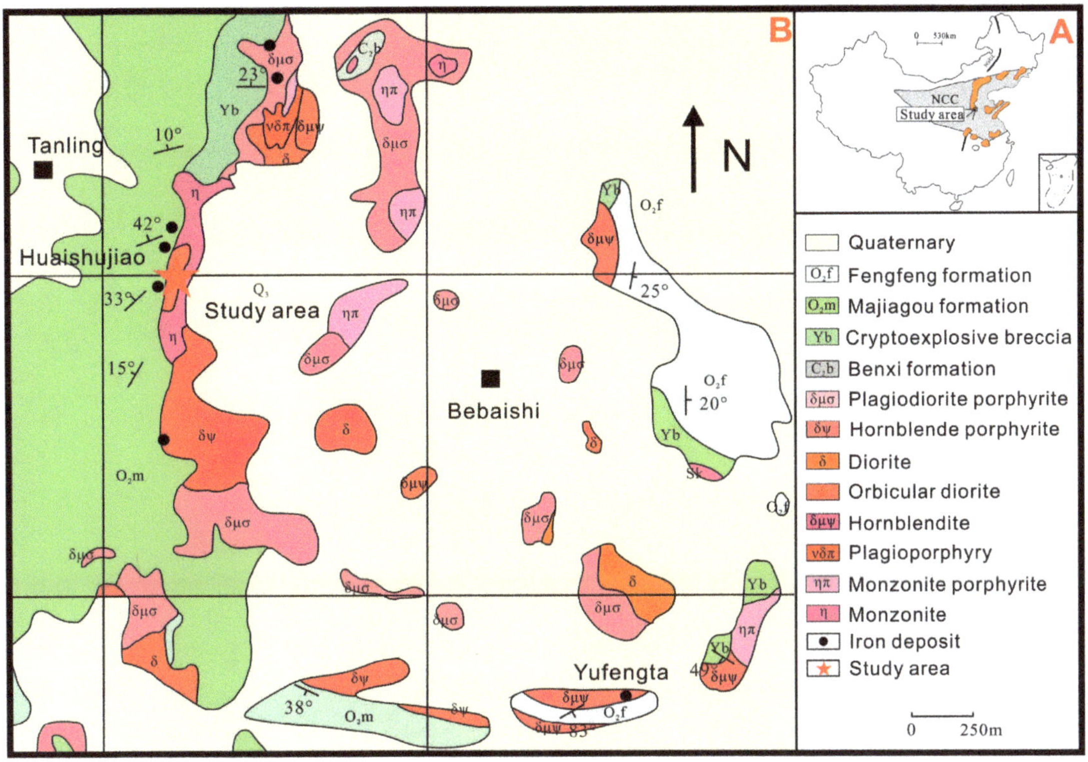

Figure 1. Regional geological map of the study area (modified after [24]).

The earliest strata exposed in this area are the Lower Zanghuang Group of the Archean [25], whose lithology mainly consists of gneiss, schist, and marble. The overlying Mesoproterozoic, Paleozoic, Mesozoic, and Cenozoic strata are in unconformable contact with it.

The region is tectonically controlled by NNE- and NEE-trending faults. The exposed strata in the Tanling area are mostly limestone and dolomite of the Ordovician Majiagou Formation, a small amount of siltstone of the Carboniferous Benxi Formation, as well as minor shale and mudstone, etc. [25].

3. Petrographic Characteristics

3.1. Orbicular Diorite

There is only one occurrence of orbicular diorite in the Wuan area. It is mainly composed of a core and a shell. The proportion varies with the degree of weathering, with the core accounting for approximately 15% of the volume of the orbicule and the outer shell for 85% (Figure 2a,b). Field observation shows that the orbicular diorite has the following characteristics: (1) the orbicular diorite can be divided into three parts: magnetite-actinolite-apatite orbicular core, epidote orbicular mantle, and single diorite orbicular shell; (2) the obvious vesicular structure is observed in the dark mass area of the orbicule's core (Figure 3b); (3) carbonate xenoliths are present in the rock mass adjacent to the orbicular diorite (Figure 2c).

Figure 2. Field and structural characteristics of the Tanling orbicular diorite: (**a**) outcrop of orbicular diorite; (**b**) internal structure of orbicular diorite; (**c**) marble xenoliths in the rock mass adjacent to orbicular diorite.

3.2. Characteristics of Magnetite in Each Part of the Orbicular Diorite

The petrographic characteristics and mineral assemblages of magnetite from the core to the rim of the orbicular diorite are as follows:

The magnetite in the orbicular core has a large particle size (approximately 1000 μm), a low degree of idiomorphism, and is mostly in the form of hypidiomorphic-heteromorphic spheroids, with curved grain edges. Apatite grains are commonly observed within both magnetite and actinolite (Figure 3b), and these inclusions are long columnar. The mineral assemblage in the core is mainly magnetite, actinolite, apatite, and titanite (Figure 3b). The core contains an extremely small number of albite and hornblende grains (Figure 4). In TIMA images, the voids are widespread within the orbicule's core, and some of them are distributed between the inclusions and magnetite grains (Figure 5).

The magnetite at the rim of the orbicule has a small particle size (<200 μm), a high degree of idiomorphism, and mostly occurs as idiomorphic to semi-idiomorphic granular grains; its magnetite particle edges are relatively flat. The degree of magnetite idiomorphism in the mantle is lower than that in the rim, and the grain boundaries are less flat than those in the rim. The mineral assemblage of the rim is magnetite, albite, hornblende, quartz, titanite, epidote, and apatite (Figure 3d).

The particle size of magnetite in the orbicule's mantle is intermediate between that of the core and the rim, and the mineral assemblage in the mantle is mainly magnetite, epidote, and plagioclase (Figure 3c). From the core to the rim of the orbicule, the pore area within the orbicule gradually decreases (Figure 5).

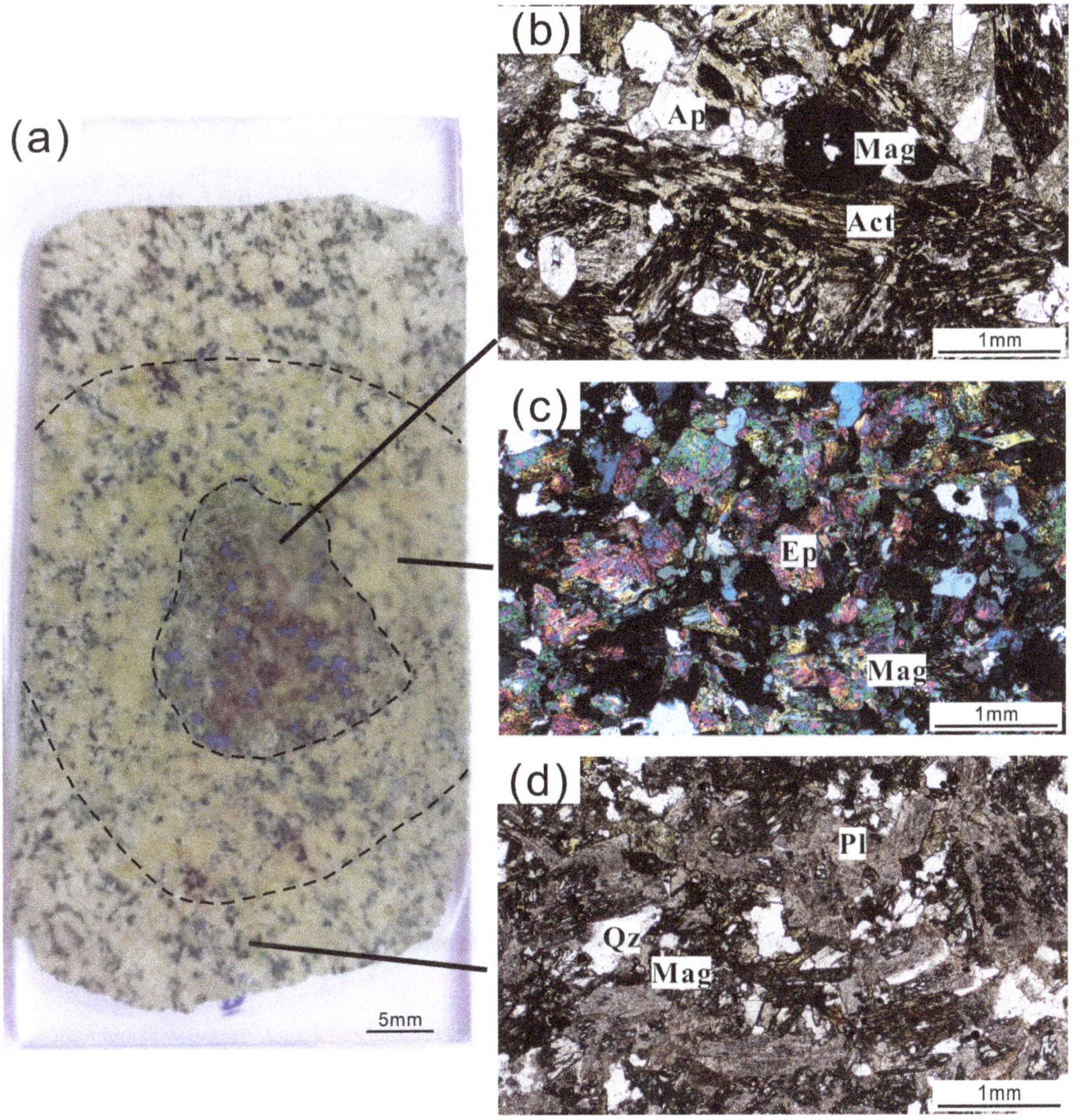

Figure 3. Microscopic characteristics of the orbicular diorite at different positions: (**a**) microphotograph of the orbicular diorite; (**b**) core mineral association (plane-polarized light, −); (**c**) mantle mineral association (cross-polarized light, +); (**d**) rim mineral association (plane-polarized light, −). Mag—magnetite; Ap—apatite; Act—actinolite; Ep—epidote; Qz—quartz; Pl—plagioclase.

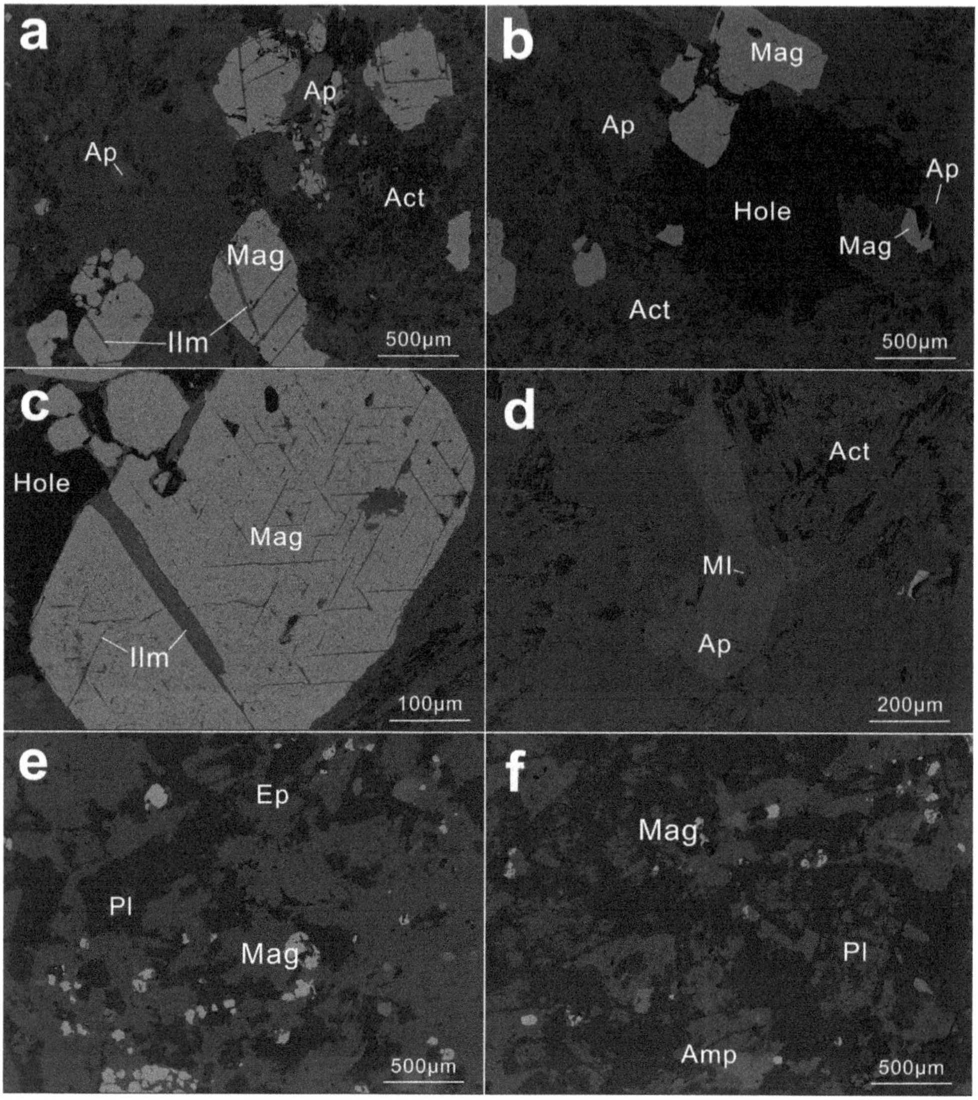

Figure 4. Magnetite BSE photos of orbicular diorite at different positions: (**a**) core mineral association; (**b**) magnetite and apatite around the voids; (**c**) exsolution of ilmenite; (**d**) magnetite inclusions in the core's apatite; (**e**) mantle mineral association; (**f**) rim mineral association. Mag—magnetite; Ap—apatite; Act—actinolite; Ilm—ilmenite; Ep—epidote; Pl—plagioclase; MI—melt inclusions.

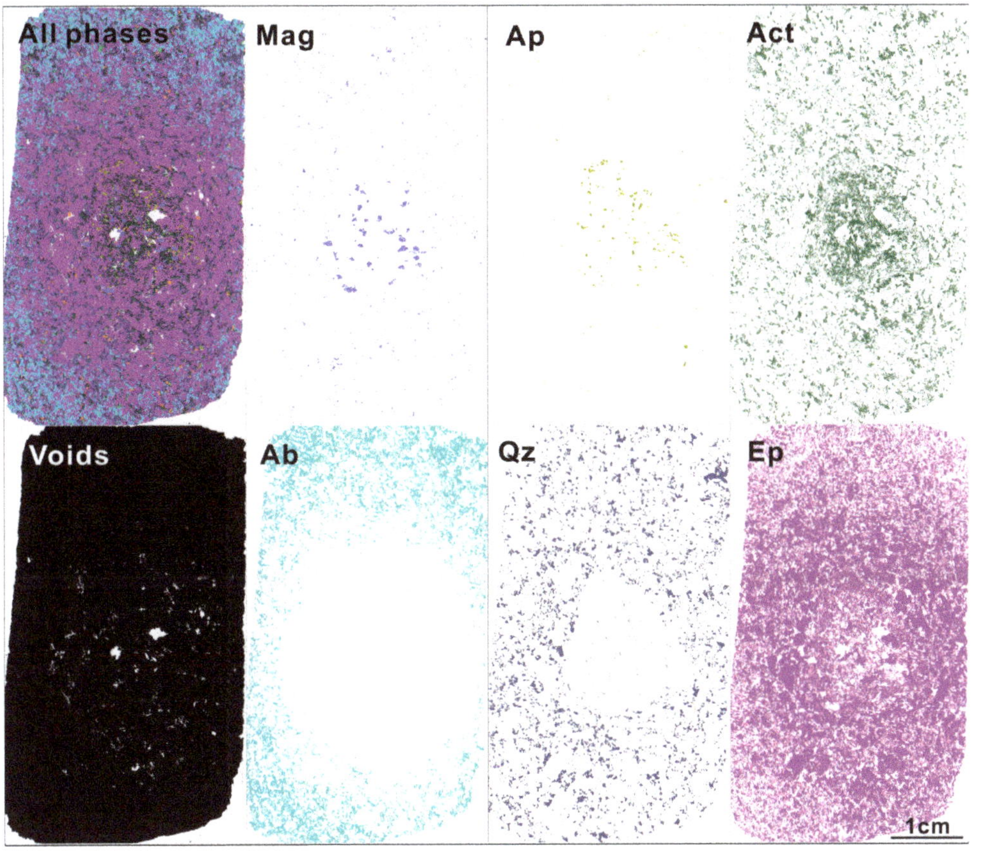

Figure 5. Distribution diagram of major mineral phases of orbicular diorite (based on TESCAN Integrated Mineral Analyzer). Mag—magnetite; Ap—apatite; Act—actinolite; Ep—epidote; Qz—quartz; Ab—albite.

4. Methods

The major element analysis of minerals was conducted using an electron microprobe in the laboratory of Hebei Institute of Regional Geology and Mineral Resources. The analytical instrument model is JXA-8230, with an accelerating voltage of 15 kV, a beam current of 1×10^{-8} A, and a beam spot size of 1–10 μm. The analytical standard samples consist of 53 mineral species from SPI Company (USA), and the analytical accuracy is better than 1%.

Double-polished thin sections (with a thickness range of 60–100 μm) were prepared for LA-ICP-MS (Laser Ablation-Inductively Coupled Plasma-Mass Spectrometry) at the State Geological Laboratory and Testing Center, Academy of Geological Sciences, Beijing China. In situ laser ablation was performed using a Coherent GeoLasPro 193 nm laser ablation system coupled with an Agilent 7700 quadrupole ICP-MS. After a 20 s gas blank monitoring step, 160 successive laser pulses (4 Hz, spot size: 44 μm) were used to ablate the sample surface for approximately 40 s. The generated aerosols were carried by a helium carrier gas, then mixed with argon make-up gas via a T-connector before entering the ICP-MS for the acquisition of ion signal intensities. For the detailed analytical process, refer to [26].

Elemental distribution maps were obtained using a TESCAN Integrated Mineral Analyzer (TIMA) at the Institute of Geology, Chinese Academy of Geological Sciences (CAGS), Beijing. Automated mineralogical analyses for phase mapping were conducted on the thin sections, with an accelerating voltage of 25 kV, a beam current of 8.47 nA, a working distance of 15 mm, and a spot size of 100.50 nm.

In situ Fe isotopic analyses were obtained using a NEPTUNE Plus MC-ICP-MS (Thermo Fisher Scientific, Bremen, Germany), which was combined with an NWR FemtoUC femtosecond system (New Wave Research, Fremont, CA, USA). For the in situ Fe isotope analyses, the laser ablation system was operated in single-spot mode, with a spot size of 40 μm, frequency of 3 Hz, and laser fluence of ~3.5 J·cm^{-2}.

For in situ Fe isotope analysis, the laser ablation system was operated in single-spot ablation mode, with a spot size of 20–40 μm, a frequency of 3–5 Hz, and a laser fluence of ~2.5–4.5 J·cm^{-2}. Ultrapure water was introduced into the ICP to form a "wet" plasma atmosphere, which effectively suppressed the matrix effects during Fe isotope analysis. LA-MC-ICP-MS Fe isotope measurements were performed at high mass resolution ($M/\Delta M \approx 7000$, defined by 5%–95% peak width) to resolve molecular interferences from argon nitrides and argon oxides on Fe isotopes. The Faraday cup detector array was set to acquire the ion beams of ^{53}Cr, ^{54}Fe, ^{56}Fe, ^{57}Fe, ^{58}Fe, and ^{60}Ni. ^{53}Cr and ^{60}Ni were monitored and used to correct for isobaric interferences on ^{54}Fe and ^{58}Fe. The SSB (Sample-Standard Bracketing) method was employed for mass fractionation correction. The international Fe isotopic standard IRMM-014 (pure iron, 99.9% purity), from the Institute for Reference Materials and Measurements (IRMM) of the European Commission, was used as an external standard, and the in-house reference material magnetite MT-8 was used as a quality control sample for in situ Fe isotope analyses. Standards and reference materials were measured before every set of 8 samples analyzed for drift monitoring. Fe isotope data reduction was performed using the Iso-Compass software [27]. Details of the LA-MC-ICP-MS operating conditions and measurement parameters are described in Feng et al. [28].

5. Results

5.1. Major Elements of Magnetite

All electron probe microanalysis (EPMA) data for the major elements of magnetite are shown in Table 1. For core magnetite (magnetite in the cores of orbicules): SiO_2 content is 0.03%–0.47%, CoO is 0.06%–0.25%, MgO is 0%–0.05%, and Cr_2O_3 is 0.03%–1.48%. For mantle magnetite (magnetite in the mantles of orbicules): SiO_2 is 0.03%–0.56%, CoO is 0.05%–0.16%, MgO is 0%–0.04%, and Cr_2O_3 is 0.01%–0.78%. For rim magnetite (magnetite in the rims of orbicules): SiO_2 is 0.02%–0.36%, CoO is 0.02%–0.17%, MgO is 0%–0.08%, and Cr_2O_3 is 0%–1.12%.

Table 1. Major composition of magnetite (wt.%).

Core Magnetite													
Sample Name	K_2O	CaO	TiO_2	CoO	P_2O_5	Na_2O	MgO	Al_2O_3	SiO_2	Cr_2O_3	MnO	FeO	Total	CaO/(Na_2O + K_2O)
Q22-A-Q1-1	0.03	0.07	0.16	0.06	0.01	0.06	0.05	0.53	0.47	1.48	0.02	86.24	89.68	0.76
Q22-A-Q1-2	0.02	0.03	0.20	0.12	n.d.	0.04	n.d.	0.42	0.22	0.50	0.08	87.56	89.72	0.43
Q22-A-Q1-3	n.d.	0.03	0.14	0.14	0.02	0.00	0.03	0.31	0.24	0.20	0.04	87.73	89.43	33.00
Q22-A-Q2-1	n.d.	0.04	0.15	0.14	0.01	0.01	0.03	0.22	0.19	0.23	n.d.	88.78	90.35	2.73
Q22-A-Q2-2	n.d.	0.01	0.11	0.13	n.d.	0.06	0.04	0.36	0.20	0.19	0.09	88.43	90.12	0.24
Q22-A-Q2-3	n.d.	0.05	0.15	0.30	0.03	n.d.	0.02	0.42	0.39	0.80	0.02	88.17	90.87	-
Q22-A-Q3-1	n.d.	0.03	0.16	0.09	n.d.	n.d.	n.d.	0.36	0.18	0.07	0.06	88.73	90.16	-
Q22-A-Q3-2	n.d.	0.04	0.18	0.11	n.d.	n.d.	0.03	0.42	0.21	0.10	0.04	88.04	89.70	-
Q22-A-Q3-3	n.d.	0.04	0.12	0.07	0.01	n.d.	0.03	0.19	0.22	0.38	0.06	88.52	90.08	-

Table 1. *Cont.*

Sample Name	K_2O	CaO	TiO_2	CoO	P_2O_5	Na_2O	MgO	Al_2O_3	SiO_2	Cr_2O_3	MnO	FeO	Total	CaO/($Na_2O + K_2O$)
Core Magnetite														
Q22-A-Q4-1	n.d.	0.03	0.09	0.06	n.d.	0.04	n.d.	0.38	0.17	0.34	0.07	87.83	89.53	0.55
Q22-A-Q4-2	n.d.	0.04	0.14	0.13	n.d.	0.01	0.02	0.30	0.05	0.14	0.05	87.58	88.93	3.00
Q22-A-Q4-3	n.d.	0.06	0.13	0.16	0.02	n.d.	0.02	0.22	0.06	0.08	0.05	87.92	89.20	-
Q22-A-Q5-1	n.d.	0.04	0.16	0.09	0.02	0.06	0.06	0.15	0.16	0.06	0.16	87.22	88.47	0.60
Q22-A-Q5-2	n.d.	0.04	0.12	n.d.	n.d.	0.15	n.d.	0.21	0.12	0.10	0.03	92.02	92.98	0.28
Q22-A-Q6-1	n.d.	0.13	0.07	n.d.	n.d.	0.01	n.d.	0.07	0.16	0.03	0.02	91.27	92.02	6.19
Q22-A-Q6-2	n.d.	0.12	0.03	n.d.	n.d.	0.01	n.d.	0.05	0.03	0.26	0.03	90.82	91.51	8.92
Q22-A-Q6-3	0.01	0.07	0.16	0.06	0.01	n.d.	0.04	0.13	0.18	0.41	0.04	87.62	89.28	5.58
Q22-A-Q7-1	n.d.	0.02	0.12	0.23	0.05	n.d.	n.d.	0.38	0.22	0.13	0.08	88.28	90.10	4.00
Q22-A-Q7-2	n.d.	0.07	0.03	n.d.	n.d.	0.01	n.d.	0.03	0.04	0.03	0.04	92.24	92.66	5.29
Q22-A-Q7-3	n.d.	0.04	0.10	0.25	n.d.	n.d.	0.08	0.46	0.29	0.46	0.06	86.77	89.12	4.30
Middle Magnetite														
Q22-B-1	n.d.	0.13	0.11	0.06	n.d.	n.d.	0.02	0.13	0.33	0.78	0.03	87.73	89.75	19.14
Q22-B-2	0.02	0.16	0.14	0.12	0.03	0.03	0.04	0.27	0.56	0.71	0.03	88.06	90.53	3.42
Q22-B-3	n.d.	0.06	0.05	n.d.	n.d.	0.06	n.d.	n.d.	0.09	0.04	n.d.	92.16	92.62	0.88
Q22-B-4	n.d.	0.22	0.06	0.08	n.d.	0.02	0.04	0.21	0.53	0.51	0.06	88.61	90.70	11.42
Q22-B-5	n.d.	0.07	0.14	0.08	n.d.	n.d.	0.03	0.18	0.25	0.56	0.03	88.34	90.16	-
Q22-B-6	n.d.	0.11	0.25	0.13	0.01	0.04	0.03	0.28	0.42	0.72	0.08	87.76	90.31	2.43
Q22-B-7	n.d.	0.10	0.04	0.09	0.05	0.05	0.01	0.05	0.11	0.04	n.d.	89.40	90.29	1.83
Q22-B-8	0.01	0.12	0.06	0.05	0.01	0.03	0.03	0.14	0.22	0.02	0.04	87.73	88.81	3.03
Q22-B-9	n.d.	0.03	0.03	0.11	0.02	n.d.	0.01	n.d.	0.07	0.03	0.03	91.71	92.35	8.50
Q22-B-10	0.01	0.07	0.05	0.11	n.d.	0.01	n.d.	0.04	0.09	0.02	0.06	91.89	92.65	3.25
Q22-B-12	n.d.	0.05	0.06	0.16	n.d.	0.05	0.03	0.02	0.05	n.d.	n.d.	91.29	92.07	1.00
Q22-B-13	0.01	0.04	0.07	0.09	n.d.	0.03	n.d.	0.04	0.06	0.05	0.03	90.22	90.99	0.97
Q22-B-14	n.d.	0.12	0.06	0.09	0.01	n.d.	0.04	0.06	0.19	0.05	0.03	88.39	89.34	-
Q22-B-15	n.d.	0.04	0.03	0.09	n.d.	0.08	n.d.	0.01	0.03	0.01	0.02	91.48	92.11	0.47
Rim Magnetite														
Q22-C-Q1-1	0.01	0.16	0.18	0.04	0.06	0.11	0.04	0.19	0.36	0.15	0.05	87.12	88.85	1.28
Q22-C-Q1-2	n.d.	0.16	0.16	0.07	n.d.	0.07	0.02	0.09	0.27	0.13	0.03	87.18	88.62	2.41
Q22-C-Q1-3	n.d.	0.07	0.12	0.06	n.d.	0.11	n.d.	0.11	0.19	0.30	0.09	88.72	90.15	0.61
Q22-C-Q1-4	0.03	0.07	0.07	0.08	n.d.	0.08	n.d.	0.08	0.39	0.06	0.03	86.56	87.83	0.63
Q22-C-Q1-5	n.d.	0.14	0.05	0.12	0.02	0.04	0.04	0.08	0.33	0.03	0.03	88.49	89.71	3.71
Q22-C-Q1-6	n.d.	0.05	0.06	n.d.	n.d.	n.d.	0.01	0.08	0.05	0.01	n.d.	93.11	93.54	7.67
Q22-C-Q1-7	0.02	0.03	0.02	0.11	0.02	0.03	0.03	0.13	0.06	0.05	0.06	91.70	92.59	0.71
Q22-C-Q2-1	0.08	0.11	0.05	0.06	0.02	0.32	0.02	0.18	0.42	0.45	n.d.	87.87	89.88	0.28
Q22-C-Q2-2	n.d.	0.06	0.04	0.17	n.d.	0.00	n.d.	n.d.	0.03	n.d.	0.02	92.01	92.55	-
Q22-C-Q2-3	0.02	0.08	0.24	0.13	n.d.	0.13	0.08	0.24	0.41	1.12	0.06	86.53	89.48	0.54
Q22-C-Q2-4	n.d.	0.09	0.06	n.d.	n.d.	0.05	0.02	0.06	0.06	0.07	n.d.	91.91	92.43	1.67
Q22-C-Q2-5	n.d.	0.05	0.07	0.10	n.d.	0.07	n.d.	0.06	0.12	0.04	0.05	88.84	89.78	0.66
Q22-C-Q2-6	0.02	0.04	0.07	0.08	0.02	0.05	0.03	0.16	0.30	0.02	0.06	87.93	89.11	0.65
Q22-C-Q3-1	n.d.	0.10	0.10	0.13	n.d.	0.11	0.03	0.07	0.17	0.10	0.02	87.57	88.78	0.87
Q22-C-Q3-2	n.d.	0.03	0.13	0.02	0.01	0.07	n.d.	0.02	0.06	0.05	0.01	93.41	94.20	0.42
Q22-C-Q3-3	0.01	0.07	0.04	0.08	0.01	n.d.	n.d.	0.01	0.04	0.01	0.02	91.86	92.51	7.20
Q22-C-Q3-4	n.d.	0.07	0.04	0.12	0.01	0.02	0.01	0.05	0.06	n.d.	0.04	85.80	86.58	3.36
Q22-C-Q3-5	n.d.	0.02	0.05	0.14	0.02	0.06	n.d.	0.07	0.02	0.04	0.03	92.76	93.60	0.28
Q22-C-Q3-6	n.d.	0.11	0.06	0.10	n.d.	0.08	0.01	0.04	0.03	0.05	0.02	88.53	89.39	1.32

Note: all Fe was recalculated as FeO in the microprobe analyses.

5.2. Trace Elements of Magnetite

Major element data for magnetite in various parts of orbicular diorite are shown in Table 1. Trace element data for magnetite in these parts are shown in Supplementary Table S1.

In the TIMA elemental distribution maps (Figure 6), the Fe and Ca elemental contents in the core of orbicular diorite are significantly higher than those in the rim. Core magnetite has higher contents of Ti (4140–17,361 ppm), Al (2163–8732 ppm), Ni (30.7–68.5 ppm), Cr (64–3147 ppm), Sc (2.14–5.19 ppm), Zn (50–1117 ppm), Co (65.7–453 ppm), Ga (34.4–82 ppm), Nb (0.46–1.29 ppm) than those in the mantle and rim (Figures 7–9). On the bulk continental crust-normalized multi-element spider diagram, core magnetite is enriched in Pb, Ge, W, Mo, Sn, Ga, Ti, Zn, Co, V, and Cr, but depleted in Si, Ca, Mg, and high field strength elements (Y, P, Nb, Ta, Zr, Hf) (Figure 10). Generally, core magnetite is more enriched in highly compatible elements with respect to magnetite.

The content range of trace elements in the rim and mantle magnetite is narrower than that in core magnetite. Most of the rim magnetite has relatively low contents of Ti (249–2522 ppm), Al (556–1266 ppm), Ni (7.9–24.6 ppm), Cr (57–195 ppm), Sc (0.19–1.89 ppm), Zn (33–258 ppm), Co (54–170 ppm), Ga (15.3–34.2 ppm), and Nb (0–0.69 ppm) (Figures 8 and 9). On the bulk continental crust-normalized multi-element spider diagram, rim magnetite is enriched in P, Pb, Ge, W, Mo, Co, and V, but depleted in Hf, Al, Nb, Ta, Sn, Mg, Ni, and Cr. (Figure 10). Mantle magnetite generally exhibits elemental characteristics intermediate between the core and rim magnetite, with Cr from 74 to 640 ppm, V from 2648 to 3685 ppm, Ti from 1197 to 4611 ppm, Al from 652 to 2859 ppm, and P from 29 to 5260 ppm. On the bulk continental crust-normalized multi-element spider diagram, mantle magnetite has similar characteristics to the rim magnetite (Figure 10). The content of Al+Mn gradually increases from the rim magnetite to the core magnetite (Figure 11).

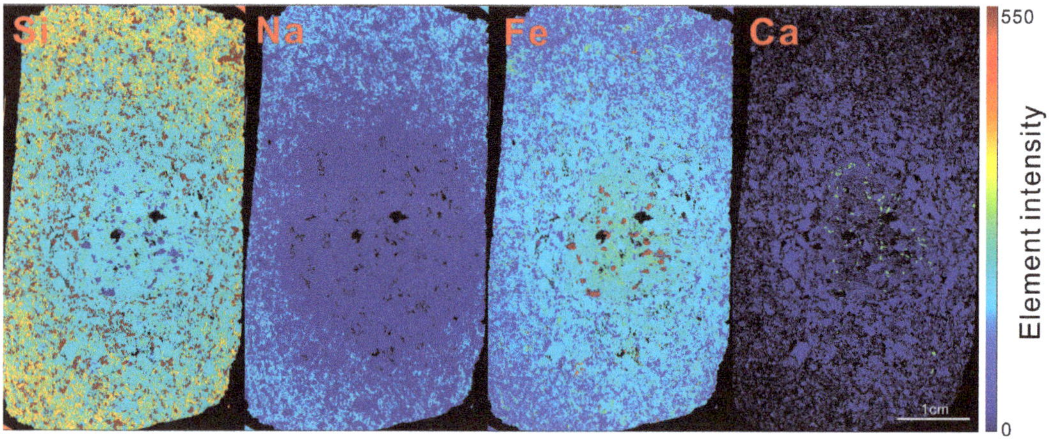

Figure 6. Elemental distribution maps of Si, Na, Fe, and Ca of orbicular diorite (acquired using TESCAN Integrated Mineral Analyzer, TIMA).

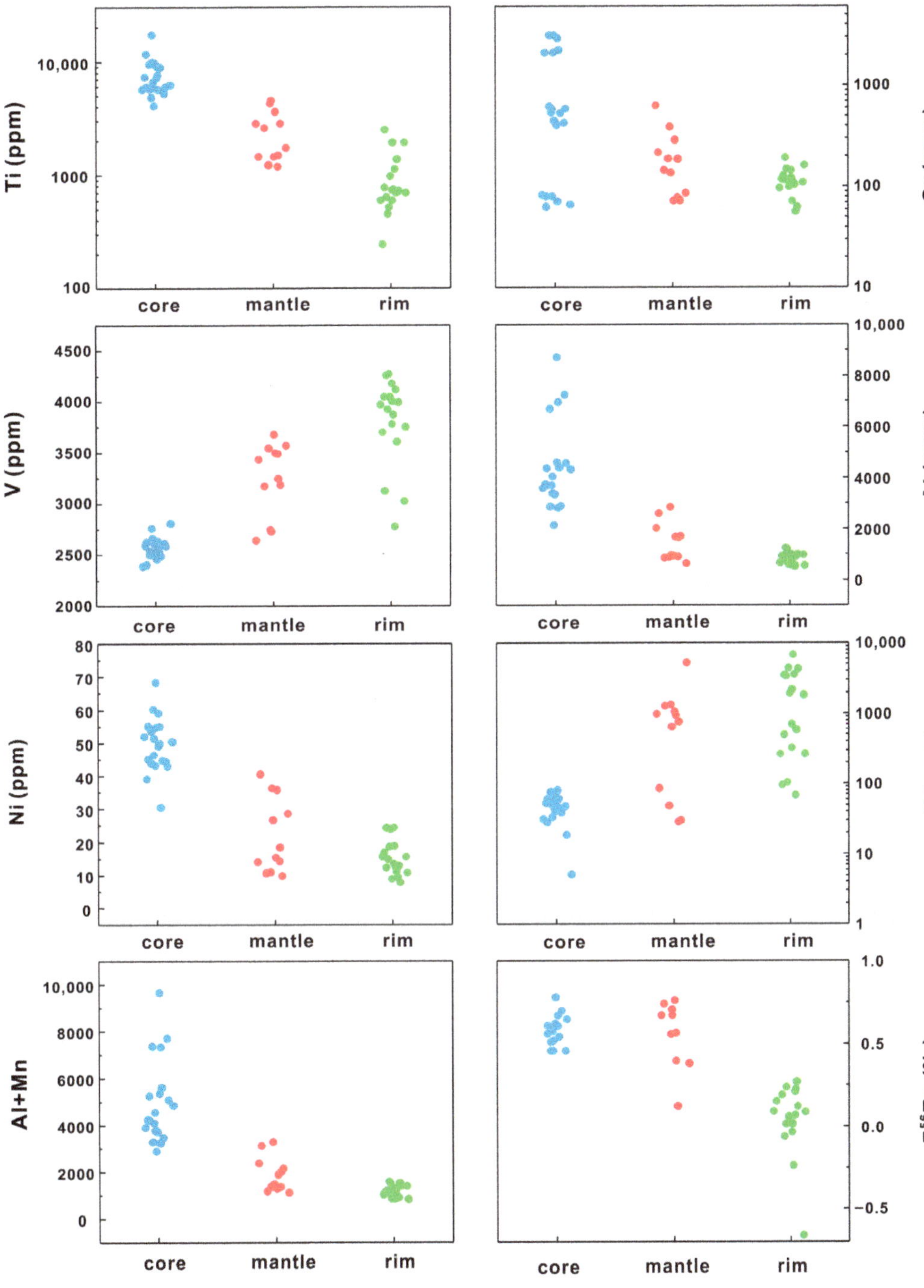

Figure 7. Element variation in magnetite from different parts of the Tanling orbicular diorite.

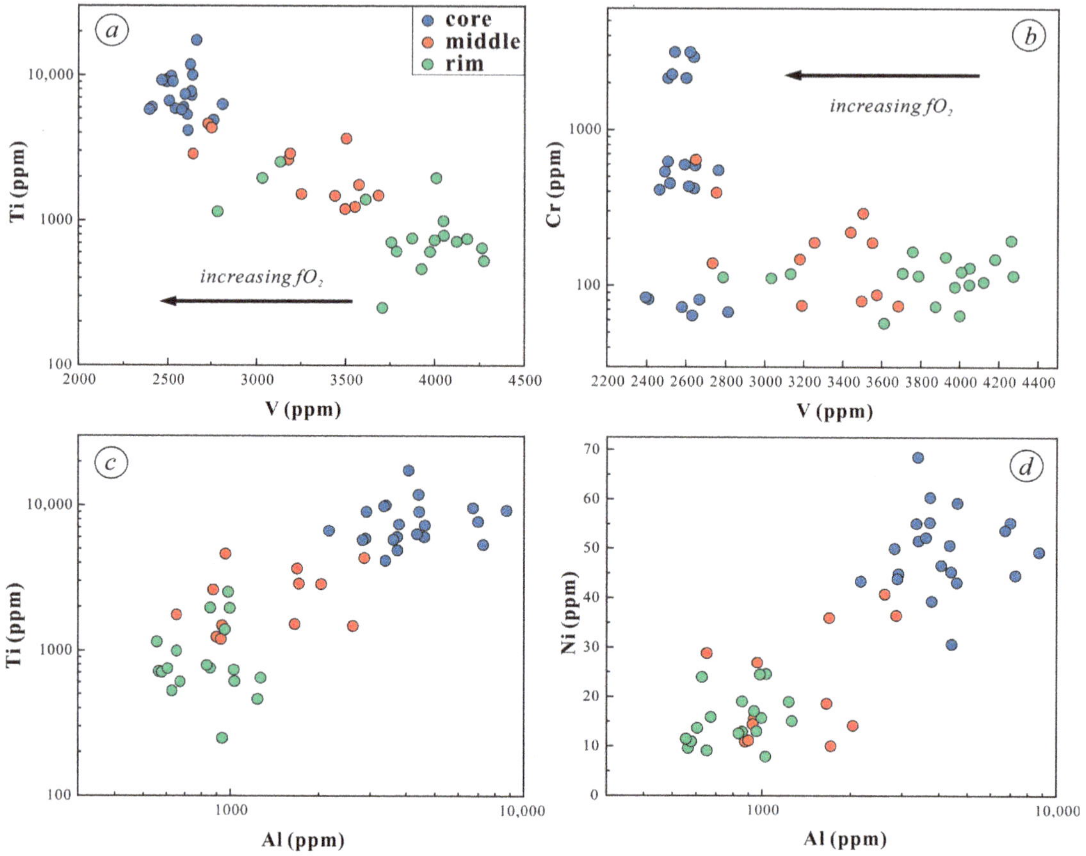

Figure 8. Binary diagrams of magnetite from different parts of the Tanling orbicular diorite: (**a**) V vs. Ti, (**b**) V vs. Cr, (**c**) Al vs. Ti, (**d**) Al vs. Ni. The direction of oxygen fugacity change is referenced from [29–32].

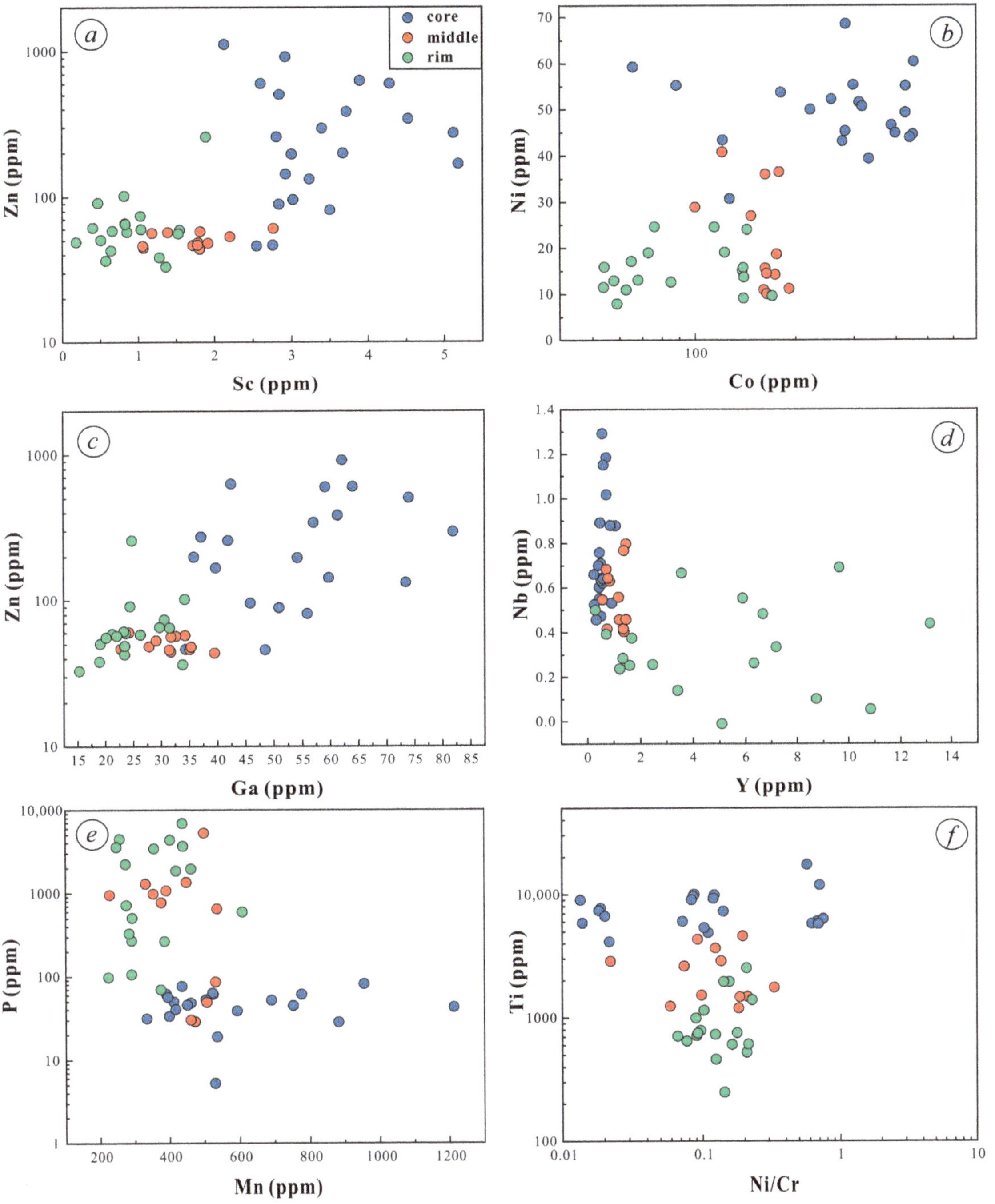

Figure 9. Binary diagrams of magnetite from different parts of the Tanling orbicular diorite: (**a**) Sc vs. Zn, (**b**) Co vs. Ni, (**c**) Ga vs. Zn, (**d**) Y vs. Nb, (**e**) Mn vs. P, (**f**) Ni/Cr vs. Ti.

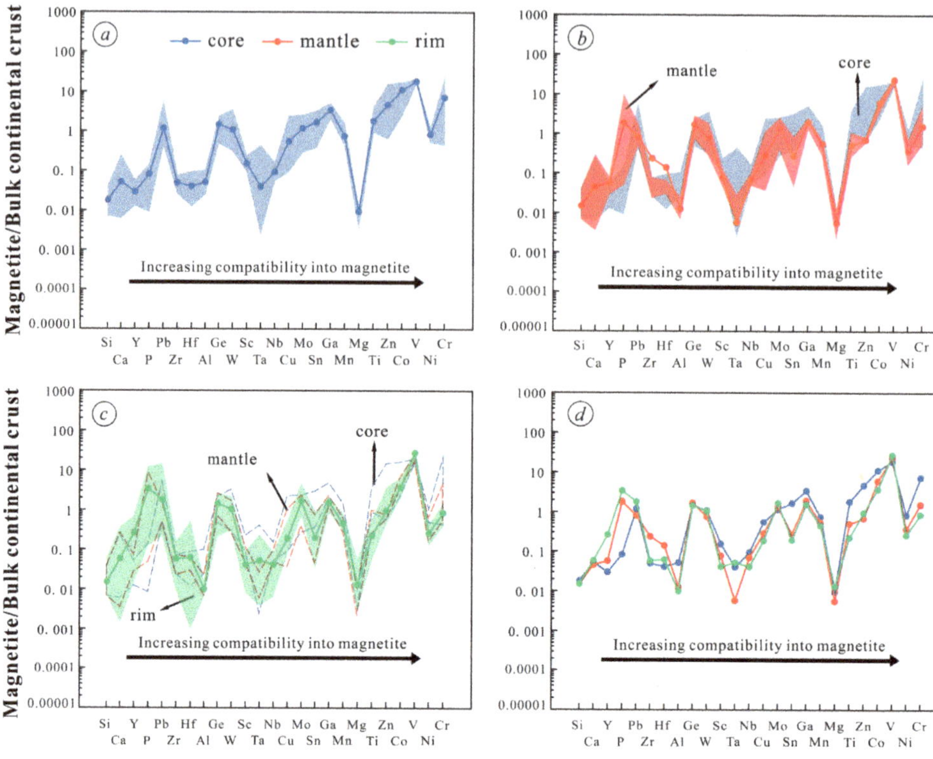

Figure 10. Multi-element spider diagrams for different types of magnetite in the Tanling orbicular diorite. All data are normalized to bulk continental crust values (after [33]). (**a**) core magnetite; (**b**) mantle magnetite; (**c**) rim magnetite; (**d**) comparison of the average values of three types of magnetite. Note: The selected elements in this figure are ordered by their increasing compatibility with magnetite, based on compiled experimental and empirical partition coefficients between magnetite and silicate magmas. More details can be found in [34] and references therein.

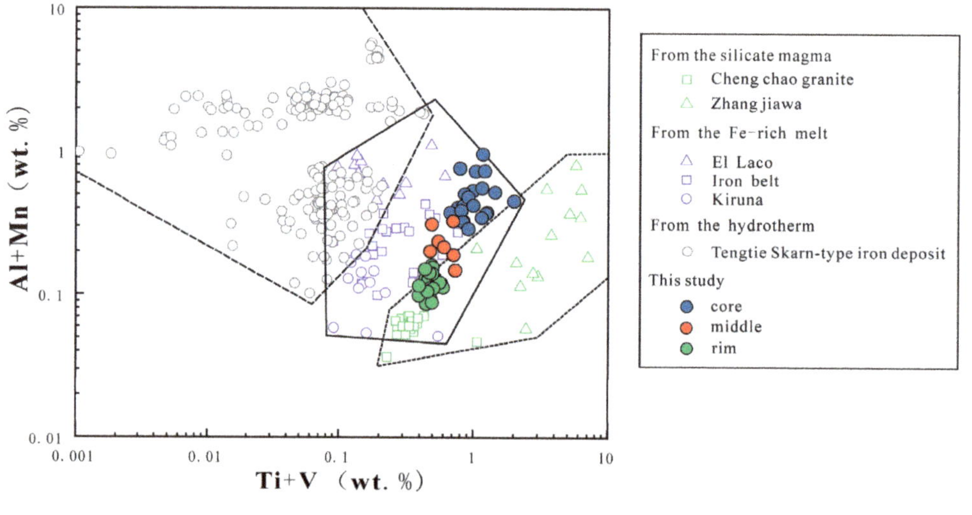

Figure 11. (Al + Mn) vs. (Ti + V) plots for discriminating among magmatic-type, iron/magma type, and skarn type magnetite (after [35]; data from: [8,36,37]).

5.3. In Situ Fe Isotopes

$\delta^{56}Fe$ test data of magnetite in different parts of the orbicular diorite are shown in Table 2.

Table 2. The analysis results of the Fe isotope for magnetite.

Sample Name	Remarks	$\delta^{56}Fe_{IRMM}$(‰)	Delta-2SE
MT-8	Standards	−0.20	0.06
MT-8	Standards	−0.19	0.06
MT-10	Standards	0.25	0.06
MT-10	Standards	0.24	0.06
MT-8	Standards	−0.08	0.06
MT-10	Standards	0.30	0.04
MT-8	Standards	−0.13	0.06
Q22-A-Q1-1	Core magnetite	0.67	0.05
Q22-A-Q1-2	Core magnetite	0.54	0.04
Q22-A-Q1-3	Core magnetite	0.58	0.04
MT-8	Standards	−0.19	0.05
Q22-A-Q2-1	Core magnetite	0.46	0.05
Q22-A-Q2-2	Core magnetite	0.46	0.05
Q22-A-Q4-1	Core magnetite	0.62	0.04
MT-8	Standards	−0.18	0.06
Q22-A-Q4-2	Core magnetite	0.61	0.04
Q22-A-Q5-1	Core magnetite	0.52	0.04
Q22-A-Q5-2	Core magnetite	0.56	0.05
MT-8	Standards	−0.11	0.05
Q22-A-Q6-1	Core magnetite	0.65	0.05
Q22-A-Q6-2	Core magnetite	0.78	0.04
Q22-A-Q6-3	Core magnetite	0.70	0.04
MT-8	Standards	−0.18	0.07
Q22-A-Q7-1	Core magnetite	0.61	0.06
Q22-A-Q7-2	Core magnetite	0.46	0.05
Q22-A-Q7-3	Core magnetite	0.51	0.05
MT-8	Standards	−0.13	0.06
MT-10	Standards	0.27	0.06
MT-8	Standards	−0.10	0.06
Q22-B-1	Middle magnetite	0.67	0.05
Q22-B-2	Middle magnetite	0.38	0.06
Q22-B-3	Middle magnetite	0.57	0.05
MT-8	Standards	−0.14	0.05
Q22-B-4	Middle magnetite	0.74	0.04
Q22-B-6	Middle magnetite	0.67	0.06
MT-8	Standards	−0.23	0.07
Q22-B-7	Middle magnetite	0.12	0.05
Q22-B-9	Middle magnetite	0.71	0.04
MT-8	Standards	−0.11	0.06
MT-8	Standards	−0.20	0.04
MT-8	Standards	−0.09	0.06
MT-10	Standards	0.31	0.06
MT-8	Standards	−0.12	0.06
Q22-B-12	Middle magnetite	0.56	0.05
MT-8	Standards	−0.10	0.04
Q22-B-13	Middle magnetite	0.76	0.06
Q22-B-15	Middle magnetite	0.40	0.05
MT-8	Standards	−0.09	0.07
MT-10	Standards	0.31	0.05
MT-10	Standards	0.30	0.05

Table 2. *Cont.*

Sample Name	Remarks	$\delta^{56}Fe_{IRMM}$(‰)	Delta-2SE
MT-8	Standards	−0.14	0.06
MT-10	Standards	0.34	0.09
MT-10	Standards	0.32	0.07
MT-8	Standards	−0.07	0.05
MT-8	Standards	−0.09	0.06
MT-10	Standards	0.23	0.07
MT-10	Standards	0.23	0.06
MT-8	Standards	−0.10	0.06
MT-10	Standards	0.28	0.05
MT-8	Standards	−0.17	0.06
Q22-C-Q1-1	Rim magnetite	0.21	0.06
Q22-C-Q1-2	Rim magnetite	0.10	0.05
Q22-C-Q1-3	Rim magnetite	0.19	0.06
MT-8	Standards	−0.11	0.08
Q22-C-Q2-1	Rim magnetite	−0.66	0.04
Q22-C-Q2-3	Rim magnetite	0.16	0.05
MT-8	Standards	−0.07	0.05
Q22-C-Q3-1	Rim magnetite	0.02	0.08
Q22-C-Q3-2	Rim magnetite	0.07	0.09
Q22-C-Q3-3	Rim magnetite	0.13	0.08
MT-8	Standards	−0.17	0.06
MT-8	Standards	−0.15	0.05
Q22-C-Q3-4	Rim magnetite	0.23	0.07
Q22-C-Q1-4	Rim magnetite	0.06	0.08
Q22-C-Q1-5	Rim magnetite	0.24	0.05
MT-8	Standards	−0.17	0.07
Q22-C-Q1-6	Rim magnetite	0.09	0.06
Q22-C-Q2-4	Rim magnetite	−0.03	0.07
MT-8	Standards	−0.18	0.07
MT-10	Standards	0.27	0.07
MT-10	Standards	0.23	0.05
MT-8	Standards	−0.11	0.06
MT-8	Standards	−0.12	0.05
Q22-C-Q1-7	Rim magnetite	0.27	0.07
Q22-C-Q2-6	Rim magnetite	−0.23	0.06
Q22-C-Q3-5	Rim magnetite	−0.06	0.07
Q22-C-Q3-6	Rim magnetite	0.02	0.10
MT-8	Standards	−0.19	0.06
MT-10	Standards	0.21	0.09
MT-8	Standards	−0.14	0.06

The overall distribution of δ^{56}Fe in the studied magnetite ranges from −0.66‰ to 0.78‰. For core magnetite, δ^{56}Fe values range from 0.46‰ to 0.78‰ with an average value of 0.58‰; for mantle magnetite, δ^{56}Fe values range from 0.12‰ to 0.76‰, with an average of 0.56‰; for rim magnetite, δ^{56}Fe values range from −0.66‰ to 0.27‰, with an average of 0.05‰. These results exhibit three characteristics:

1. The δ^{56}Fe values of magnetite in the orbicular diorite tend to decrease from the core to the rim (Figure 12).
2. The Fe isotopic composition of rim magnetite falls within the range of Fe isotopic composition of magnetite from magmatic deposits [12] reported in previous studies, but the rim magnetite is slightly enriched in light Fe isotopes (Figure 13).

3. Compared with magnetite in skarn hydrothermal deposits and magmatic deposits [16,38], the core and mantle magnetite are more enriched in heavier Fe isotopes, with maximum δ^{56}Fe values reaching 0.78‰ (Figure 13).

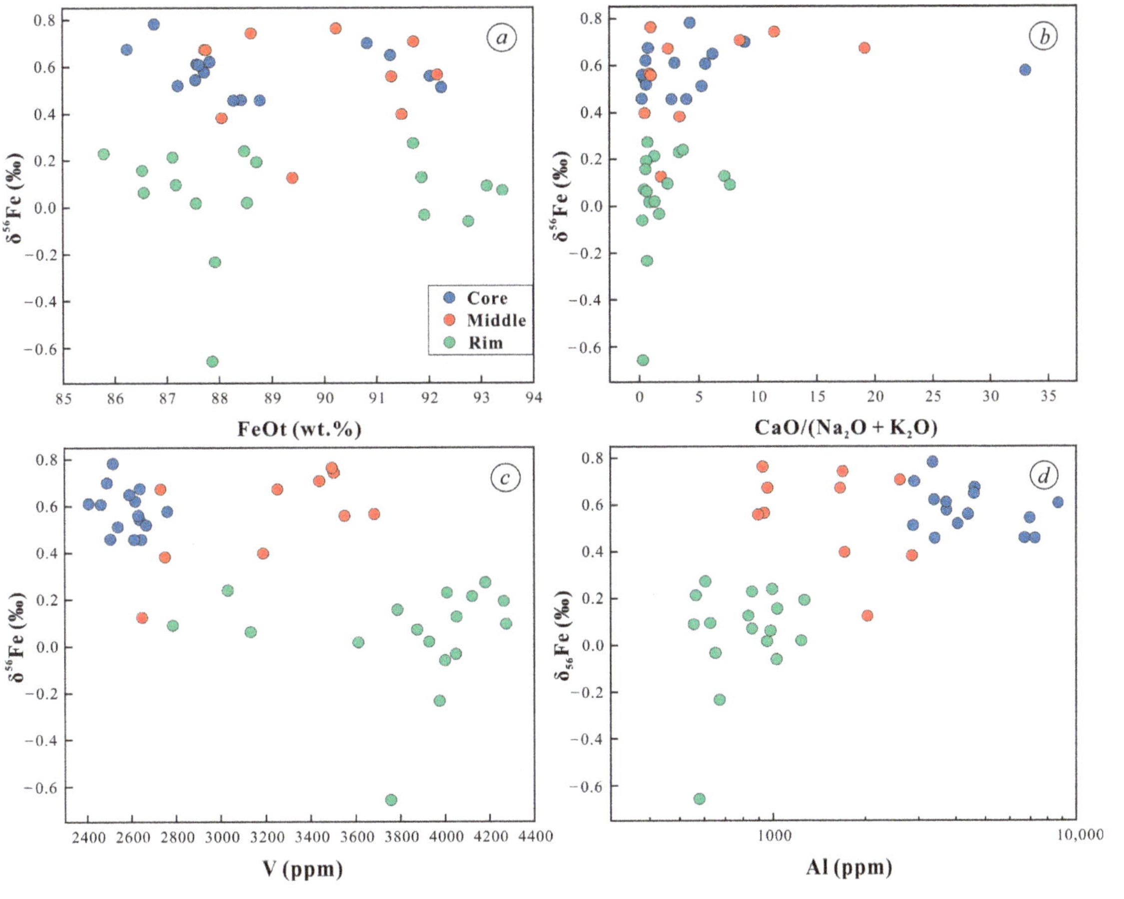

Figure 12. Binary diagrams of magnetite from orbicule in diorite: (**a**) FeO_t vs. δ^{56}Fe, (**b**) CaO/(Na_2O + K_2O) vs. δ^{56}Fe, (**c**) V vs. δ^{56}Fe, (**d**) Al vs. δ^{56}Fe.

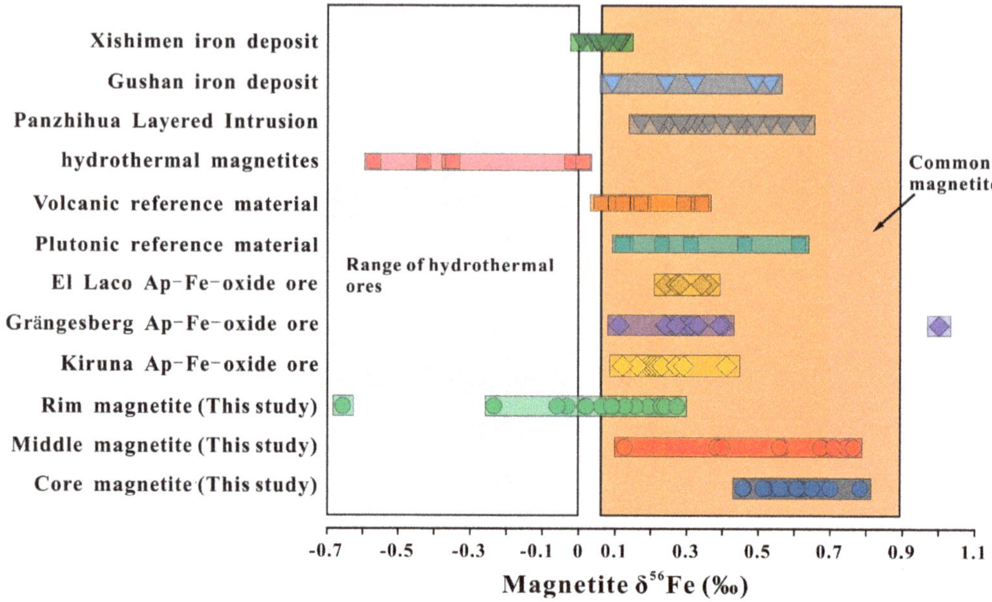

Figure 13. Iron isotope distribution diagrams of magnetite from different parts of Tanling orbicular diorite. Shown is the distribution of iron isotopes (δ^{56}Fe) in magnetite, incorporating data from this study and the available literature [12,14,15,39–44]. Reference fields for common hydrothermal magnetites [12,13,45] and magmatic magnetites [14,15,41,46] are included for comparison (literature iron isotope data are provided in Table 3). The orange area denotes the range of common igneous magnetites, the light grey area denotes the range of hydrothermal ores.

Table 3. Iron isotope analysis for magnetite from reference materials.

Sample	Sample Description	Sample Provenance	δ^{56}Fe in ‰	2σ
Apatite-iron oxide ore				
Kiruna Mining District (KMD), Northern Sweden				
Kiruna-17NY28	Banded massive (Ap-)magnetite ore	Kiirunavaara mine, Kiruna	0.19	±0.03
K-Mt-1 (907/75)	Massive magnetite ore	Kiirunavaara mine, Kiruna	0.2	±0.03
Ki-Mi-2a (1079/251)	Massive magnetite ore	Kiirunavaara mine, Kiruna	0.21	±0.02
Ki-Mi-2b (1079/251)	Massive magnetite ore	Kiirunavaara mine, Kiruna	0.22	±0.04
K-Mt-1079/303	Massive magnetite ore	Kiirunavaara mine, Kiruna	0.27	±0.04
K-Mt-1079/437	Massive magnetite ore	Kiirunavaara mine, Kiruna	0.16	±0.02
M1931	Massive magnetite ore	Kiirunavaara mine, Kiruna	-	-
M1937	Skeletal magnetite ore	Kiirunavaara mine, Kiruna	-	-
LVA-3	Massive magnetite ore	Luossavaara mine, Kiruna	0.23	±0.02
LVA-FW-1	Massive magnetite ore	Luossavaara mine, Kiruna	0.12	±0.04
LVA-FW-2	Massive magnetite ore	Luossavaara mine, Kiruna	0.27	±0.03
K-Mt-3	Massive magnetite ore	Mertainen mine, Kiruna	0.41	±0.03
K-Mt-4	Massive magnetite ore	Mertainen mine, Kiruna	0.29	±0.03
M7557*	Massive magnetite ore	Rektorn mine, Kiruna	-	-
Grängesberg Mining District (GMD), Central Sweden				
DC717-KES090068	Massive (Ap-)magnetite ore	Grängesberg mine, Grängesberg	0.4	±0.03
DC717-KES090070	Massive (Ap-)magnetite ore	Grängesberg mine, Grängesberg	0.24	±0.03
DC717-KES090072	Massive (Ap-)magnetite ore	Grängesberg mine, Grängesberg	0.33	±0.03
DC717-KES090084	Magnetite vein in intermediate volcanic rock	Grängesberg mine, Grängesberg	0.11	±0.03

Table 3. *Cont.*

Sample	Sample Description	Sample Provenance	δ^{56}Fe in ‰	2σ
DC690-KES090012	Massive (Ap-)magnetite ore	Grängesberg mine, Grängesberg	0.31	±0.04
DC690-KES090020	Massive Ap-veined magnetite ore	Grängesberg mine, Grängesberg	0.3	±0.04
DC690-KES090024	Massive (Ap-)magnetite ore	Grängesberg mine, Grängesberg	0.26	±0.03
DC690-KES090027	Ap-veined/banded massive magnetite ore	Grängesberg mine, Grängesberg	0.29	±0.03
DC690-KES090030	Silicate-spotted massive (Ap-)magnetite ore	Grängesberg mine, Grängesberg	0.39	±0.04
DC690-KES090034	Coarse-grained Ap-spotted massive magnetite ore	Grängesberg mine, Grängesberg	0.27	±0.03
DC690-KES090044	Disseminated magnetite in intermediate volcanic rock	Grängesberg mine, Grängesberg	0.24	±0.03
DC575-KES103011	Magnetite-dominated massive ore	Grängesberg mine, Grängesberg	0.31	±0.03
DC575-KES103016	Coarse, massive (Ap-)magnetite ore	Grängesberg mine, Grängesberg	0.27	±0.04
DC575-KES103003	Magnetite-dominated massive ore	Grängesberg mine, Grängesberg	1	±0.03
KES091013b	Massive (Ap-)magnetite ore	Blötberget mine, Blötberget	0.33	±0.03
El Laco Ap-Fe-oxide deposit, Chile				
EJ-LS-11-1	Massive magnetite ore	Laco Sur, El Laco	0.28	±0.03
EJ-LS-11-2	Massive magnetite ore	Laco Sur, El Laco	0.24	±0.03
EJ-LS-11-3	Massive magnetite ore	Laco Sur, El Laco	0.36	±0.03
EJ-LS-11-4	Massive magnetite ore	Laco Sur, El Laco	0.34	±0.03
LS-2	Massive magnetite ore	Laco Sur, El Laco	0.27	±0.03
LS-52	Massive magnetite ore	Laco Sur, El Laco	0.28	±0.03
Plutonic reference material				
Ruoutevare	Ti-magnetite, layered igneous intrusion	Kvikjokk, Norrbotten, Sweden	0.31	±0.03
Ulvön	Ti-magnetite, layered igneous intrusion	Ulvön island, Ångermanland, Sweden	0.13	±0.03
Taberg	Ti-magnetite, layered igneous intrusion	Iron mine, Taberg, Småland, Sweden	0.23	±0.04
EM419	Massive Fe-Ti magnetite ore	Northern pit, Panzhihua, China	0.61	±0.05
EM424	Massive Fe-Ti magnetite ore	Nalahe, Panzhihua, China	0.12	±0.04
Bushveld	Massive magnetite ore	Upper Zone, Bushveld Complex, South Africa	-	-
Gabbrobomb	Magnetite from a gabbro xenolith	NW-Flank, Skjaldbreiður, Iceland	0.46	±0.03
Volcanic reference material				
TEF-NER-18	Magnetite from an ankaramite dyke	NE Rift Zone, Tenerife, Spain	0.07	±0.05
TEF-NER-57B	Magnetite from an ankaramite dyke	NE Rift Zone, Tenerife, Spain	0.16	±0.02
TEF-NER-70	Magnetite from a pyroxene phyric dyke	NE Rift Zone, Tenerife, Spain	0.1	±0.02
MG-07	Igneous magnetite from dacite	S-Flank, Mt. Ruapehu, New Zealand	0.32	±0.03
MG-09	Igneous magnetite from dacite	S-Flank, Mt. Ruapehu, New Zealand	0.29	±0.03
Kelut A1	Igneous magnetite from basaltic andesite	Mt. Kelut, Java, Indonesia	0.1	±0.04
GD-D-2	Igneous magnetite from basaltic andesite	Gede Dome, Java, Indonesia	0.12	±0.03
AK-B1	Igneous magnetite from basaltic andesite	SE-Flank, Anak Krakatau, Indonesia	0.06	±0.03
AK-B3	Igneous magnetite from basaltic andesite	SE-Flank, Anak Krakatau, Indonesia	0.16	±0.03

Table 3. *Cont.*

Sample	Sample Description	Sample Provenance	δ^{56}Fe in ‰	2σ
A-BA-1	Igneous magnetite from basaltic andesite	Mt. Agung, Bali, Indonesia	0.18	±0.05
M-BA06-KA-3	Igneous magnetite from basaltic andesite	Mt. Merapi, Java, Indonesia	0.17	±0.03
83/CRS/6	Igneous magnetite from a dolerite dyke	Agros, Troodos Massif, Cyprus	0.34	±0.03
Low-temperature or hydrothermal magnetites				
KES091007B	Calcite bearing-magnetite ore	Björnberget, Sweden	−0.02	±0.03
DM-1	Iron-skarn magnetite ore	Botenhäll, Dannemora, Sweden	−0.36	±0.03
DM-2	Iron-skarn magnetite ore	Norrnäs 3, Dannemora, Sweden	0.01	±0.03
DM-3	Iron-skarn magnetite ore	Konstäng, Dannemora, Sweden	−0.43	±0.03
DM-4	Iron-skarn magnetite ore	Strömsmalmen, Dannemora, Sweden	−0.35	±0.03
EJ092008	Magnetite from a banded iron formation deposit	Striberg, Bergslagen, Sweden	−0.57	±0.03
Panzhihua Layered Intrusion				
LJ1482	Leuco-gb	the Lanjiahuoshan open pit, Panzhihua, China	0.34	0.05
LJ1487a	Leuco-gb	the Lanjiahuoshan open pit, Panzhihua, China	0.36	0.03
LJ1487	Leuco-gb	the Lanjiahuoshan open pit, Panzhihua, China	0.32	0.01
LJ1434	Ap-gb	the Lanjiahuoshan open pit, Panzhihua, China	0.29	0.03
LJ1438	Ap-gb	the Lanjiahuoshan open pit, Panzhihua, China	0.4	0.04
LJ1453	Ap-gb	the Lanjiahuoshan open pit, Panzhihua, China	0.57	0.01
LJ1457	Ap-gb	the Lanjiahuoshan open pit, Panzhihua, China	0.3	0.03
LJ1458	Ap-gb	the Lanjiahuoshan open pit, Panzhihua, China	0.62	0.02
LJ1461	Mela-gb	the Lanjiahuoshan open pit, Panzhihua, China	0.53	0.04
LJ1462	Mela-gb	the Lanjiahuoshan open pit, Panzhihua, China	0.46	0.04
LJ1463	Mela-gb	the Lanjiahuoshan open pit, Panzhihua, China	0.49	0.03
LJ1430a	Mela-gb	the Lanjiahuoshan open pit, Panzhihua, China	0.23	0.03
LJ1430	Mela-gb	the Lanjiahuoshan open pit, Panzhihua, China	0.25	0.04
LJ1427	Mela-gb	the Lanjiahuoshan open pit, Panzhihua, China	0.28	0.01
LJ1426	Oxide ore	the Lanjiahuoshan open pit, Panzhihua, China	0.22	0.01
LJ1424	Oxide ore	the Lanjiahuoshan open pit, Panzhihua, China	0.22	0.02
LJ1420	Oxide ore	the Lanjiahuoshan open pit, Panzhihua, China	0.17	0.05
LJ1410	Micro-gb	the Lanjiahuoshan open pit, Panzhihua, China	0.28	0.05
LJ1407	Micro-gb	the Lanjiahuoshan open pit, Panzhihua, China	0.44	0.03
Gushan iron deposit, China				
10GS_4_Hm	Massive ore(vesicular structure)	Gushan iron deposit, China	0.24	
10GS_5_Hm	Massive ore (vesicular structure)	Gushan iron deposit, China	0.32	
10GS_6_Hm	Massive ore (vesicular structure)	Gushan iron deposit, China	0.53	

Table 3. Cont.

Sample	Sample Description	Sample Provenance	δ^{56}Fe in ‰	2σ
10GS_17_Hm	Massive ore (vesicular structure)	Gushan iron deposit, China	0.24	
10GS_19_Hm	Massive ore (vesicular structure)	Gushan iron deposit, China	0.09	
10GS_20_Hm	Massive ore (vesicular structure)	Gushan iron deposit, China	0.49	
Xishimen iron deposit				
XSM2-130503-01	Massive magnetite ore	Xishimen iron deposit, China	0.008	0.04
XSM3-130504-06	Massive magnetite ore	Xishimen iron deposit, China	0.034	0.04
XSM3-130504-09	Massive magnetite ore	Xishimen iron deposit, China	0.115	0.04
XSM3-130504-14	Massive magnetite ore	Xishimen iron deposit, China	0.047	0.04
XSM3-130504-16	Massive magnetite ore	Xishimen iron deposit, China	0.054	0.04
XSM3-130504-18	Massive magnetite ore	Xishimen iron deposit, China	0.111	0.04
XSM3-130504-33	Massive magnetite ore	Xishimen iron deposit, China	0.113	0.04
XSM4-130506-22	Massive magnetite ore	Xishimen iron deposit, China	0.105	0.052
XSM4-130506-23	Massive magnetite ore	Xishimen iron deposit, China	0.079	0.052
XSM4-130506-26	Massive magnetite ore	Xishimen iron deposit, China	0.06	0.052
XSM4-130506-29	Massive magnetite ore	Xishimen iron deposit, China	0.049	0.04
XSM4-130506-34	Massive magnetite ore	Xishimen iron deposit, China	0.067	0.052

6. Discussion

6.1. Controls on Compositions of Magnetite

Factors controlling the trace element contents in magmatic magnetite include silicate melt composition, magnetite crystallography, coexisting minerals, oxygen fugacity, temperature, cooling rate, sulfur fugacity, and silica activity [29,47–50]. Magmatic magnetite crystallizes from evolved silicate melts, sulfide melts, or Fe-Ti-(P)-rich melts and thus contains various trace elements. Coexisting minerals, magma composition, and oxygen fugacity are the dominant controlling factors. Next, we will discuss these major factors that may influence the trace element distribution of magnetite in the Tanling orbicular diorite.

6.1.1. Magma Composition and Coexisting Minerals

The minerals associated with magnetite can reflect the crystallization environment and conditions of the magnetite. In the Tanling orbicular diorite, rim magnetite has a small particle size (<200 μm), and its associated minerals are mostly plagioclase and hornblende, indicating that it crystallized from normal diorite magma. Core magnetite has a large particle size (about 1000 μm) and is associated with apatite and actinolite. This mineral association is similar to that in Chile ore magma flow [36,51], and the P content of magnetite in the core and apatite is 1–2 orders of magnitude lower than that in the mantle and rim (Figure 9e). These observations suggest that core magnetite may have crystallized from an evolved iron-rich melt [36,51–53].

The trace elements of magnetite may carry important information about magmatic composition [8–15,30,35,47,49,54–56]. Magnetites in the cores of orbicules have higher Ti, Al, Ni, Cr, Sc, Zn, Co, Ga, and Nb than those in the mantle and rim (Figures 8 and 9), indicating that the core magnetite is significantly different from rim magnetite and is not directly crystallized from normal diorite magma. The Cr content of the core magnetite shows a multi-peaked distribution (Figure 8b), which is consistent with the data reported by Wang and Zhou [57]. The highly compatible element Cr (D_{Cr} magnetite/melt = 100–600); [58] is inconsistent with the assumption of fractional crystallization of magnetite in diorite magma, where its content is expected to decline sharply and continuously. In addition, Detao He et al. noted that Fe-rich melts formed under high oxygen fugacity may have relatively high Al_2O_3 contents [59,60]. The Al content of magnetites in the cores of orbicules is an order of magnitude higher than that of the rim and mantle magnetites (Figure 8c,d), suggesting that the magnetites in the cores of orbicules may crystallize from iron-rich melts with high oxygen fugacity. The enriched characteristics of compatible elements in the magnetites in the cores of orbicules of the Tanling orbicular diorite in Wuan, Hebei Province (Figure 10) indicate that they crystallize from a homogeneous iron-rich melt, which is consistent with the latest model proposed by Liu et al. [61] based on evidence from multiphase solid inclusions.

To gain a more concise and intuitive understanding of the genesis of magnetite, many researchers have addressed this need by establishing genetic discrimination diagrams for magnetite. Dupuis and Beaudoin [35] first proposed the identification diagram of magnetite origins. Based on the compositions of characteristic elements of magnetite from different deposits, they constructed the (Al + Mn) vs. (Ti + V) diagram to identify different types of magnetite. While collecting data on magnetite in our study, we found a new trend in the (Al + Mn) vs. (Ti + V) diagram (Figure 11) of magmatic origin towards iron-rich magma origin, and the changes from rim to core of magnetite in orbicular diorite are consistent with this trend. Therefore, we propose that the (Al + Mn) vs. (Ti + V) chart can be used to distinguish among iron-magma-type, magmatic-type, and skarn-type magnetite. Additional data will be required to verify the broad applicability of this chart.

6.1.2. Oxygen Fugacity

Vanadium in magnetite can be used to trace melt-fluid evolution and identify magma evolution processes [62]. Published papers [27,30–32] have shown that only V^{3+} is easily incorporated into the lattice of magnetite, and the content of V^{3+} in magnetite is negatively correlated with fO_2. In the Wuan orbicular diorite in Hebei Province, the V content varies as follows: it decreases from the rim magnetite (average: 653 ppm) to the mantle magnetite (average: 0.30 ppm), then slightly increases in the core magnetite (average: 2.98 ppm) (Figure 8a,b, Supplementary Table S1). This variation indicates that the core magnetite formed under high fO_2 conditions. This is consistent with the phenomenon of the dissolution of ilmenite lamellae in the large-grained magnetite of the core. In addition, tremolite-actinolite aggregates occur in the core center. Yang Et Al. [63] calculated the $Mg/(Mg + Fe^{2+})$ ratio using electron probe microanalysis (EPMA) data of calcium amphibole, and the results were concentrated in the range of 0.88~0.9, further confirming the relatively high oxygen fugacity of the core.

The above observations show that the oxygen fugacity increases gradually from the rim of the orbicule to the core, suggesting that the core part must involve the participation of oxidizing agents. However, what causes the oxygen fugacity of the core to be higher than that of the rim? According to previous studies [64], when the gas phase consists of CO_2-CO, H_2O-H_2, S_2-H_2S-SO_2, the following reaction occurs when CO_2 is injected into the silicate melt-gas system containing FeO:

$$CO_2 + 2FeO = CO + Fe_2O_3$$

This reaction affects the distribution of Fe^{3+}/Fe^{2+} in the melt and modifies the oxygen fugacity. Meanwhile, researchers have also found that the total amount of CO_2 involved was sufficient to significantly increase oxygen fugacity by several log units [64,65]. They proposed that the interaction between basic magma and carbonate rocks forms calcium silicate rocks and releases large amounts of CO_2, and that the interaction between the basic magma and surrounding carbonate rocks triggers mineralization. The CO_2-rich fluid released by rock decarbonization of carbonate rocks oxidizes the invading magma and promotes the crystallization of iron and titanium oxides, thereby leading to the formation of ore deposits [66].

Therefore, the crystallization of large-grained magnetite in the core of the orbicular diorite occurs under high fO_2 conditions, followed by that of magnetite in the mantle, whereas magnetite at the rim of the orbicule forms under relatively low fO_2 conditions.

Therefore, we propose that the accumulation of large-grained magnetite in the core is caused by the addition of CO_2, which increases the oxygen fugacity of the magma. This indicates that the oxygen fugacity of the magma may play an important role in controlling the crystallization of magnetite in the orbicular diorite.

6.2. Reasons for High $\delta^{56}Fe$ in Magnetites in the Cores of Orbicules

The δ^{56}Fe values of magnetite in the orbicular diorite are mostly >0‰, exhibiting igneous rock characteristics. The δ^{56}Fe of the core and mantle magnetite are much higher than those of other related magmatic deposits except the marginal magnetite (Figure 13). The high δ^{56}Fe of magnetites in the cores of orbicules may be the result of crystallization of light ^{54}Fe silicate minerals [15,16,38], high oxygen fugacity [67,68], magma differentiation/immiscibility [16,39,40,46,69], exsolution of ilmenite [26,41,70,71]. These issues are discussed below in the context of Tanling orbicular diorite magnetite.

6.2.1. Crystallization of Light δ^{56}Fe Silicate Minerals

The fractional crystallization of minerals in magma can lead to changes in the isotopic composition of iron. Minerals enriched in light δ^{56}Fe values—such as olivine and pyroxene—crystallize, a process that leads to higher δ^{56}Fe values in the residual magma [15,38,72]. In contrast, magnetite tends to enrich in heavy iron isotopes during its crystallization [16,46]. For magnetite in different parts of the Tanling orbicular diorite, there is little variation in the Fe_{total}, and no correlation is observed between Fe_{total} and δ^{56}Fe (Figure 12a). This indicates that the fractional crystallization of minerals enriched in the light ^{54}Fe isotope is not the main mechanism driving Fe isotopic fractionation in magnetite from the Tanling orbicular diorite.

6.2.2. Hydrothermal Alteration

Hydrothermal alteration is another mechanism for iron isotope fractionation [40,41,70]. The δ^{56}Fe values of iron ores associated with hydrothermal fluids show significant variations, with a maximum range of 5‰ [73–77]. For the orbicular diorite, no strong correlation was observed between δ^{56}Fe and the $CaO/(Na_2O + K_2O)$ ratio, and the δ^{56}Fe values of magnetites in the orbicule cores are more concentrated (Figures 12c and 13). It can be concluded that the high δ^{56}Fe values of core magnetites in the orbicules are unlikely to result from hydrothermal alteration.

6.2.3. High Oxygen Fugacity

As mentioned above, the oxygen fugacity (fO_2) of magnetite in different parts of the orbicular diorite varies greatly during its formation. The crystallization of large-grained magnetite in the core of orbicular diorite occurs under high fO_2 conditions, followed by that of magnetite in the mantle and then the rim. Since the partitioning of vanadium (V) in magnetite depends on fO_2 [31], the V content of magnetite decreases from the rim to the core, while its δ^{56}Fe value gradually increases (Figure 12c), which suggests that the higher δ^{56}Fe values of core magnetites (relative to the low values of the rim magnetites) may result from crystallization under high oxygen fugacity. Therefore, fO_2 may play an important role in controlling the iron isotopic system of the orbicular diorite and the crystallization of magnetite.

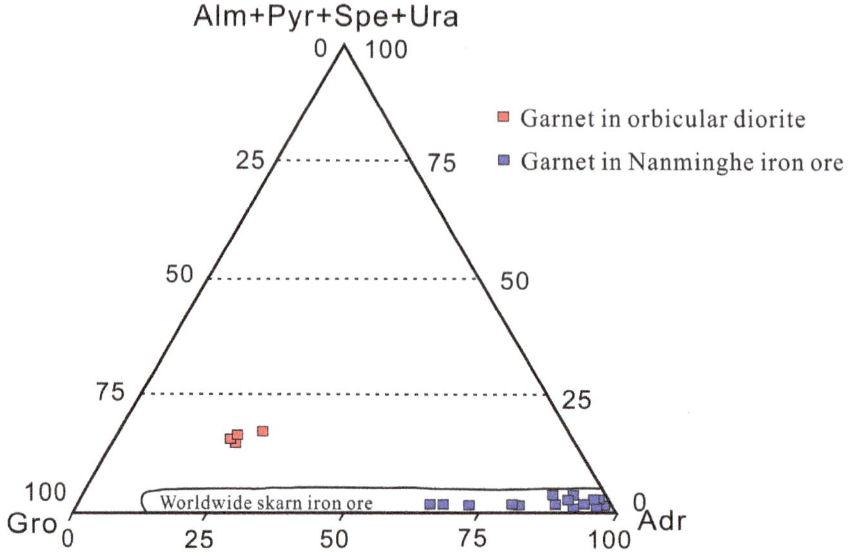

Figure 14. Triangular diagram of the end-member content of garnet in orbicular diorite (Nanminghe data from [78]). Garnet abbreviations: Gro—grossular, And—andradite, Alm—almandine.

6.2.4. Magma Differentiation/Immiscibility

Yue Wang et al. [12,40] analyzed extensive Fe isotope data and showed that iron isotopic fractionation occurs not only during hydrothermal processes, but also during magma differentiation and liquid immiscibility. Magma immiscibility results in the enrichment of heavy Fe isotopes in the immiscible Fe-rich melt fraction and light Fe isotopes in the Si-rich melt fraction. Yonghua Cao et al. [39] proved via their Fe isotope data that magnetite in the Panzhihua deposit was crystallized from iron-rich melts formed by liquid immiscibility from parent magma under high oxygen fugacity conditions. When the melt evolves toward a Fe-rich trend, the residual melt becomes enriched in light Fe isotopes [69], and the more Fe-rich melt had higher Fe^{3+}, more heavy iron enrichment, and higher fO_2 [64]. As noted earlier, Fe-rich melts formed at high fO_2 conditions may have relatively high Al_2O_3 contents [59,60]. Consistent with this, the Al content of core magnetite (2163–8732 ppm) is one order of magnitude higher than that in the rim (556–1266 ppm) and mantle (652–2859 ppm) magnetites (Figure 8c,d). There is a significant positive correlation between $\delta^{56}Fe$ and Al content of magnetite (Figures 12d and 13). The $\delta^{56}Fe$ values of different types of magnetite commonly found in the world are summarized. The high $\delta^{56}Fe$ values of core magnetite in the Tanling orbicular diorite in Wuan, Hebei Province, are similar to those in the Panzhihua Pluton. This evidence shows that the core magnetites were already enriched in heavy Fe isotopes when they formed, which is called Fe-rich magma. This Fe enrichment may be driven by magmatic degassing (a common process in many volcanic systems) or more likely by Fe oxide-rich melts formed via Fe precipitation [79,80].

6.2.5. Exsolution Lamellae of Ilmenite

Ilmenite generally occurs within the magnetite in the core of orbicules (Figure 4c), and the $\delta^{56}Fe$ values in ilmenite are indeed very low [38]. The magnetites in the orbicule cores have ilmenite exsolution lamellae, while mantle magnetites do not have such lamellae, but there is no significant difference in $\delta^{56}Fe$ values between the two (average $\delta^{56}Fe$ of the magnetites in the core magnetites: 0.58‰; the average $\delta^{56}Fe$ of mantle magnetites: 0.56‰).

Although the exsolution of ilmenite (enriched in light ^{54}Fe) would cause an increase of $\delta^{56}Fe$ values in the host magnetite, instead, the two remain in iron isotope equilibrium. There is no significant difference in $\delta^{56}Fe$ values between core and mantle magnetites, indicating that ilmenite exsolution has little effect on the high $\delta^{56}Fe$ values of the magnetites.

6.3. Origin of Skarn Iron Ore

6.3.1. Melting of Carbonate Xenoliths by Magma

The petrographic observations show that the minerals in the core of the orbicule are obviously different from those in the rim. The main minerals in the rim include plagioclase, hornblende, quartz, apatite, and magnetite. The core minerals are mainly magnetite, apatite, tremolite-actinolite, epidote, and garnet (mainly grossular). The minerals in the core are more enriched in Ca than those in the rim (Figure 5). A large amount of skarn minerals and skarn alteration features were discovered in the WuanIron Mine, as well as typical skarn minerals garnet and epidote developed in the mantle and edge of the orbicules. This leads us to believe that the formation of minerals in the orbicules has a type similar to that of skarn deposits. However, magnetite within orbicules differs from typical skarn-type magnetite in that it contains internal vesicles. The presence of vesicles in magnetite indicates that the magnetite particles formed under conditions of rapid magma crystallization, and the gas was unable to escape, becoming trapped in their lattice. Previous studies have shown that pores in magnetite are important evidence for iron-rich magma mineralization [81], and the development of vesicles within or between minerals is consistent with the characteristics of viscous melt containing a large amount of volatile matter (i.e., ore-bearing melt-fluid flows [44,82]). Therefore, the formation process of magnetite in orbicular diorite differs from that of skarn-type magnetite.

The composition and its variation characteristics of garnet can usually reflect the changes in physicochemical conditions such as temperature, pressure, oxygen fugacity, and pH during mineralization [83–85]. Calculations of the elemental composition of garnet (Table 4) show that it consists mainly of grossular (Gro), andradite (And), and almandine (Alm) (with an average elemental formula is $Alm_{16.04}And_{23.68}Gro_{59.88}Pyr_{0.10}Spe_{0.50}Ura_{0.06}$). The garnet in the orbicular diorite is mainly calcium-aluminum garnet (Figure 14), which differs from the garnet in typical iron skarns and that in the Nanminghe iron ore of this area. This indicates that this garnet does not have a hydrothermal origin typical of traditional skarns, and when combined with the previous analysis, it is inferred that this garnet may have formed during the evolution stage of iron-rich melts. The garnet associated with magnetite is mainly calcium-aluminum garnet, indicating that the magnetites in the cores of orbicules formed in advance, consumed Fe^{3+}, inhibited the formation of calcium-iron garnet, and made it become calcium-aluminum garnet. The formation of this calcium-aluminum-rich garnet may be related to the formation of carbonate rocks [86]. At relatively low temperatures, decarbonation reactions occur in the system to form iron-rich garnet ($Alm_{75}Prp_{17}Grs_8$), magnesiowüstite ($Mg^{\#} \leq 0.13$), and CO_2-rich fluids. However, Mg-bearing wüstite is unstable under high-pressure CO_2-rich fluids and in the presence of a carbonate-silicate melt: It is either completely oxidized or is dissolved in the melt or fluid phase, leading to the formation of carbonate-silicate melts enriched in Fe^{2+} and Fe^{3+}. These melts thus act as potential metasomatic agents in the mantle [87].

Table 4. Major composition of Grt (wt.%).

Sample Name	SiO_2	TiO_2	Al_2O_3	Cr_2O_3	P_2O_5	V_2O_3	FeO	MnO	MgO	CaO	NiO	Na_2O	K_2O	Total
Grt1	40.20	n.d.	26.02	n.d.	n.d.	n.d.	10.17	0.22	0.02	23.28	0.02	0.06	0.02	100
Grt2	40.01	n.d.	23.88	n.d.	n.d.	n.d.	12.94	0.06	0.02	23.04	0.01	0.04	n.d.	100
Grt3	40.14	0.04	25.54	0.01	n.d.	n.d.	11.02	0.02	n.d.	23.16	n.d.	0.03	0.03	99.99
Grt4	39.59	0.02	26.20	0.05	n.d.	n.d.	10.68	0.42	0.02	23.03	0.01	n.d.	n.d.	99.95

Based on the field distribution characteristics of the strata, orbicular diorite is found to occur at the contact area between diorite intrusions and limestone, and carbonate xenoliths are present in the adjacent rock masses (Figure 2b). Detao He [59] recorded the recycling of sedimentary carbonates to the deep mantle by studying the abundant carbonate xenoliths in the Dali Lake basalts. Qiu Jiaxiang (1985) showed that intermediate-acidic magmas have a strong ability to assimilate surrounding rocks, while assimilative mixing with carbonates is the most likely to occur, resulting in an increase in the contents of Ca, Mg, and Fe, while a decrease in the contents of Si, Na, and K in the magma [88]. According to Yang's [63] whole-rock principal weight data and TIMA element distribution map (Figure 6), orbicular calcium content increased significantly, while silicon, sodium, and potassium decreased significantly. It is hypothesized that the increase in calcium content in the core is attributed to the assimilation of the diorite intrusion and mixing of some limestone fragments, and this process alters the magma composition.

By studying the isotopic composition of rock masses and ores in this area, numerous researchers have found that: ① the εNd values in this area show characteristics of mixing between Archaean TTGs (εNd = −30) and Mesozoic intrusions (εNd = −8~−15) [89]; ② the Sr and Nd isotopic compositions are roughly negatively correlated, which also implies significant mixing of lower crustal components [90]; ③ there is little difference in Pb isotopic composition between the Pb isotopic composition of ores and that of the diorite pluton in this area, and these isotopic compositions exhibit significant characteristics of mixed Pb—with Pb mainly derived from the crust beneath the lead source area, and a small amount of mantle-derived Pb incorporated [91]. The magnetite in the cores of orbicules in the orbicular diorite of this area exhibits a strong positive Pb anomaly (Figure 10a), indicating that its formation is driven by the addition of crust-derived components. This might indicate that the mineralization process or iron enrichment occurred near the carbonate formations.

Therefore, both petrographic and geochemical characteristics show that there is evidence of carbonate xenoliths mixing in the formation of orbicular diorite. During the upwelling process of magma, the carbonate formations were softened. Carbonate xenoliths fell into the magma and were melted into orbicules, eventually forming orbicular diorite. A possible reaction is:

$$\text{Diorite magma} + \text{carbonate} \rightarrow \text{iron-rich magma (Magnetite + actinolite + apatite + garnet + epidote + feldspar)} + \text{hornblende} + CO_2\uparrow$$

6.3.2. Formation Mechanism of Fe-Rich Melt

The ore magma concept was first proposed by J.E. Spurr in 1923. R. Fischer conducted experiments on the immiscibility of high-iron oxides, silicates, and phosphates, while A. R. Philpotts carried out experiments on the immiscibility of apatite-rich liquids, magnetite-apatite melts, and silicate melts [92–94]. These achievements enabled the iron ore magma concept to be firmly established and promoted its solid development in mineral deposit science. At the same time, the Gushan iron deposit, the Meishan main orebody, and the Hanxing area are also considered as ore magma type iron ore deposits, rather than skarn type iron ore deposits in the traditional sense [24,44].

Orbicular diorite occurs within diorite around the Wu'an iron ore. This special structure may represent a small processing plant and reflect the mineralization process of the Wu'an Iron Mine. Orbicular diorite has the pseudomorph of carbonate xenoliths, reflecting the process of its interaction with magma. A part of the carbonate xenoliths reacted completely and not recorded, while another part demonstrated the process of their mutual reaction, namely the presence of iron-rich melts. As for large-grained magnetite, it is mainly enriched in the center of the orbicules (Figure 5). We believe that the magma melted the carbonate xenoliths. During this process, an iron-calcium exchange occurred. When the reaction was complete, the center of the orbicules was the original location where the carbonate xenoliths existed, and also the center where iron was enriched. The mantle and rim of the orbicules are the contact points between the carbonate xenoliths and the magma, eventually forming orbicular diorite. Actinolite is one of the important minerals in magmatic ore deposits. Many previous studies have often directly assumed that actinolite in ores is a product of hydrothermal alteration, similar to actinolite formed during the wet skarn stage in skarn deposits, and thus inferred that this actinolite originated from diopside alteration, leading to the conclusion that these deposits are of hydrothermal origin. However, the metasomatism of diopside by actinolite has not been observed within the mining areas of some magmatic ore deposits in Chile; instead, idiomorphic actinolite phenocrysts or aggregates are present in fresh volcanic lavas and clastic rocks [95]. Similarly, in the Tanling orbicular diorite of Wuan, Hebei Province, actinolite often occurs in columnar or fibrous forms and is associated with magnetite and apatite. The evidence suggests that at least some actinolite, especially the part of the massive ore associated with magnetite and apatite, is likely to have crystallized directly from magma.

Regarding the exsolution mechanism of Fe-rich melt, previous studies suggested that the addition of phosphorus in the orbicule's core caused the exsolution of Fe-rich melts when the magma evolved to the diorite stage. Philpotts' experiments on the diorite-magnetite-apatite system demonstrated that iron can be relatively concentrated in a liquid phase to form iron-rich magma. The $FeO\text{-}Ca_5(PO_4)_3F\text{-}NaAlSiO_4\text{-}CaMgSiO_6$ experiment conducted by Su Lianghe and the $FeO\text{-}CaMgSi_2O_6\text{-}KMg_3(AlSi_3O_{10})F_2$ experiment conducted by Yu Xuehui both proved that Fe-rich melt can be formed by separation with the participation of volatile components [96–98]. During the intrusion process of magma, it reacts with carbonate rocks to generate components such as CO_2. During this process, physical and chemical conditions (e.g., oxygen fugacity) change, resulting in immiscibility between iron-rich melts and the magma. After multiple enrichments, iron-rich magma is formed.

The results of this study show that core magnetite in the Wuan orbicular diorite (Hebei Province, China) crystallized from iron-rich melts. Variations in magnetite composition indicate that the genesis of the Hanxing-type iron deposits is the result of a series of reactions between magma and carbonate rocks. Therefore, the following conceptual model is proposed to explain this step-by-step process (Figure 15):

① Iron primarily exists as Fe^{2+} in normal silicate magma. During magma intrusion, the hot magma reacts with carbonate xenoliths [(Ca, Mg) CO_3], generating large amounts of CO_2, which greatly increases the oxygen fugacity of the magmatic system.

② Under the action of volatile components (e.g., CO_2), liquid immiscibility occurs between the iron-rich magma and the silicate melt, iron-rich oxide melts are formed via melting and subsequent enrichment, and iron-rich magma is finally formed after multiple rounds of enrichment [6,94,99].

③ Iron oxides (Fe_3O_4/Fe_2O_3) will crystallize close to the liquidus due to high oxygen fugacity [64,99], accompanied by the formation of numerous iron-poor silicate minerals (e.g., tremolite and actinolite).

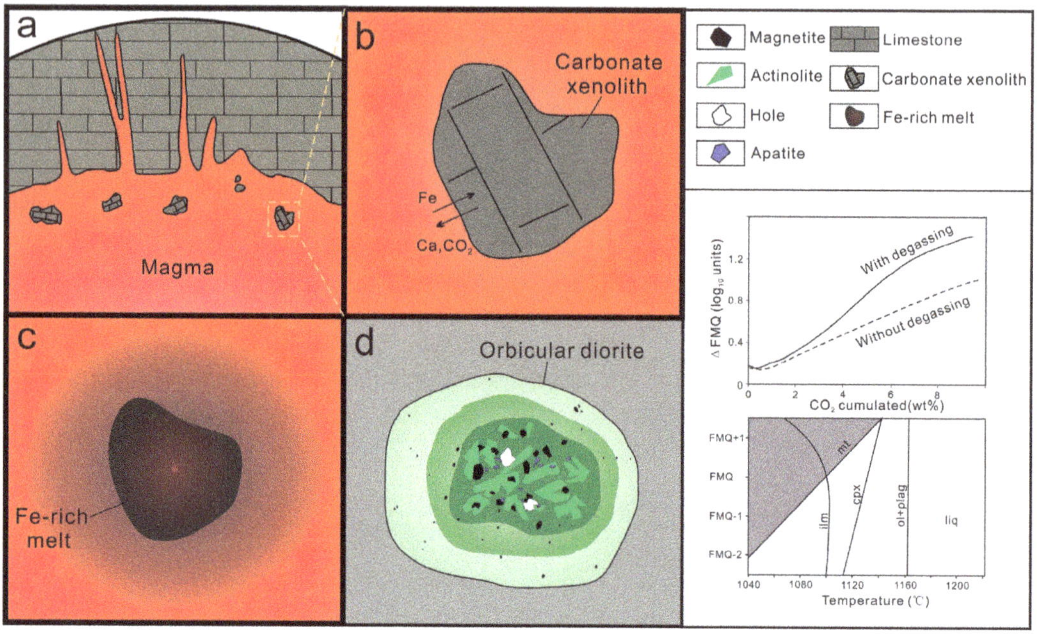

Figure 15. Metallogenic model of orbicular diorite in Wuan (phase diagram from [64]). During (**a**) upward intrusion of magma, (**b**) hot magma reacts with the carbonate xenoliths to gen-erate abundant CO_2, which greatly increases the oxygen fugacity of the magmatic system; (**c**) liquid immiscibility occurs in the magma chamber, and the Fe-rich oxide melt is formed by melting away. Iron oxides will crystallize close to the liquidus due to high oxygen fugacity; Finally, (**d**) it cools and crystallizes to form orbicular diorite.

7. Conclusions

Comprehensive analysis of petrographic and geochemical characteristics of magnetite in the Tanling orbicular diorite yields the following conclusions:

1. Magnetite composition varies significantly across the orbicule: rim magnetite has a small particle size (about 200 μm) and is associated mainly with plagioclase and hornblende, indicating that it crystallizes from normal diorite magma. The magnetite in the core has a large particle size (>1000 μm), and is associated with apatite and actinolite, with apatite inclusions and a large number of vesicles in the core. The magnetites in the cores of orbicules have higher Ti, Al, Ni, Cr, Sc, Zn, Co, Ga, and Nb than those in the rim. The multi-peak distribution of Cr and the (Al + Mn) vs. (Ti + V) diagram of the magnetites in the cores of orbicules indicate that they crystallized from iron-rich melts.
2. The $\delta^{56}Fe$ value of the core magnetite (0.46‰–0.78‰) is much higher than that of the mantle and rim magnetite in orbicules. The $\delta^{56}Fe$ values of magnetite in different parts have no correlation with Fetotal and $CaO/(Na_2O + K_2O)$, but they increase as the V content of magnetite gradually decreases. This large fractionation of iron isotopes is not caused by mineral fractionation, crystallization, or fluid alteration,

but may be driven by liquid immiscibility that forms iron-rich melts under high oxygen fugacity.

3. The iron enrichment mechanism of the Wuan orbicular diorite is as follows: During upward intrusion of magma, hot magma reacts with the carbonate xenoliths [(Ca, Mg) CO_3] to generate abundant CO_2, which greatly increases the oxygen fugacity of the magmatic system; Under the action of CO_2 and other volatile components, liquid immiscibility occurs in the magma chamber, and the Fe-rich oxide melt is formed by melting away. Iron oxides (Fe_3O_4/Fe_2O_3) will crystallize close to the liquidus due to high oxygen fugacity.
4. The aforementioned characteristics of magnetite in the Tanling orbicular diorite (Wuan, Hebei Province) indicate that diorite magma reacts with carbonate xenoliths to form "Fe-rich melts"; furthermore, skarn iron ore deposits are probably formed by the reaction of intermediate-to-basic magma with carbonate rocks to generate such "Fe-rich melts". A possible reaction is as follows:

$$\text{Diorite magma + carbonate} \rightarrow \text{Fe-rich melt (Magnetite + actinolite + apatite) + garnet + epidote} \\ \text{+ feldspar + hornblende + } CO_2\uparrow$$

Supplementary Materials: The following supporting information can be downloaded at: https://www.mdpi.com/article/10.3390/min15101061/s1. Table S1. Trace elemental compositions of magnetite (ppm).

Author Contributions: Conceptualization, R.L. and S.S.; Methodology, S.S.; Software, P.W.; Validation, R.L., S.S. and P.W.; Formal analysis, R.L.; Investigation, R.L.; Resources, S.S.; Data curation, P.W.; Writing—original draft preparation, R.L.; Writing—review and editing, R.L.; Visualization, P.W.; Supervision, R.L.; Project administration, S.S.; Funding acquisition, S.S. All authors have read and agreed to the published version of the manuscript.

Funding: This research was funded by the Strategic Key Metals Supernormal Enrichment Kinetics (92162213) Project of the Major Research Program.

Data Availability Statement: Data are contained within the article and Supplementary Materials.

Conflicts of Interest: The authors declare no conflicts of interest.

References

1. Edwards, R.; Atkinson, K. *Ore Deposit Geology and Its Influence on Mineral Exploration*; Springer: Dordrecht, The Netherlands, 1986; pp. 143–174.
2. Hall, D.L.; Cohen, L.H.; Schiffman, P. Hydrothermal alteration associated with the iron hat iron skarn deposit, eastern Mojave Desert, San Bernardino County, California. *Econ. Geol.* **1988**, *83*, 568–587. [CrossRef]
3. Zheng, J.M.; Mao, J.W.; Chen, M.H.; Li, G.D.; Ban, C.Y. Geological characteristics and metallogenic model of skarn iron deposits in the Handan- Xingtai area, southern Hebei, China. *Geol. Bull. China* **2007**, *26*, 150–154. (In Chinese)
4. Shi, Z.L.; Jin, Z.M.; Xiong, P.F.; Wang, D.Y.; Huang, K.K. A Preliminary Discussion on Problems of ore-magma mineralization of "Daye-Type" iron deposits at Tieshan, Hubei. *Earth Sci.* **1981**, *2*, 145–154. (In Chinese)
5. Liu, P.P.; Zhou, M.F.; Chen, W.T.; Gao, J.F.; Huang, X.W. In-situ LA-ICP-MS trace elemental analyses of magnetite: Fe-Ti-(V) oxide-bearing mafic-ultramafic layered intrusions of the Emeishan Large Igneous Province, SW China. *Ore Geol. Rev.* **2015**, *65*, 853–871. [CrossRef]
6. Li, Y.H.; Xie, G.Q.; Duan, C.; Han, D.; Wang, C.Y. Effect of Sulfate Evaporate Salt Layer over the Formation of Skarn-Type Iron Ore. *Acta Geol. Sin.* **2013**, *87*, 1324–1334. (In Chinese)
7. Wang, W.; Su, S.; Santosh, M.; Lu, X.; Hou, X.; Li, X. Formation mechanism of skarn iron deposit: Evidence from Fe-rich enclave in porphyritic monzonite. *Ore Geol. Rev.* **2024**, *165*, 105902. [CrossRef]
8. Maia, M.; Barrulas, P.; Nogueira, P.; Mirao, J.; Noronha, F. In-situ LA-ICP-MS trace element analysis of magnetite as a vector towards mineral exploration: A comparative case study of Fe-skarn deposits from SW Iberia (Ossa-Morena Zone). *J. Geochem. Explor.* **2022**, *234*, 106941. [CrossRef]

9. Chen, W.; Zhou, M.; Gao, J.; Hu, R. Geochemistry of magnetite from Proterozoic Fe-Cu deposits in the Kangdian metallogenic province, SW China. *Miner. Depos.* **2015**, *50*, 795–809.
10. Huang, X.W.; Gao, J.F.; Qi, L.; Zhou, M.F. In-situ LA-ICP-MS trace elemental analyses of magnetite and Re-Os dating of pyrite: The Tianhu hydrothermally remobilized sedimentary Fe deposit, NW China. *Ore Geol. Rev.* **2015**, *65*, 900–916. [CrossRef]
11. Stergiou, C.L.; Melfos, V.; Papadopoulou, L.; Ladas, A.D.; Aidona, E. Volcanic Rocks from Western Limnos Island, Greece: Petrography, Magnetite Geochemistry, and Magnetic Susceptibility Constraints. *Minerals* **2025**, *15*, 673. [CrossRef]
12. Wang, Y.; Zhu, X.K.; Mao, J.W.; Li, Z.H.; Cheng, Y.B. Iron isotope fractionation during skarn-type metallogeny: A case study of Xinqiao Cu-S-Fe-Au deposit in the Middle-Lower Yangtze valley. *Ore Geol. Rev.* **2011**, *43*, 194–202. [CrossRef]
13. Anbar, A.D. Iron stable isotopes: Beyond biosignatures. *Earth Planet. Sci. Lett.* **2004**, *217*, 223–236. [CrossRef]
14. Dziony, W.; Horn, I.; Lattard, D. In-situ Fe isotope ratio determination in Fe–Ti oxides and sulfides from drilled gabbros and basalt from the IODP Hole 1256D in the eastern equatorial Pacific. *Chem. Geol.* **2014**, *363*, 101–113.
15. Bilenker, L.D.; VanTongeren, J.A.; Lundstrom, C.C.; Simon, A.C. Iron isotopic evolution during fractional crystallization of the uppermost Bushveld Complex layered mafic intrusion. *Geochem. Geophys. Geosyst.* **2017**, *18*, 956–972.
16. Liu, P.P.; Zhou, M.F.; Luais, B.; Cividini, D.; Rollion-Bard, C. Disequilibrium iron isotopic fractionation during the high-temperature magmatic differentiation of the Baima Fe–Ti oxide-bearing mafic intrusion, SW China. *Earth Planet. Sci. Lett.* **2014**, *399*, 21–29.
17. Luo, Z.H.; Deng, J.F.; Zhao, G.C.; Cao, Y.Q. Characteristics of magmatic activities and orogenic process of Taihangshan intraplate orogen. *Earth Sci.* **1997**, *22*, 57–62. (In Chinese)
18. Chen, B.; Chen, C.J.; He, J.B.; Liu, A.K. Origin of Mesozoic high-Mg adakitic rocks from northeastern China:Petrological and Nd-Sr-Os isotopic constraints. *Chin. Sci. Bull.* **2013**, *58*, 1941–1953. (In Chinese)
19. Li, Y.H.; Duan, C.; Dan, H.; Chen, X.; Wang, C.; Yang, B.; Zhang, C.; Liu, F. Effect of sulfate evaporate salt layer for formation of porphyrite iron ores in the Middle-Lower Yangtze River area. *Acta Petrol. Sin.* **2014**, *30*, 1355–1368.
20. Shen, J.; Santosh, M.; Li, S.; Zhang, H.; Yin, N.; Dong, G.; Wang, Y.; Ma, G.; Yu, H. The Beiminghe skarn iron deposit, eastern China: Geochronology, isotope geochemistry and implications for the destruction of the North China Craton. *Lithos* **2013**, *156*, 218–229. [CrossRef]
21. Sun, Y.; Xiao, L.; Zhu, D.; Wu, T.; Deng, X.; Bai, M.; Wen, G. Geochemical, geochronological, and sr-nd-hf isotopic constraints on the petrogenesis of the qicun intrusive complex from the handan-xingtai district: Implications for the mechanism of lithospheric thinning of the north china craton. *Ore Geol. Rev.* **2014**, *57*, 363–374. [CrossRef]
22. Menzies, M.A.; Xu, Y.G.; Zhang, H.F.; Fan, W.M. Integration of geology, geophysics and geochemistry: A key to understanding the North China Craton. *Lithos* **2007**, *96*, 1–21. [CrossRef]
23. Zhang, K.J. Destruction of the North China Craton: Lithosphere folding-induced removal of lithospheric mantle? *J. Geodyn.* **2012**, *53*, 8–17. [CrossRef]
24. Su, S.G.; Hou, J.G.; Luo, Z.H.; Wu, G.Y.; Chen, Y.J. Genetic mechanism of skarn iron deposit: Evidence from Xishimen Iron deposit in Wu'an, Hebei Province. *J. Jilin Univ. (Earth Sci. Ed.)* **2015**, *201*, 346–347. (In Chinese)
25. Liu, L.L.; Su, S.G.; Wang, N.; Wang, W.B. Deep magmatic processes and shallow responses during the lithospheric thinning of North China Craton: Taking Tanling intrusive complex as an example. *Acta Petrol. Sin.* **2019**, *35*, 2873–2892. (In Chinese)
26. McFarlane, C.R.M.; Luo, Y. U-Pb Geochronology Using 193 nm Excimer LA-ICP-MS Optimized for In Situ Accessory Mineral Dating in Thin Sections. *Geosci. Can.* **2012**, *39*, 158–172.
27. Zhang, W.; Hu, Z.C.; Liu, Y.S. Iso-Compass: New freeware software for isotopic data reduction of LA-MC-ICP-MS. *J. Anal. At. Spectrom.* **2020**, *35*, 1087–1096. [CrossRef]
28. Feng, Y.T.; Zhang, W.; Hu, Z.C.; Luo, T.; Li, M.; Liu, Y.S.; Liu, H.; Li, Q.L. A new synthesis program of pyrite and chalcopyrite reference materials for in situ iron and sulfur isotope analysis using LA-MC-ICP-MS. *J. Anal. At. Spectrom.* **2022**, *37*, 551–562.
29. Toplis, M.J.; Corgne, A. An experimental study of element partitioning between magnetite, clinopyroxene and iron-bearing silicate liquids with particular emphasis on vanadium. *Contrib. Mineral. Petrol.* **2002**, *144*, 22–37.
30. Nadoll, P.; Angerer, T.; Mauk, J.L.; French, D.; Walshe, J. The chemistry of hydrothermal magnetite: A review. *Ore Geol. Rev.* **2014**, *61*, 1–32. [CrossRef]
31. Bordage, A.; Balan, E.; Villiers, J.P.R.d.; Cromarty, R.; Juhin, A.; Carvallo, C.; Calas, G.; Raju, P.V.S.; Glatzel, P. V oxidation state in Fe–Ti oxides by high-energy resolution fluorescence-detected X-ray absorption spectroscopy. *Phys. Chem. Miner.* **2011**, *38*, 449–458. [CrossRef]
32. Barnes, S.J.; Godel, B.; Gürer, D.; Brenan, J.M.; Robertson, J.; Paterson, D. Sulfide-Olivine Fe-Ni Exchange and the Origin of Anomalously Ni Rich Magmatic Sulfides. *Econ. Geol.* **2013**, *108*, 1971–1982. [CrossRef]
33. Rudnick, R.L.; Gao, S. Composition of the Continental Crust. *Treatise Geochem.* **2003**, *3*, 1–64.
34. Dare, S.A.S.; Barnes, S.J.; Beaudoin, G.; Méric, J.; Boutroy, E.; Potvin-Doucet, C. Trace elements in magnetite as petrogenetic indicators. *Miner. Depos.* **2014**, *49*, 785–796. [CrossRef]

35. Dupuis, C.; Beaudoin, G. Discriminant diagrams for iron oxide trace element fingerprinting of mineral deposit types. *Miner. Depos.* **2011**, *46*, 319–335. [CrossRef]
36. Nystroem, J.O.; Henriquez, F. Magmatic features of iron ores of the Kiruna type in Chile and Sweden; ore textures and mag-netite geochemistry. *Econ. Geol.* **1994**, *89*, 820–839. [CrossRef]
37. Zhao, W.W.; Zhou, M.F. In-situ LA-ICP-MS trace elemental analyses of magnetite: The Mesozoic Tengtie skarn Fe deposit in the Nanling Range, South China. *Ore Geol. Rev.* **2015**, *65*, 872–883. [CrossRef]
38. Chen, L.; Song, X.; Zhu, X.; Zhang, X.; Yu, S.; Yi, J. Iron isotope fractionation during crystallization and sub-solidus re-equilibration: Constraints from the Baima mafic layered intrusion, SW China. *Chem. Geol.* **2014**, *380*, 97–109. [CrossRef]
39. Cao, Y.; Wang, C.Y.; Huang, F.; Zhang, Z. Iron Isotope Systematics of the Panzhihua Mafic Layered Intrusion Associated With Giant Fe-Ti Oxide Deposit in the Emeishan Large Igneous Province, SW China. *J. Geophys. Res.* **2019**, *124*, 358–375.
40. Wang, Y.; Zhu, X.; Tang, C.; Mao, J.; Chang, Z. Discriminate between magmatic- and magmatic-hydrothermal ore deposits using Fe isotopes. *Ore Geol. Rev.* **2021**, *130*, 103946. [CrossRef]
41. Heimann, A.E.A.; Beard, B.L.; Johnson, C.M. View Correspondence The role of volatile exsolution and sub-solidus fluid/rock interactions in producing high (Heimann)Fe/(Heimann)Fe ratios in siliceous igneous rocks. *Geochim. Cosmochim. Acta* **2009**, *72*, 4379–4396. [CrossRef]
42. Bilenker, L.D.; Simon, A.C.; Reich, M.; Lundstrom, C.C.; Gajos, N.; Bindeman, I.; Barra, F.; Munizaga, R. Fe-O stable isotope pairs elucidate a high-temperature origin of Chilean iron oxide-apatite deposits. *Geochim. Cosmochim. Acta* **2016**, *177*, 94–104. [CrossRef]
43. Weis, F. Oxygen and Iron Isotope Systematics of the Grängesberg Mining District (GMD), Central Sweden. Ph.D. Thesis, Uppsala University, Uppsala City, Sweden, 2017.
44. Su, S.G.; Lu, X.; Santosh, M.; Hou, J.G.; Cui, Y.; Cui, X.L. Geochemical and Fe-isotope characteristics of the largest Mesozoic skarn deposit in China: Implications for the mechanism of Fe skarn formation. *Ore Geol. Rev.* **2021**, *138*, 104400. [CrossRef]
45. Severmann, S.; Anbar, A.D. Reconstructing Paleoredox Conditions through a Multitracer Approach: The Key to the Past Is the Present. *Elements* **2009**, *5*, 359–364. [CrossRef]
46. Sossi, P.A.; Foden, J.D.; Halverson, G.P. Redox-controlled iron isotope fractionation during magmatic differentiation: An example from the Red Hill intrusion, S. Tasmania. *Contrib. Mineral. Petrol.* **2012**, *164*, 757–772. [CrossRef]
47. Dare, S.A.S.; Barnes, S.J.; Beaudoin, G. Variation in trace element content of magnetite crystallized from a fractionating sulfide liquid, Sudbury, Canada: Implications for provenance discrimination. *Geochim. Cosmochim. Acta* **2012**, *88*, 27–50. [CrossRef]
48. Buddington, A.F.; Lindsley, D.H. Iron-titanium oxide minerals and synthetic equivalents. *J. Petrol.* **1964**, *5*, 310–357. [CrossRef]
49. Frost, B.R.; Lindsley, D.H. Occurrence of iron-titanium oxides in igneous rocks. *Rev. Miner. Geochem.* **1991**, *25*, 433–468.
50. Ghiorso, M.S.; Sack, O. Iron-titanium oxide geothermometry: Thermodynamic formulation and the estimation of intensive variables in silicic magmas. *Contrib. Mineral. Petrol.* **1991**, *108*, 485–510. [CrossRef]
51. Keller, T.; Tornos, F.; Hanchar, J.M.; Pietruszka, D.K.; Soldati, A.; Dingwell, D.B.; Suckale, J. Genetic model of the El Laco magnetite-apatite deposits by extrusion of iron-rich melt. *Nat. Commun.* **2022**, *13*, 6114. [CrossRef]
52. Wang, S.; Hou, T.; Xie, Q. Formation of orthopyroxene-magnetite symplectites by reactive melt flow: Insights into the Shangzhuang layered intrusion in Beijing, China. *J. Asian Earth Sci.* **2024**, *273*, 106262. [CrossRef]
53. Xie, Q.; Zhang, Z.; Hou, T.; Cheng, Z.; Campos, E.; Wang, Z.; Fei, X. New insights for the formation of Kiruna-type iron deposits by immiscible hydrous Fe-P melt and high-temperature hydrothermal processes: Evidence from El Laco deposit. *Econ. Geol.* **2019**, *114*, 35–46. [CrossRef]
54. Gao, J.; Zhou, M.; Lightfoot, P.; Wang, C.; Qi, L.; Sun, M. Sulfide Saturation and Magma Emplacement in the Formation of the Permian Huangshandong Ni-Cu Sulfide Deposit, Xinjiang, Northwestern China. *Econ. Geol.* **2013**, *108*, 1833–1848. [CrossRef]
55. Huang, X.; Zhou, M.; Qi, L.; Gao, J.; Wang, Y. Re–Os isotopic ages of pyrite and chemical composition of magnetite from the Cihai magmatic–hydrothermal Fe deposit, NW China. *Miner. Depos.* **2013**, *48*, 925–946. [CrossRef]
56. McCarthy, T.S.; Cawthorn, R.G. The geochemistry of vanadiferous magnetite in the bushveld complex: Implications for crystallization mechanisms in layered complexes. *Miner. Depos.* **1983**, *18*, 505–518. [CrossRef]
57. Wang, C.; Zhou, M. New textural and mineralogical constraints on the origin of the Hongge Fe-Ti-V oxide deposit, SW China. *Miner. Depos.* **2013**, *48*, 787–798. [CrossRef]
58. McCarthy, T.S.; Cawthorn, R.G.; Wright, C.J.; McIver, J.R. Mineral layering in the Bushveld Complex: Implications of chromium abundances in magnetite from closely spaced magnetitite and intervening silicate-rich layers. *Econ. Geol.* **1985**, *80*, 1062–1074. [CrossRef]
59. He, D.T.; Liu, Y.; Chen, C.; Foley, S.F.; Ducea, M.N. Oxidization of the mantle caused by sediment recycling may contribute to the formation of iron-rich mantle melts. *Sci. Bull.* **2020**, *65*, 519–521. [CrossRef]
60. Goldstein, S.B.; Francis, D. The Petrogenesis and Mantle Source of Archaean Ferropicrites from the Western Superior Province, Ontario, Canada. *J. Petrol.* **2008**, *49*, 1729–1753. [CrossRef]

61. Liu, P.P.; Zhou, M.; Chen, W.; Boone, M.; Cnudde, V. Using Multiphase Solid Inclusions to Constrain the Origin of the Baima Fe-Ti-(V) Oxide Deposit, SW China. *J. Petrol.* **2014**, *55*, 951–976. [CrossRef]
62. Zhao, L.; Chen, H.; Zhang, L.; Li, D.; Zhang, W.; Wang, C.; Yang, J.; Yan, X. Magnetite geochemistry of the Heijianshan Fe–Cu (–Au) deposit in Eastern Tianshan: Metallogenic implications for submarine volcanic-hosted Fe–Cu deposits in NW China. *Ore Geol. Rev.* **2018**, *100*, 422–440. [CrossRef]
63. Yang, R.N. Genesis of Orbicular Diorite in Tanling, Wuan, Hebei. Master's Thesis, China University of Geosciences (Beijing), Beijing, China, 2017. (In Chinese).
64. Ganino, C.; Arndt, N.; Zhou, M.F.; Gaillard, F.; Chauvel, C. Interaction of magma with sedimentary wall rock and magnetite ore genesis in the Panzhihua mafic intrusion, SW China. *Miner. Depos.* **2008**, *43*, 677–694. [CrossRef]
65. Ganino, C.; Harris, C.; Arndt, N.T. Assimilation of carbonate country rock by the parent magma of the Panzhihua Fe-Ti-V deposit (SW China): Evidence from stable isotopes. *Geosci. Front.* **2013**, *4*, 547–554. [CrossRef]
66. Pecher, A.; Arndt, N.; Jean, A. Structure of the Panzhihua intrusion and its Fe-Ti-V deposit, China. *Geosci. Front.* **2013**, *4*, 571–581. [CrossRef]
67. Dauphas, N.; Roskosz, M.; Alp, E.E.; Neuville, D.R.; Hu, M.Y.; Sio, C.K.; Tissot, F.L.; Zhao, J.; Tissandier, L.; Médard, E. Magma redox and structural controls on iron isotope variations in Earth's mantle and crust. *Earth Planet. Sci. Lett.* **2014**, *398*, 127–140. [CrossRef]
68. Prissel, K.B.; Krawczynski, M.J.; Nie, N.X.; Dauphas, N.; Couvy, H.; Hu, M.Y.; Alp, E.E.; Roskosz, M. Experimentally determined effects of olivine crystallization and melt titanium content on iron isotopic fractionation in planetary basalts. *Geochim. Cosmochim. Acta* **2018**, *238*, 580–598. [CrossRef]
69. Shahar, A.; Young, E.D.; Manning, C.E. Equilibrium high-temperature Fe isotope fractionation between fayalite and magnetite: An experimental calibration. *Earth Planet. Sci. Lett.* **2008**, *268*, 330–338. [CrossRef]
70. Poitrasson, F.; Freydier, R. Heavy iron isotope composition of granites determined by high resolution MC-ICP-MS. *Chem. Geol.* **2005**, *222*, 132–147. [CrossRef]
71. Wang, Y.J.; Hu, Y.Y.; Shen, J.F.; Qu, K.; Yin, N.; Yu, H.J.; Ma, G.G. Sulfur and lead isotope composition and tracing for sources of ore-forming materials in Beiming river iron deposits, southern Taihang mountains. *Nat. Geosci.* **2011**, *25*, 846–852.
72. Tang, D.; Qin, K.; Mao, Y.; Evans, N.J. Magnetite geochemistry and iron isotope signature of disseminated and massive mineralization in the Kalatongke magmatic Cu-Ni sulfide deposit, northwest China. *Chem. Geol.* **2022**, *605*, 120965. [CrossRef]
73. Dauphas, N.; Zuilen, M.V.; Wadhwa, M.; Davis, A.M.; Marty, B.; Janney, P.E. Clues from Fe Isotope Variations on the Origin of Early Archean BIFs from Greenland. *Science* **2004**, *306*, 2077–2080. [CrossRef]
74. Dauphas, N.; Cates, N.L.; Mojzsis, S.J.; Busigny, V. Identification of chemical sedimentary protoliths using iron isotopes in the >3750 Ma Nuvvuagittuq supracrustal belt, Canada. *Earth Planet. Sci. Lett.* **2007**, *254*, 358–376. [CrossRef]
75. Dauphas, N.; van Zuilen, M.; Busigny, V.; Lepland, A.; Wadhwa, M.; Janney, P.E. View Correspondence Iron isotope, major and trace element characterization of early Archean supracrustal rocks from SW Greenland: Protolith identification and metamorphic overprint. *Geochim. Cosmochim. Acta* **2008**, *71*, 4745–4770. [CrossRef]
76. Frost, C.D.; Blanckenburg, F.v.; Schoenberg, R.; Frost, B.R.; Swapp, S.M. Preservation of Fe isotope heterogeneities during diagenesis and metamorphism of banded iron formation. *Contrib. Mineral. Petrol.* **2007**, *153*, 211–235. [CrossRef]
77. Hyslop, E.; Valley, J.W.; Johnson, C.M.; Beard, B.L. The effects of metamorphism on O and Fe isotope compositions in the Biwabik Iron Formation, northern Minnesota. *Contrib. Mineral. Petrol.* **2008**, *155*, 313–328. [CrossRef]
78. Ruan, C.T. The Gem-Mineralogy Characteristics of Garnet in the Nanminghe Skarn Iron Deposit in Wuan, Hebei. Master's Thesis, China University of Geosciences (Beijing), Beijing, China, 2020. (In Chinese).
79. Troll, V.R.; Weis, F.A.; Jonsson, E.; Andersson, U.B.; Majidi, S.A.; Högdahl, K.; Harris, C.; Millet, M.-A.; Chinnasamy, S.S.; Kooijman, E.; et al. Global Fe–O isotope correlation reveals magmatic origin of Kiruna-type apatite-iron-oxide ores. *Nat. Commun.* **2019**, *10*, 1712. [CrossRef]
80. Telus, M.; Dauphas, N.; Moynier, F.; Tissot, F.L.H.; Teng, F.Z.; Nabelek, P.I.; Craddock, P.R.; Groat, L.A. Iron, zinc, magnesium and uranium isotopic fractionation during continental crust differentiation: The tale from migmatites, granitoids, and pegmatites. *Geochim. Cosmochim. Acta* **2012**, *97*, 247–265. [CrossRef]
81. Mungall, J.E.; Brenan, J.M.; Godel, B.; Barnes, S.J.; Gaillard, F. Transport of metals and sulphur in magmas by flotation of sulphide melt on vapour bubbles. *Nat. Geosci.* **2015**, *8*, 216–219. [CrossRef]
82. Su, S.G.; Tang, Z.L.; Luo, Z.H.; Deng, J.F.; Wu, G.Y.; Zhou, M.F.; Song, C.; Xiao, Q.H. Magmatic Conduit Metallogenic System. *Acta Petrol. Sin.* **2014**, *11*, 3120–3130. (In Chinese)
83. Clechenko, C.C.; Valley, J.W. Oscillatory zoning in garnet from the Willsboro Wollastonite Skarn, Adirondack Mts, New York: A record of shallow hydrothermal processes preserved in a granulite facies terrane. *J. Metamorph. Geol.* **2003**, *21*, 771–784. [CrossRef]
84. Zhai, D.; Liu, J.; Zhang, H.; Wang, J.; Su, L.; Yang, X.; Wu, S. Origin of oscillatory zoned garnets from the Xieertala Fe–Zn skarn deposit, northern China: In situ LA–ICP-MS evidence. *Lithos* **2014**, *190*, 279–291. [CrossRef]

85. Mirnejad, H.; Hasannejad, M.; Miller, N.; Hassanzadeh, J.; Bocchio, R.; Modabberi, S. Origin and Evolution of Oscillatory Zoned Garnet from Kasva Skarn, Northeast Tafresh, Iran. *Can. Mineral.* **2018**, *56*, 15–37. [CrossRef]
86. Zhu, Q.Q.; Xie, G.Q.; Li, W. In-situ analysis of garnets from the Jingshandian iron skarn deposit, Hubei Province, and its geological implications. *Geol. China* **2014**, *41*, 1944–1963. (In Chinese)
87. Bataleva, Y.V.; Palyanov, Y.N.; Sokol, A.G.; Borzdov, Y.M.; Bayukov, O.A. Wüstite stability in the presence of a CO_2-fluid and a carbonate-silicate melt: Implications for the graphite/diamond formation and generation of Fe-rich mantle metasomatic agents. *Lithos* **2016**, *244*, 20–29. [CrossRef]
88. Qiu, J.X. *Magmatic Petrology*; Geological Publishing House: Beijing, China, 1985; p. 340. (In Chinese)
89. Fan, Q.C. Zircon chronology and rare earth geochemistry of xenolith granulites in Hanuoba. *Chin. Sci. Bull.* **1998**, *2*, 133–137. (In Chinese)
90. Chen, B.; Zhai, M.G.; Tian, W.; Jiang, B.M. Petrogenesis of the Mesozoic Intrusive Complexes from the Southern Taihang Orogen, North China Craton: Elemental and Sr-Nd-Pb Isotopic Constraints. *Bull. Mineral. Petrol. Geochem.* **2005**, *24*, 10. (In Chinese) [CrossRef]
91. Zheng, J.M.; Xie, G.Q.; Liu, J.; Chen, M.H.; Wang, S.M.; Guo, S.F.; Gao, X.; Li, G.D. ^{40}Ar-^{39}Ar dating of phologopite from the Xishimen skarn iron deposit in the Handan-Xingtai area, southern Hebei, and its implications. *Acta Petrol. Sinca* **2007**, *23*, 2513–2518. (In Chinese)
92. Förster, H.; Bachtiar, I.; Borumandi, H. Petrographische Detailuntersuchungen im Bereich der Eisenerzlagerstätten von Bafq/Zentraliran. *Z. Dtsch. Geol. Ges.* **1973**, *124*, 121–134.
93. Park, C.F. A magnetite "flow" in northern Chile. *Econ. Geol.* **1961**, *56*, 431–436. [CrossRef]
94. Philpotts, A.R. Origin of certain iron-titanium oxide and apatite rocks. *Econ. Geol.* **1967**, *62*, 303–315. [CrossRef]
95. Lledo, H.L.; Jenkins, D.M. Experimental Investigation of the Upper Thermal Stability of Mg-rich Actinolite; Implications for Kiruna-Type Iron Deposits. *J. Petrol.* **2008**, *49*, 225–238. [CrossRef]
96. Philpotts, A.R. Compositions of immiscible liquids in volcanic rocks. *Contrib. Mineral. Petrol.* **1982**, *80*, 201–218. [CrossRef]
97. Su, L.H. The importance of liquid immiscibility in petrology and mineral deposits. *J. Earth Sci.* **1984**, *1*, 1–12. (In Chinese)
98. Yu, X.H. The geological significance and the phase equilibrium experiments of wustite-fluor-phlogopite-diopsied melt system atonebar and high temperature. *J. Earth Sci.* **1984**, *1*, 13–18+135–136. (In Chinese)
99. Li, Y.H.; Duan, C.; Han, D.; Liu, F.; Wan, D.F.; Wang, C.Y. Oxygen isotopic discriminant marker of magmatic iron deposits: Ningwu porphyrite iron ore as an example. *Acta Petrol. Sin.* **2017**, *33*, 3411–3421. (In Chinese)

minerals

MDPI

Article

Ore Genesis Based on Microtextural and Geochemical Evidence from the Hydrothermal As–Sb Mineralization of the Matra Deposit (Alpine Corsica, France)

Danis Ionut Filimon [1], John A. Groff [2], Emilio Saccani [3,*] and Maria Di Rosa [1]

[1] Dipartimento di Scienze della Terra, Università di Pisa, Via Santa Maria 53, 56126 Pisa, Italy; danis.filimon@phd.unipi.it (D.I.F.); maria.dirosa@unipi.it (M.D.R.)
[2] Physical Sciences Department, Tuscaloosa Academy, Tuscaloosa, AL 35406, USA; jgroff@tuscaloosaacademy.org
[3] Dipartimento di Fisica e Scienze della Terra, Università di Ferrara, Via Ludovico Ariosto 35, 44121 Ferrara, Italy
* Correspondence: sac@unife.it

Abstract: The Matra As–Sb deposit (Alpine Corsica, France) is hosted in the normal N–S trending Matra Fault. Sulfide minerals in ore consist of realgar, stibnite, and pyrite with minor orpiment and hörnesite. The gangue includes quartz, dolomite, and calcite. In this study, the microstructural analysis of selected ore samples has been combined with the geochemical characterization of the sulfides. The results depict a succession of events that record the evolution of the ore deposit related to fault movement. In the pre–ore stage, plumose, crustiform, jigsaw, and feathery textures of quartz testify to a short–lived boiling event. The mineral assemblage of the main–ore stage includes an Fe(–Zn) substage dominated by the formation of different textures of pyrite. In general, pyrite samples contain significant concentrations of As ($\leq$32,231 ppm) and Sb ($\leq$10,684 ppm), with lesser amounts of by Tl ($\leq$1257 ppm) and Ni ($\leq$174 ppm). This is followed by an Sb–As–Fe substage of pyrite–stibnite–realgar $\pm$orpiment. The precipitation of the sulfides was mainly driven by changes in fS_2. The increasing level of oxidation is attributed to a progressive influx of meteoric water resulting from reactivation of the Matra Fault.

Keywords: As–Sb–Fe mineralization; trace element mineral chemistry; LA–ICP–MS on pyrite and stibnite; Matra deposit; Alpine Corsica

1. Introduction

The Matra deposit (MD), located in eastern Corsica, France, produced a total of ~30,000 tons of ore containing 30% As until 1945 [1]. Realgar is the main ore mineral with stibnite and minor orpiment. The deposit was discovered between 1880 and 1890 as a zone of green siliceous clay–like material containing realgar and orpiment exposed in a stream bank [2–4]. The mineralization occurs along the post–Alpine normal N–S–trending Matra Fault, which juxtaposes the Jurassic–Cretaceous(?) Upper Castagniccia and Santo Pietro di Tenda Units, both belonging to the Schistes Lustrés Complex [5]. The Matra Fault is part of a larger network of normal faults that accommodated the extension of the Tyrrhenian Sea [6].

The Early Oligocene in the Corsica–Northern Apennine–Tuscany sector marked an important transition from compression to extensional tectonics [7]. This shift was associated with igneous activity and the formation of mineralization at a large scale in the Tuscan Archipelago and the Tuscany Magmatic Province. Ore deposits in the Tuscany Magmatic

Province include carbonate–hosted Sb (±Au) mineralization that is characterized by the presence of invisible gold in pyrite, a trace element signature of As–Hg–Tl–Ba, jasperoid alteration, and dilute low–temperature fluids [8–10]. Sillitoe and Brogi [10] proposed that carbonate–hosted Sb (±Au) mineralization, related to the Tuscany Magmatic Province, was Carlin–style based on ore characteristics and the evolution of plutons emplaced at depths of 5–7 km. Ore produced from the MD has a similar mineral assemblage and geological setting, formed during the transition from compressional to extensions tectonics along the Corsica–Tuscany transect.

The objectives of this study are therefore to establish the ore paragenesis of the MD and to document the geochemistry of the different sulfides. Data obtained using various analytical techniques such as scanning electron microscopy/energy dispersive X-ray spectroscopy (SEM–EDS), electron probe microanalyses (EPMA), and laser ablation–inductively coupled plasma–mass spectrometry (LA–ICP–MS) are used to produce a genetic model, including the tectonic setting, for the MD As–Sb mineralization. Finally, as both As and Sb are included in the list of European Critical Raw Materials [11], this contribution may provide new insights that will benefit future exploration for this type of ore deposit.

2. Geological Setting

According to Durand–Delga [12], Corsica can be divided into two distinct geological domains (Figure 1): the Autochthonous Hercynian Domain (to the west) and Alpine Corsica (to the east). The Hercynian Domain is primarily composed of Carboniferous to Permian granitoids intruding a Paleozoic basement, which is unconformably overlain by Jurassic to Middle Eocene sedimentary units (e.g., [13]). Alpine Corsica comprises a complex stack of tectonic units derived from both oceanic and continental domains. The oceanic units represent remnants of the Ligure–Piemontese oceanic basin, which existed between the European and Adria continental margins during the Middle to Late Jurassic (e.g., [12–20]). The oceanic–derived units include an ophiolitic sequence covered by associated sedimentary covers, and based on their structural position and the degree of deformation and metamorphism, they compose the Upper Units and the Schistes Lustrés Complex [12]. The tectonic units of Alpine Corsica are unconformably overlain by Miocene (Burdigalian–Langhian) sedimentary rocks hosted in basins that are exposed in the Saint–Florent and Francardo regions (e.g., [21]).

The Upper Units (e.g., Balagne Nappe) occupy the highest structural level of Alpine Corsica [22]. They are primarily composed of basalts with associated sedimentary cover rocks, marked by a significant input of continental–derived sediments [23], ranging in age from the Dogger to the Cretaceous [12]. Basalts within the Balagne Nappe exhibit ocean–floor metamorphism [23–25].

The Schistes Lustrés Complex is widely exposed in eastern Corsica and Cap Corse and lies below the Upper Units [26]. It is composed of several tectonic units having oceanic affinity and includes serpentinized ultramafic rocks, troctolites, gabbros, ferrogabbros, plagiogranites, dolerites, and basalts—most of which display N–MORB–type magmatic affinities [24,27,28]. The associated ophiolitic sedimentary cover rocks range in age from the Late Jurassic to the Mid–Cretaceous and are composed of pelagic deposits (radiolarian cherts, shales, and micritic limestones) that are typical of oceanic sequences formed in the deeper parts of the Ligure–Piemontese oceanic basin [29]. Metamorphism of the Schistes Lustrés Complex varies from low blueschist to the eclogite facies [30–32].

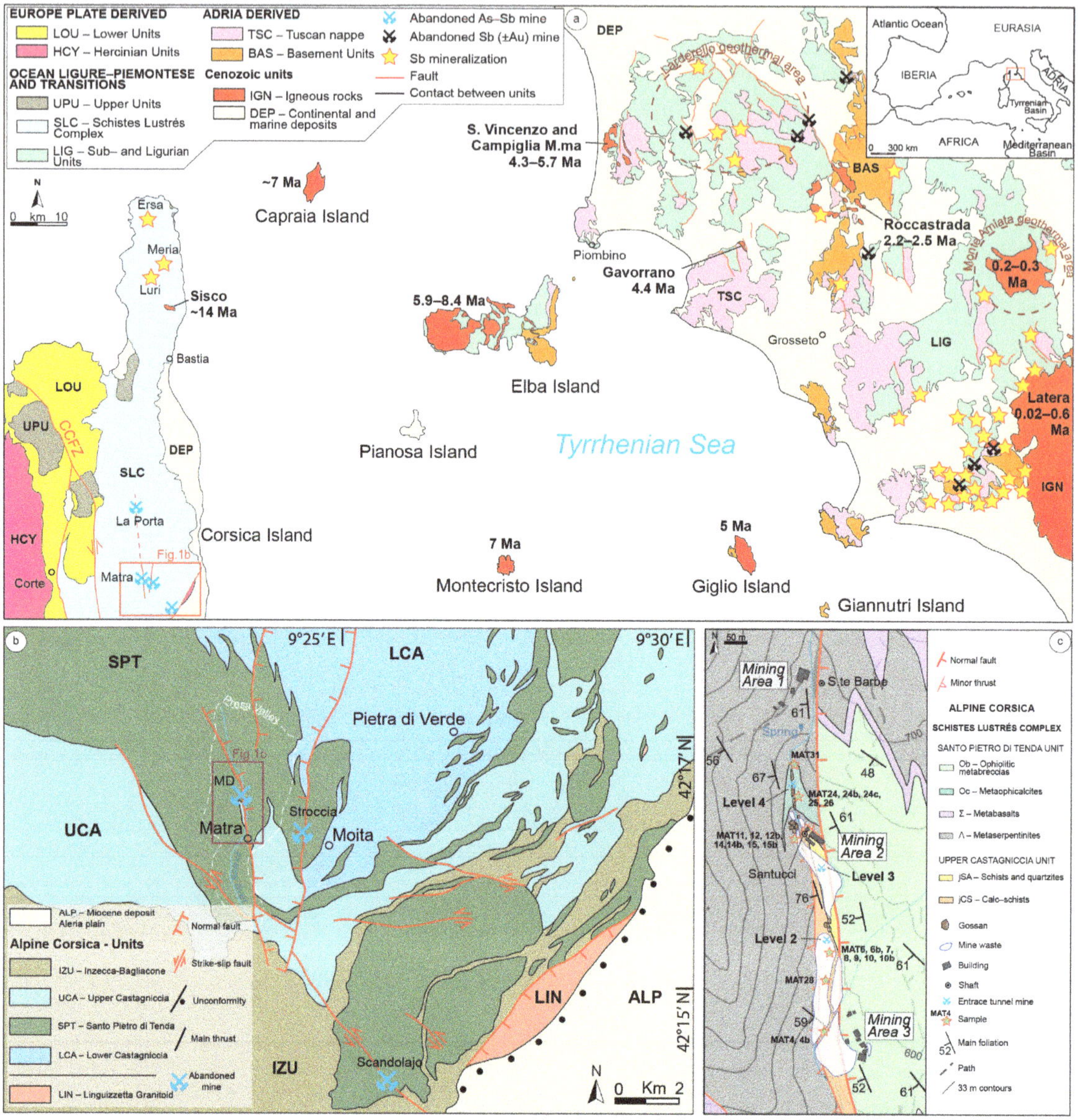

Figure 1. (**a**) Regional geological map including the studied mine area within a small section of northwestern Italy. The inset panel at the upper right shows the location of the study area relative to a larger section of the western Mediterranean area (modified after [10,17]). (**b**) Geological setting of the Matra area (modified after [5]). (**c**) Geological map of the area of the Matra deposit (modified after [5]).

Between the Late Cretaceous and Late Eocene, convergence between the European and Adria plates led to the progressive closure of the Ligure–Piemontese oceanic basin [33–35]. This tectonic evolution began with intra–oceanic subduction, which was followed by the subduction of the continental margins and subsequent collision–related processes (e.g., [36]). Plate convergence ceased in the Early Oligocene and was replaced by widespread extensional tectonics (e.g., [37]). Evidence of these subduction–related events is well–preserved in the Schistes Lustrés Complex, which reached metamorphic peak conditions represented by the eclogite facies between ~83 Ma and 35 Ma [16,38].

During the Early Oligocene, a shift in plate kinematics transferred the convergence zone east of Alpine Corsica, initiating a new westward–dipping subduction system [7,39,40]. This change generated a broad extensional regime that caused the collapse of the previously formed orogenic wedge in Corsica and led to the opening of the Liguro–Provençal back–arc basin [37,41,42].

This was followed by Late Miocene extension that triggered the opening of the Tyrrhenian back–arc basin, effectively separating Corsica from the Apennine orogenic system [43–45]. In Alpine Corsica, this episode of extension was accompanied by magmatic activity (~14 Ma), based on the age of lamproitic rocks at Sisco, which are attributed to the Tuscany Magmatic Province ([46]; Figure 1a). The Tuscany Magmatic Province comprises a suite of mafic to silicic intrusive and extrusive rocks that become progressively younger toward the east, culminating in ages of 0.2–0.3 Ma at Mt. Amiata [46,47].

The Alpine Corsican tectonic units in the Matra area include the Santo Pietro di Tenda and the Upper Castagniccia Units, which are part of the Schistes Lustrés Complex [5,48]. The Santo Pietro di Tenda Unit comprises an ophiolitic sequence consisting of metaserpentinites, metagabbros, metaophicalcites, metabasalts, and metaophiolitic breccias overlain by metacarbonates and schists (e.g., [28,29]), whereas the Upper Castagniccia Unit is mainly composed of calc–schists, schists, and quartzites [49]. Both of these units were affected by Alpine ductile deformation and metamorphism that reached the eclogite facies [5]. In the post–Alpine deformation, these units were juxtaposed along the Matra normal fault, a brittle structure trending N170° and dipping 75° eastward (Figure 1b; [5]). The Matra Fault developed during Miocene extension in association with the opening of the Tyrrhenian Sea and hosts the MD composed of Sb–As–Fe sulfides and the related dolomite and minor quartz gangue. Given the kinematic indicators (e.g., slickenfibers) formed within both the sulfides and the dolomite gangue, Filimon et al. [5] demonstrated that the faulting activity and the formation of MD and related gangue were coeval.

3. Sampling and Analytical Methods

Forty–one samples of gossan and metamorphic rocks hosting mineralization were collected from mine waste and outcrops surrounding the MD in the Presa Valley (Figures 1c and 2, Table 1). The mine was inaccessible to sampling due to rock collapse at its entrance. Twenty–three samples representing different ore types, host rocks, and gossan were studied using optical microscopy (e.g., transmitted and reflected light) and X-ray fluorescence spectroscopy (XRF) at the Ferrara University (Supplementary Table S1).

These samples were then studied using a combination of SEM, EPMA, and LA–ICP–MS. The evaluation of these data allowed us to define mineral assemblages, microtextures, and the geochemistry of the sulfide and gangue minerals (Supplementary Tables S2–S4). Scanning electron microscope analyses were undertaken using a Hitachi TM3030Plus Tabletop Electron Scanning Microscopy laboratory coupled with a TM3030 Swift ED3000 microanalysis system (EDS) in the Dipartimento di Scienze della Terra di Pisa (Pisa, Italy). The WDS analyses (spot and map modes) of ore minerals (e.g., realgar, stibnite, and pyrite) were performed using a JEOL JXA–8800 microprobe in the Dipartimento di Scienze della Terra "A. Desio" at the Università di Milano Statale.

Figure 2. (**a**) Landscape around the Matra deposit showing mine waste along the Presa valley. (**b**) Field occurrence of gossan proximal to the Matra Fault and hosted by metaserpentinites of the Santo Pietro di Tenda Unit. (**c–l**) Photographs illustrating the different rock types and ore located along the Matra Fault. (**c**) Example of gossan that contain a dolomite vein displaced by a microfault that developed within the larger normal fault network. Relics of unaltered ophicalcite are also shown. (**d**) Ore breccia associated with altered metaserpentinite of the Santo Pietro di Tenda Unit, sample MAT24. (**e**) Realgar (Rlg) and stibnite (Sbn–II) veins in the calc–schist of the Upper Castagniccia Unit, sample MAT4. (**f**) Ore breccia characterized by vugs filled with realgar (Rlg) and orpiment (Orp), sample MAT11. (**g**) Ore breccia with pyrite (Py–I), sample MAT6. (**h**) Dolomite (Dol) and realgar (Rlg) ore breccia, sample MAT24c. (**i**) Brittle deformation in the ore breccia, sample MAT9. (**j**) Ore sample cut by realgar (Rlg) and stibnite (Sbn–II) veins, sample MAT10. (**k**) Ore sample with stibnite (Sbn–I) and quartz (Qz) characterized by vugs filled with realgar (Rlg), sample MAT12. (**l**) Euhedral crystals of realgar (Rlg) and hörnesite (Hns) filling a vug in sample MAT6b. Mineral abbreviations are from Warr [50].

Table 1. Summary and description of samples collected around the Matra mine that were used in this study.

Sample	Type	Description
MAT 4, 4b	Host rock	Foliated calc–schist of Upper Castagniccia Unit, characterized by different veins of realgar and stibnite.
MAT 6, 6b	Breccia	Breccia with quartz and dolomite clasts with the matrix containing traces of pyrite, realgar, and orpiment.
MAT 7	Ore	Mineralized sample with dolomite gangue, showing veins of realgar and disseminated stibnite and pyrite.
MAT 8	Ore	Framework of calcite veins with massive realgar.
MAT 9	Ore	Dolomite gangue with interstitial stibnite, realgar, and crosscutting calcite veins.
MAT 10, 10b	Ore	Dolomite gangue with stibnite, realgar, and pyrite in veinlets or disseminations.
MAT 11	Breccia	Breccia with quartz and dolomite clasts in a matrix containing traces of mineralization, including pyrite, realgar, and hörnesite.
MAT 12, 12b	Ore	Quartz gangue with disseminated stibnite and pyrite, along with nodules of realgar.
MAT 14, 14b	Gossan	Non–cohesive altered and mineralized rock, showing iron–oxide coloration with traces of green zones enriched with nickel oxides.
MAT 15, 15b	Host rock	Serpentinite sample of the Santo Pietro di Tenda Unit, with dolomite and realgar veins cutting the main foliation.
MAT 24, 24b, 24c	Ore breccia	Mineralized breccia composed of drusy dolomite gangue with clasts of pyrite, stibnite, and realgar, featuring slickensides and slickenfibers on quartz and stibnite.
MAT 25	Gossan	Altered sample with iron–oxide coloration, containing drusy calcite veins.
MAT 26	Gossan	Altered sample with iron–oxide coloration and quartz veins.
MAT 28	Ore	Mineralized sample with dolomite and calcite, containing massive realgar and disseminated stibnite and pyrite.
MAT 31	Gossan	Weathered, altered, with a reddish–brown iron–oxide coating. Contains residual quartz and calcite veinlets.

Spot analyses were performed at 15 kV and 2 nA with a beam size diameter of 1 μm. The standards used for sulfides were Ni_2Si (Ni), pyrite (Fe, S), cuprite (Cu), stibnite (Sb), and GaAs (As). Back–scattered electron mapping conditions were 15 kV and 1 nA, with a counting time of 20 ms per grid point. Mineral phases documented in the maps were identified using the spot analysis mode of XMapTools [51]. The EPMA of carbonate gangue was performed using a beam diameter of 3 μm, with a 15 kV voltage and a 2 nA sample current at the ISTerre of Grenoble (Grenoble, France). The standards used included calcite (Ca), dolomite (Mg), siderite (Fe, Mn), and strontianite (Sr). Additional analytical information is given in the Supplementary Materials, Table S3.

The trace element contents of pyrite and stibnite were determined using a PerkinElmer NexION 2000 ICP–MS coupled with a New Wave Research–193 Ar–F 193 nm excimer laser in the Centre for Instrument Sharing at the University of Pisa (CISUP). The laser operated at a repetition rate of 10 Hz with a spot size between 35 and 40 μm, depending on the mineral size, and 5 J/cm^2 of energy density. The trace elements analyzed included ^{49}Ti, ^{53}Cr, ^{55}Mn, ^{57}Fe, ^{59}Co, ^{60}Ni, ^{65}Cu, ^{66}Zn, ^{71}Ga, ^{72}Ge, ^{75}As, ^{77}Se, ^{95}Mo, ^{107}Ag, ^{111}Cd, ^{115}In, ^{118}Sn, ^{121}Sb, ^{125}Te, ^{182}W, ^{195}Pt, ^{197}Au, ^{205}Tl, ^{208}Pb, and ^{209}Bi. Data reduction was performed with the IOLITE v4.10.7 software [52]. Iron and sulfur from EPMA were used as internal standards in pyrite and stibnite, respectively. The sulfide analyses were calibrated using the standard STDGL3 [53], and the data quality was verified using UQAC5.

X-ray diffraction (XRD) was used to identify minor mineral phases (e.g., hörnesite and orpiment). Owing to the small amount of material, X-ray powder diffraction data were collected using a Bruker D8 Venture single–crystal diffractometer in the Dipartimento di Scienze della Terra di Pisa (Pisa, Italy), equipped with a Photon III CCD area detector (microfocus Cu$K\alpha$ radiation) simulating a Gandolfi–like geometry (Supplementary Table S5).

4. Mineralogy and Geochemistry of Ore Minerals

4.1. Pyrite

In mineralized samples from the MD, pyrite was categorized as either Py–I, Py–II, or Py–III based on textural features and mineralogical associations (Figure 3). Py–I occurs as large euhedral crystals ≤1 cm in size and is associated with sphalerite (Figure 3a,b or Figure 4a). The cubic shape is a characteristic of both single crystals and aggregates disseminated in the quartz and dolomite gangue. No zoning was identified in Py–I, as compositional variations are limited ($Fe_{0.780–0.989}As_{0.020–0.380}Sb_{0–0.001}Ni_{0–0.035}Cu_{0–0.001}S_2$). The most abundant trace elements were As, Sb, and Tl. Note that Ni and Zn were also detected (Table 2; Figure 5).

Py–II crystals are anhedral and range in size from 100 μm to 2 mm. These occur in the dolomite gangue, often in association with Py–I. Visible banding was evident (Figure 3c,d), and EPMA identified rhythmic zoning in some crystals, while others displayed a mottled texture (Figure 4b). This is reflected in the compositional range of Fe ($Fe_{0.934–1.005}As_{0.005–0.191}Sb_{0–0.042}Ni_{0–0.052}Cu_{0–0.002}S_2$; Supplementary Table S4). Analyses by LA–ICP–MS revealed high concentrations of As and Sb (Table 2). Other elements such as Tl, Ni, and Zn were also detected (Figure 5). This type of pyrite occurs either as crystal aggregates in veins or disseminated within the dolomite gangue.

The most common type of pyrite identified in ore samples is Py–III. It is characterized by highly fractured crystals ranging in size from 50 μm to 1 mm (Figure 3e,f). The occurrence of these crystals is confined to veins that are distinguished by the presence of dolomite–calcite, stibnite, realgar, and, to a lesser extent, quartz. This pyrite formed through the fracturing of Py–I and subsequent displacement of these fragments. The EPMA micro–maps show that Py–III crystals have rims enriched in As and Sb. Although Ni shows a uniform distribution, the back–scattered electron (BSE) images for As show inner zones that are lighter in color and have high As concentrations relative to the darker zones. The LA–ICP–MS analyses also recorded significant variations in several elements, including As, Sb, Ni, and Tl (Table 2). The highest recorded concentrations of Cu and Co were 441 ppm and 2868 ppm, respectively (Figure 5). Although Py–III has a wider range of element concentrations, there is significant overlap between the compositions of Py–I and Py–II (Table 2; Supplementary Materials Table S2). What readily distinguishes Py–II is that the intense fracturing of pre–existing pyrite and displacement of the small fragments (≤15 μm) formed parallel layers within the dolomite gangue (Figure 3g,h or Figure 4d).

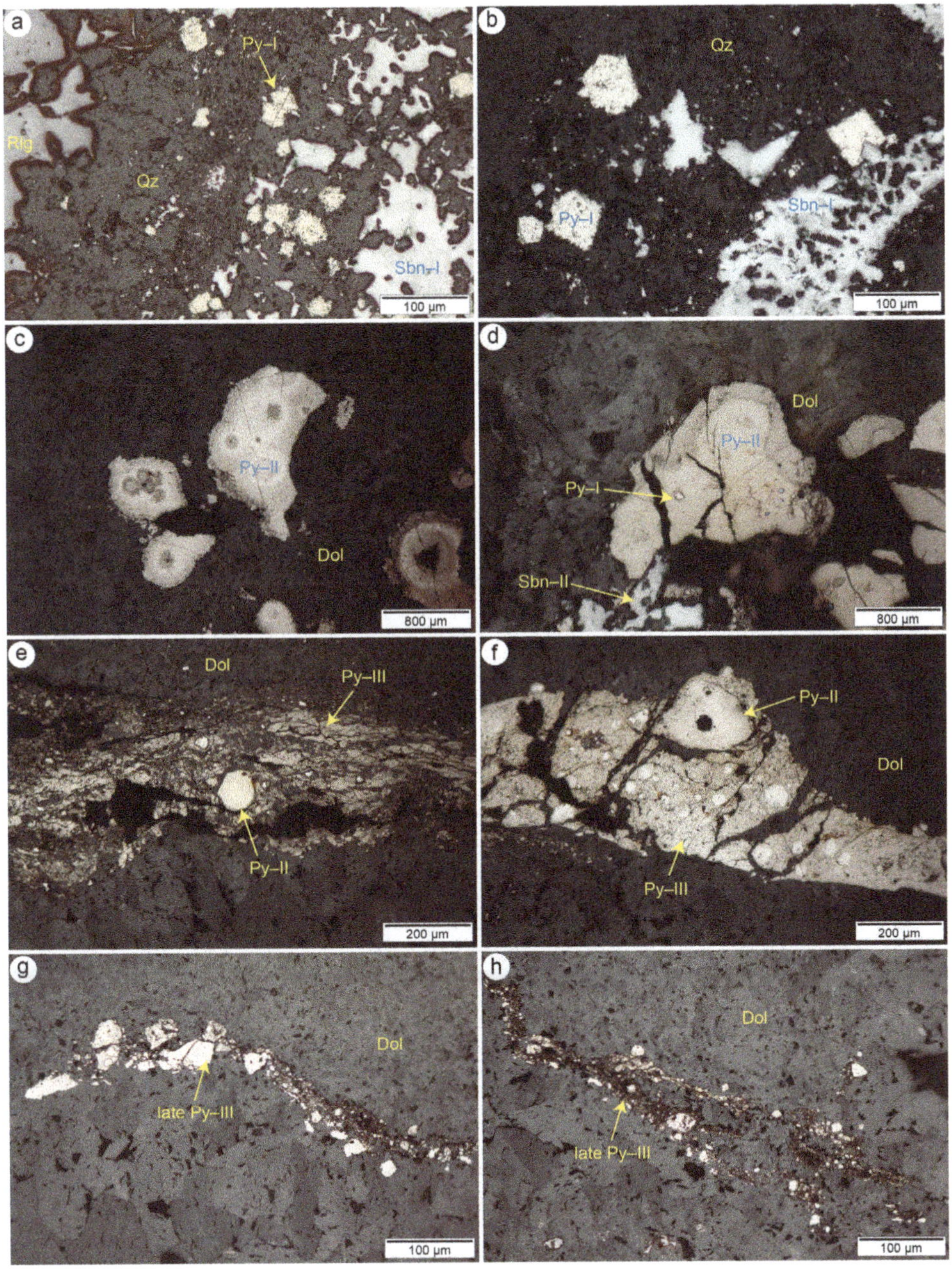

Figure 3. Photomicrographs (reflected light) showing the different pyrite types (e.g., Py–I, Py—-II, Py–III, and late Py–III). (**a**) Pyrite (Py–I), stibnite (Sbn–I), and realgar in quartz gangue, sample MAT12. (**b**) Pyrite (Py–I) and stibnite (Sbn–I) in quartz gangue, sample MAT12. (**c**) Zoned Pyrite (Py–II) in dolomite gangue, sample MAT10. (**d**) Pyrite (Py–I and Py–II) with stibnite (Sbn–II) in dolomite gangue, sample MAT10. (**e**) Transition from pyrite (Py–II to Py–III) in dolomite gangue, sample MAT10. (**f**) Disjointed pyrite (Py–II with Py–III) and dolomite gangue, sample MAT10. (**g**) Pyrite (Py–III and late Py–III) in dolomite gangue, sample MAT10. (**h**) Pyrite (late Py–III) in dolomite gangue, sample MAT10.

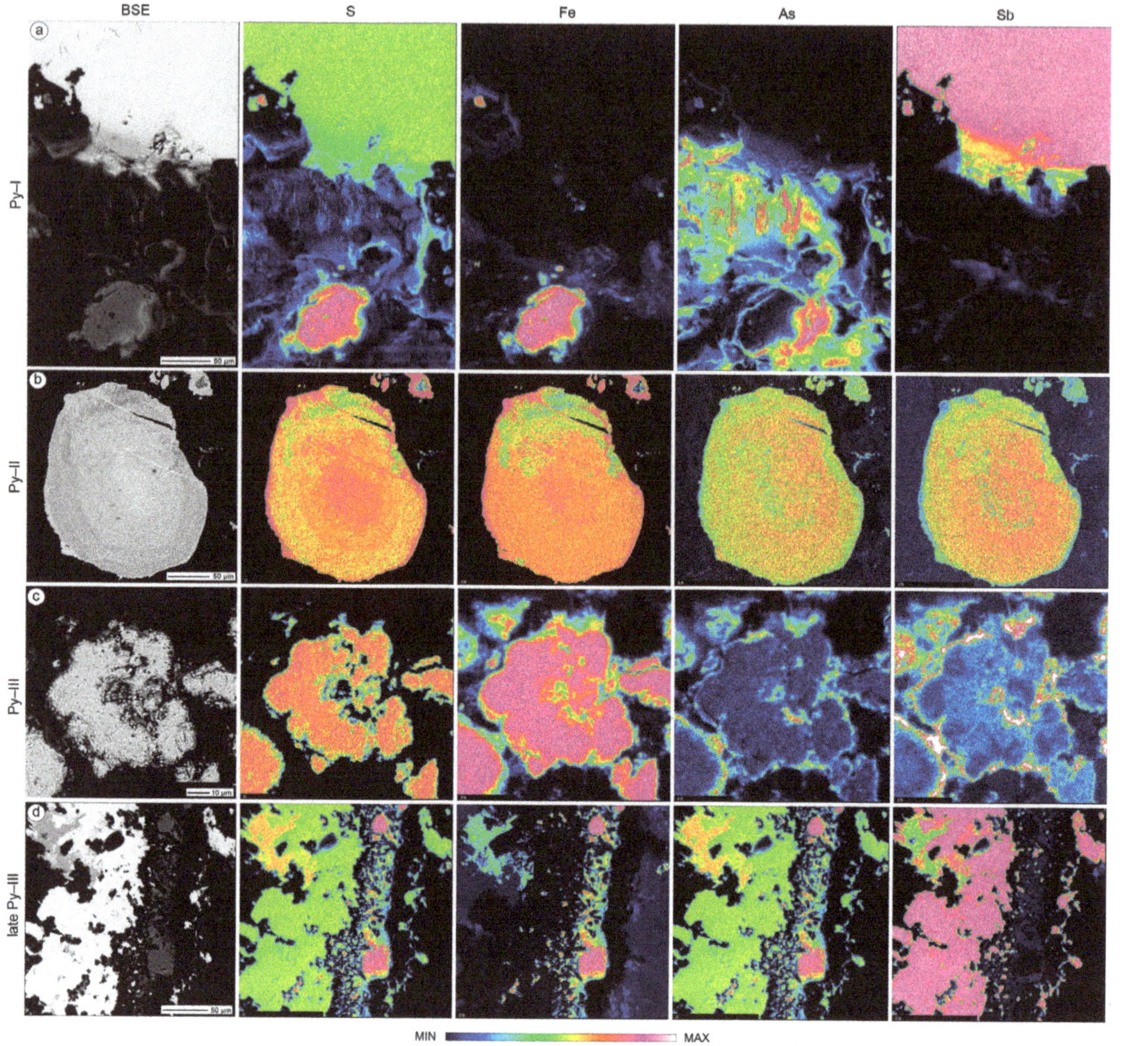

Figure 4. EPMA micro–maps showing the contents of major elements in pyrite (Py–I, Py–II, Py–III, and late Py–III). The distributions expressed in counts per second (cps, see the bar in the figure) range between a minimum (MIN) and a maximum (MAX) value, which are as follows: (**a**) Py–I, S: 0–536 cps; Fe: 0–671 cps; As: 49–788 cps; and Sb: 266–1421 cps; (**b**) Py–II, S: 140–514 cps; Fe: 253–636 cps; As: 0–88 cps; and Sb: 0–116 cps; (**c**) Py–III, S: 244–546 cps; Fe: 0–687 cps; As: 0–169 cps; and Sb: 0–101 cps; (**d**) late Py–III, S: 0–530 cps; Fe: 0–678 cps; As: 0–89 cps; and Sb: 0–1437 cps.

Table 2. Summary of trace element data (%adl: percentage of data above the detection limit) for pyrite (Py–I, Py–II, and Py–III) and stibnite (Sbn–I, Sbn–II, and Sbn–III) using LA–ICP–MS.

Minerals	N		Ti	Cr	Mn	Fe	Co	Ni	Cu	Zn	Ga	Ge	As	Se	Mo	Ag	Cd	In	Sn	Sb	Te	W	Pt	Au	Tl	Pb	Bi
Py–I	20	max	3.83	n.d.	1.63	n.d.	3.17	219	1.46	386	n.d.	n.d.	49,417	n.d.	5.59	n.d.	n.d.	0.10	0.39	26,876	n.d.	0.12	n.d.	n.d.	1795	0.10	n.d.
		min	2.00	n.d.	0.85	n.d.	0.18	11.5	0.34	1.74	n.d.	n.d.	11,601	n.d.	0.63	n.d.	n.d.	0.03	0.26	8999	n.d.	0.12	n.d.	n.d.	905	0.06	n.d.
		median	3.00	n.c.	1.25	n.c.	0.53	35.5	0.51	2.91	n.c.	n.c.	19,793	n.c.	2.91	n.c.	n.c.	0.05	0.33	12,699	n.c.	0.12	n.c.	n.c.	1075	0.10	n.c.
		mean	2.86	n.c.	1.23	n.c.	0.88	58.2	0.62	22.5	n.c.	n.c.	23,286	n.c.	2.92	n.c.	n.c.	0.05	0.33	13,261	n.c.	0.12	n.c.	n.c.	1165	0.09	n.c.
		stdev	0.55	n.c.	0.21	n.c.	0.92	52.4	0.34	85.6	n.c.	n.c.	10,443	n.c.	1.32	n.c.	n.c.	0.02	0.09	4458	n.c.	n.c.	n.c.	n.c.	237	0.02	n.c.
		%adl	100	n.c.	100	n.c.	85.0	100	45.0	100	n.c.	n.c.	100	n.c.	100	n.c.	n.c.	100	10.0	100	n.c.	5.00	n.c.	n.c.	100	15.0	n.c.
Py–II	20	max	12.7	n.d.	627	n.d.	55.0	2413	6.79	257	n.d.	n.d.	154,785	n.d.	1160	n.d.	0.60	1.16	11.9	76,967	n.d.	7.28	n.d.	n.d.	13,604	0.93	n.d.
		min	1.93	n.d.	15.1	n.d.	1.24	0.58	0.34	4.73	n.d.	n.d.	26,996	n.d.	3.99	n.d.	0.60	0.14	0.41	2285	n.d.	0.11	n.d.	n.d.	780	0.07	n.d.
		median	5.84	n.c.	154	n.c.	2.55	34.4	1.11	46.2	n.c.	n.c.	46,566	n.c.	33.8	n.c.	0.60	0.34	1.28	15,055	n.c.	1.32	n.c.	n.c.	3662	0.46	n.c.
		mean	5.64	n.c.	224	n.c.	13.0	340	1.90	72.8	n.c.	n.c.	63,066	n.c.	115	n.c.	0.60	0.40	2.11	23,539	n.c.	2.90	n.c.	n.c.	4589	0.49	n.c.
		stdev	2.50	n.c.	199	n.c.	20.0	710	2.07	70.7	n.c.	n.c.	39,703	n.c.	263	n.c.	n.c.	0.23	2.64	20,007	n.c.	3.44	n.c.	n.c.	3751	0.43	n.c.
		%adl	100	n.c.	100	n.c.	40.0	75.0	50.0	100	n.c.	n.c.	100	n.c.	100	n.c.	5.00	95.0	90.0	100	n.c.	30.0	n.c.	n.c.	100	15.0	n.c.
Py–III	37	max	768	n.d.	780	n.d.	2869	11,405	442	907	n.d.	n.d.	320,187	3.70	2580	0.23	1.85	1.56	2.06	251,513	1.07	145	n.d.	n.d.	47,443	5.50	n.d.
		min	1.98	n.d.	0.68	n.d.	0.47	10.3	0.43	1.87	n.d.	n.d.	4454	3.70	0.57	0.23	0.67	0.03	0.49	389	1.07	0.07	n.d.	n.d.	9.67	0.11	n.d.
		median	7.82	n.c.	40.9	n.c.	68.0	2381	43.0	12.2	n.c.	n.c.	32,232	3.70	6.59	0.23	1.01	0.16	1.24	2319	1.07	1.69	n.c.	n.c.	928	0.59	n.c.
		mean	40.0	n.c.	87.2	n.c.	153	3068	79.4	52.8	n.c.	n.c.	76,138	3.70	303	0.23	1.07	0.28	1.22	28,707	1.07	8.17	n.c.	n.c.	6620	1.32	n.c.
		stdev	130	n.c.	144	n.c.	466	3217	114	151	n.c.	n.c.	90,896	n.c.	619	n.c.	0.42	0.30	0.52	61,702	n.c.	26.1	n.c.	n.c.	11,842	1.56	n.c.
		%adl	91.9	n.c.	100	n.c.	100	100	86.5	100	n.c.	n.c.	100	0.00	81.1	2.70	18.9	97.3	16.2	100	2.70	81.1	n.c.	n.c.	100	73.0	n.c.
Sbn–I	19	max	n.d.	3.33	0.19	n.d.	n.d.	n.d.	3.04	8.87	0.26	1.30	44,078	28.6	0.76	n.d.	n.d.	0.05	9.30	n.d.	n.d.	0.04	n.d.	n.d.	1606	0.06	n.d.
		min	n.d.	1.04	0.19	n.d.	n.d.	n.d.	0.26	1.55	0.03	0.69	12,247	5.30	0.24	n.d.	n.d.	0.01	1.15	n.d.	n.d.	0.04	n.d.	n.d.	126	0.03	n.d.
		median	n.c.	1.75	0.19	n.c.	n.c.	n.c.	1.20	4.08	0.05	1.16	15,548	8.73	0.42	n.c.	n.c.	0.03	4.62	n.c.	n.c.	0.04	n.c.	n.c.	432	0.04	n.c.
		mean	n.c.	2.13	0.19	n.c.	n.c.	n.c.	1.36	4.07	0.08	1.06	19,681	12.8	0.46	n.c.	n.c.	0.03	4.92	n.c.	n.c.	0.04	n.c.	n.c.	575	0.04	n.c.
		stdev	n.c.	0.96	n.c.	n.c.	n.c.	n.c.	0.73	1.80	0.06	0.18	9755	10.6	0.24	n.c.	n.c.	0.01	3.65	n.c.	n.c.	n.c.	n.c.	n.c.	421	0.02	n.c.
		%adl	n.c.	36.8	5.26	n.c.	n.c.	n.c.	100	100	84.2	68.4	100	21.1	21.1	n.c.	n.c.	52.6	31.6	n.c.	n.c.	5.26	n.c.	n.c.	100	10.5	n.c.
Sbn–II	18	max	n.d.	0.70	1.01	148	0.03	8.71	7.19	4.69	0.02	1.20	34,885	7.09	6.80	0.02	0.23	0.04	10.5	n.d.	0.29	0.03	n.d.	n.d.	898	0.13	0.005
		min	n.d.	0.42	0.09	19.0	0.03	1.94	0.10	0.58	0.01	0.65	4508	0.78	0.69	0.02	0.14	0.01	0.19	n.d.	0.29	0.02	n.d.	n.d.	75.6	0.02	0.005
		median	n.c.	0.56	0.31	63.0	0.03	4.06	1.56	3.35	0.01	0.98	14,180	1.53	1.40	0.02	0.14	0.02	2.25	n.c.	0.29	0.03	n.c.	n.c.	302	0.05	0.005
		mean	n.c.	0.56	0.40	69.3	0.03	5.01	1.85	2.98	0.01	0.93	14,005	2.36	2.34	0.02	0.16	0.02	2.98	n.c.	0.29	0.03	n.c.	n.c.	378	0.06	0.005
		stdev	n.c.	0.20	0.32	n.c.	n.c.	2.91	2.14	1.43	0.01	0.16	8774	2.67	1.98	n.c.	0.05	0.01	3.20	n.c.	n.c.	0.01	n.c.	n.c.	233	0.05	n.c.
		%adl	n.c.	11.1	50.0	38.9	5.56	27.8	88.9	100	33.3	83.3	100	27.8	100	5.56	22.2	50.0	44.4	n.c.	5.56	11.1	n.c.	n.c.	100	22.2	5.56
Sbn–III	19	max	n.d.	21.3	273	2584	0.59	20.5	1.78	4.53	0.05	1.54	21,237	1.52	5.72	n.d.	n.d.	0.04	8.02	n.d.	0.25	4.32	n.d.	n.d.	681	3.82	0.01
		min	n.d.	1.10	0.30	64.5	0.09	3.63	0.20	0.95	0.03	0.77	4383	1.52	0.38	n.d.	n.d.	0.01	0.57	n.d.	0.11	0.02	n.d.	n.d.	0.08	0.03	0.01
		median	n.c.	2.78	0.63	276	0.35	12.8	0.28	2.06	0.05	1.12	13,775	1.52	1.90	n.c.	n.c.	0.02	4.13	n.c.	0.18	0.06	n.c.	n.c.	19.18	0.18	0.01
		mean	n.c.	5.89	30.6	655	0.34	12.3	0.43	2.10	0.05	1.16	13,403	1.52	2.40	n.c.	n.c.	0.02	4.21	n.c.	0.18	0.82	n.c.	n.c.	113	0.66	0.01
		stdev	n.c.	6.66	81.8	841	0.21	8.44	0.51	0.86	0.01	0.24	4122	n.c.	1.70	n.c.	n.c.	0.01	3.34	n.c.	0.10	1.72	n.c.	n.c.	194	1.29	n.c.
		%adl	n.c.	52.6	57.9	63.2	21.05	15.8	0.00	94.7	26.3	80.0	100	5.26	73.7	n.c.	n.c.	31.6	21.1	n.c.	10.5	31.6	n.c.	n.c.	100	42.1	5.26

N = number of analyses; n.d. = not detected; n.c. = not calculated.

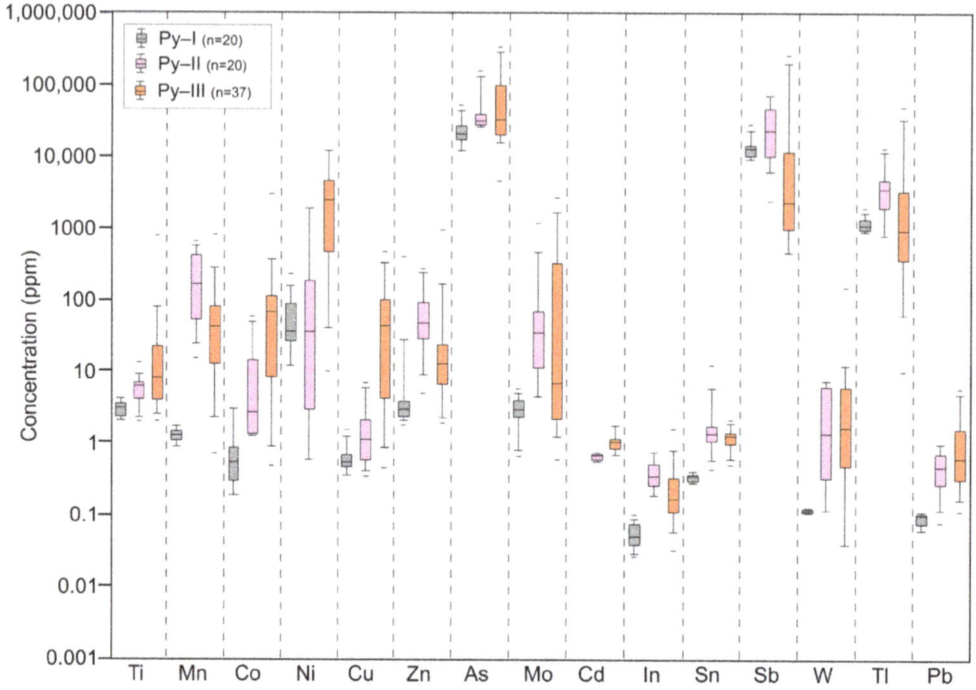

Figure 5. Whisker diagram showing composition ranges of selected trace elements in different pyrite types (Py–I, Py–II, and Py–III) using LA–ICP–MS data. The boxes represent data from the 25th–75th percentiles, the whiskers indicate the minimal and maximum values, the horizontal lines represent the medians, and the lines represent outlier values (Supplementary Materials, Table S2).

4.2. Stibnite

Stibnite (Sb_2S_3) is a common mineral that is spatially associated with other sulfides in Matra ore samples. As stibnite formed in association with distinct mineral assemblages, this was used as a basis to distinguish three types of stibnite (e.g., Sbn–I, Sbn–II, and Sbn–III). Sbn–I occurs either as large (≤1 cm) anhedral crystals or small (≤100 µm) acicular crystals (Figure 6a,b). It is associated with Py–I, realgar, sphalerite, and quartz. The results of the LA–ICP–MS analyses show significant ranges in concentrations of As, Tl, and Zn (Figure 7).

Sbn–II occurs as interstitial fillings (≤100 µm; Figure 6c,d) and veinlets (≤3 cm thick; Figure 6c,d). It is associated with dolomitic gangue and locally encapsulated by realgar. Sbn–II consists of 70.8 wt.% Sb and 28.1 wt.% S (compositional range of $Sb_{1.897–2.068}As_{0.032–0.281}Ni_{0–0.004}Fe_{0–0.004}Cu_{0–0.004}S_3$). The trace element composition of Sbn–II is dominated by As, Tl, and Ni (Figure 7).

Crystals of Sbn–III range in size from 50 µm to 1 cm and occur in veins also containing Py–III and quartz that cut the dolomite gangue (Figure 6e,f). Positive anomalies of As and Fe were recorded by EPMA analyses, but no compositional zoning was identified. Median concentrations for Sb and S were 71.32 wt.% and 28.47 wt.%, respectively (compositional range of $Sb_{1.748–2.030}As_{0.041–0.282}Ni_{0–0.003}Fe_{0.001–0.012}Cu_{0–0.004}S_3$). There was significant variation in the concentrations of As, Tl, Fe, and Ni (Figure 7).

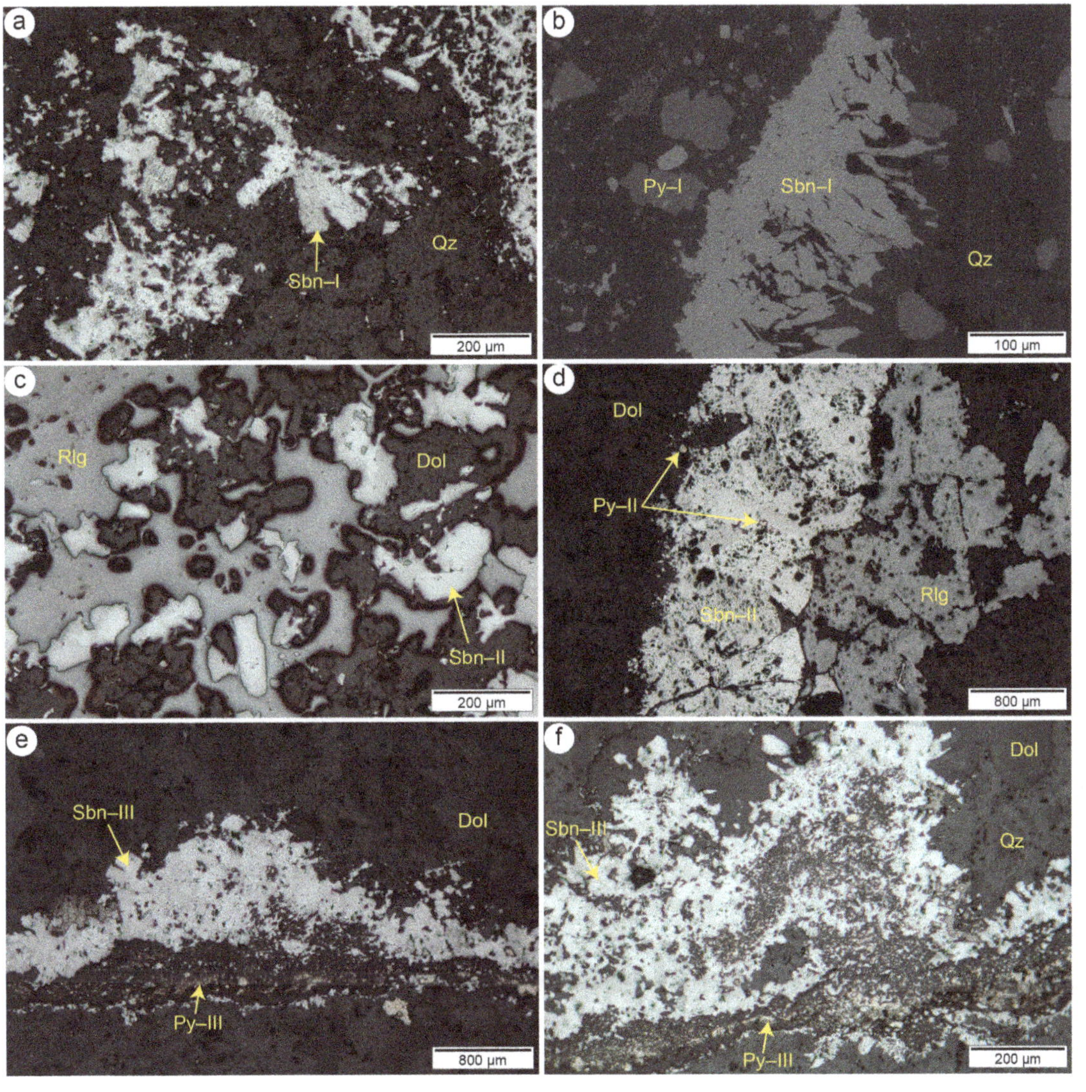

Figure 6. Photomicrographs showing the different stibnite types (e.g., Sbn–I, Sbn–II, and Sbn–III). (**a**) Stibnite (Sbn–I) in dolomite–quartz gangue, sample MAT12 (reflected light). (**b**) Stibnite (Sbn–I) and pyrite (Py–I) in quartz gangue (Qz), sample MAT12 (BSE image). (**c**) Stibnite (Sbn–II) and realgar (Rlg) intergrown in dolomite gangue (Dol), sample MAT9 (reflected light). (**d**) Sbn–II and realgar (rlg) vein associated with Py–II in dolomite gangue (Dol), sample MAT10 (reflected light). (**e**) Stibnite (Sbn–III) and pyrite (Py–III) in dolomite gangue (Dol), sample MAT10 (reflected light). (**f**) Stibnite (Sbn–III) and pyrite (Py–III) in dolomite (Dol) and quartz gangue (Qz), sample MAT10 (reflected light).

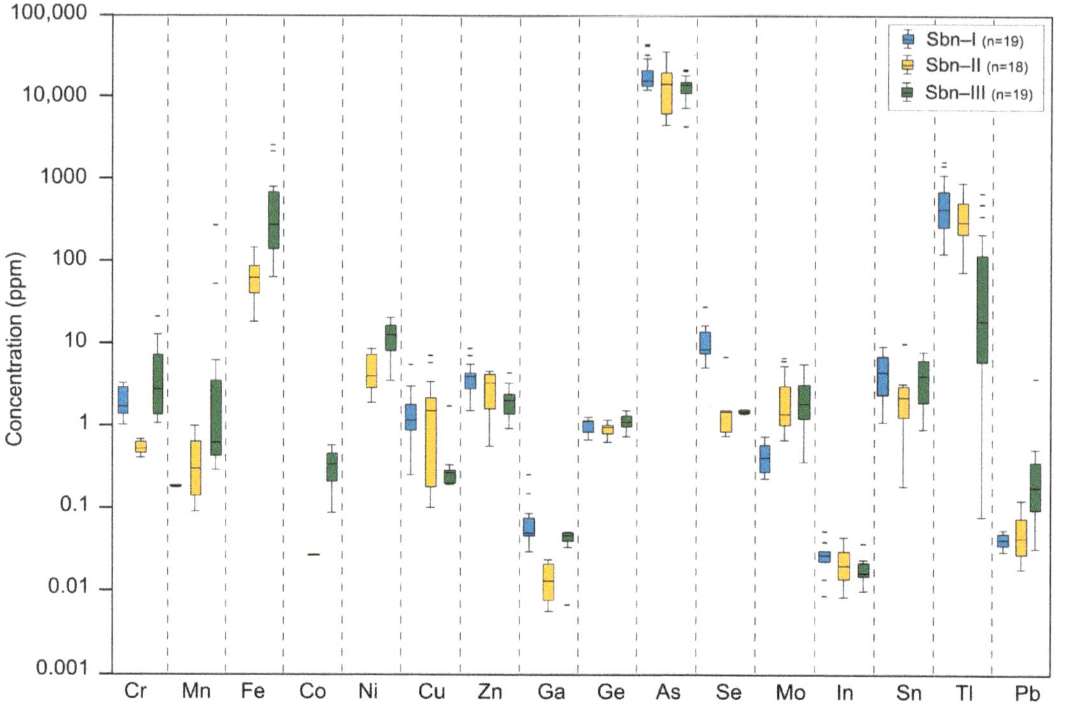

Figure 7. Whisker diagram showing composition ranges of selected trace elements in different stibnite types (Sbn–I, Sbn–II, and Sbn–III) using LA–ICP–MS data. The boxes represent data from the 25th to 75th percentiles, the whiskers represent the minimal and maximum values, the horizontal lines represent the medians, and the lines indicate outlier values (Supplementary Materials, Table S3).

4.3. Arsenic Minerals

The most abundant arsenic sulfide phase is realgar (As_4S_4). It occurs in veinlets ≤1 cm wide (Figure 2d–f,h–l) as nodule–like masses ≤5 cm across (Figure 2f) or as interstitial fillings (Figure 2i). Petrographic studies indicate that realgar grew on different types of pyrite and is intergrown with dolomite and quartz (Figure 3a). EPMA data for realgar indicate 68.1 wt.% for As and 30.6 wt.% for S, with an Fe content of ≤0.18 wt.% (compositional range of $As_{3.682–3.813}Sb_{0–0.012}Fe_{0–0.005}Ni_{0–0.004}S_4$). These data record a range of 0.09–0.17 wt.% for Sb and Ni values ≤0.05 wt% (Supplementary Materials, Table S3).

Another primary arsenic–bearing minerals, orpiment (As_2S_3), was identified in a limited number of samples. The breccia samples contain yellow prismatic orpiment crystals ≤80 µm in length, which have a vitreous luster and partially fill vugs. The occurrence of orpiment was verified by XRD (Figure 2f). Pristine realgar crystals occur in some vugs in contact with orpiment (Supplementary Materials, Table S5). A similar relationship was identified in one sample between pristine realgar crystals and hörnesite filling a vug.

4.4. Gangue

Ore samples contain gangue consisting of dolomite and quartz. Dolomite is more abundant and occurs in crystal aggregates or veins (Figure 2h,i). The SEM–EDS analyses recorded peaks of Ca, Mg, and Fe. EPMA analyses confirmed elevated Fe concentrations of 3.35–5.73 wt.% to support its classification as ferroan dolomite ([54] Supplementary Materials, Table S3).

Macroscopic and petrographic studies of hydrothermal quartz veinlets identified distinct textures. Plumose and crustiform textures (Figure 8a,b) were evident at a macroscopic scale compared with mosaic and jigsaw textures at a microscopic scale (Figure 8c,d). Veins containing Sbn–I, Py–II, and Py–III, quartz grains tend to be small (<100 μm).

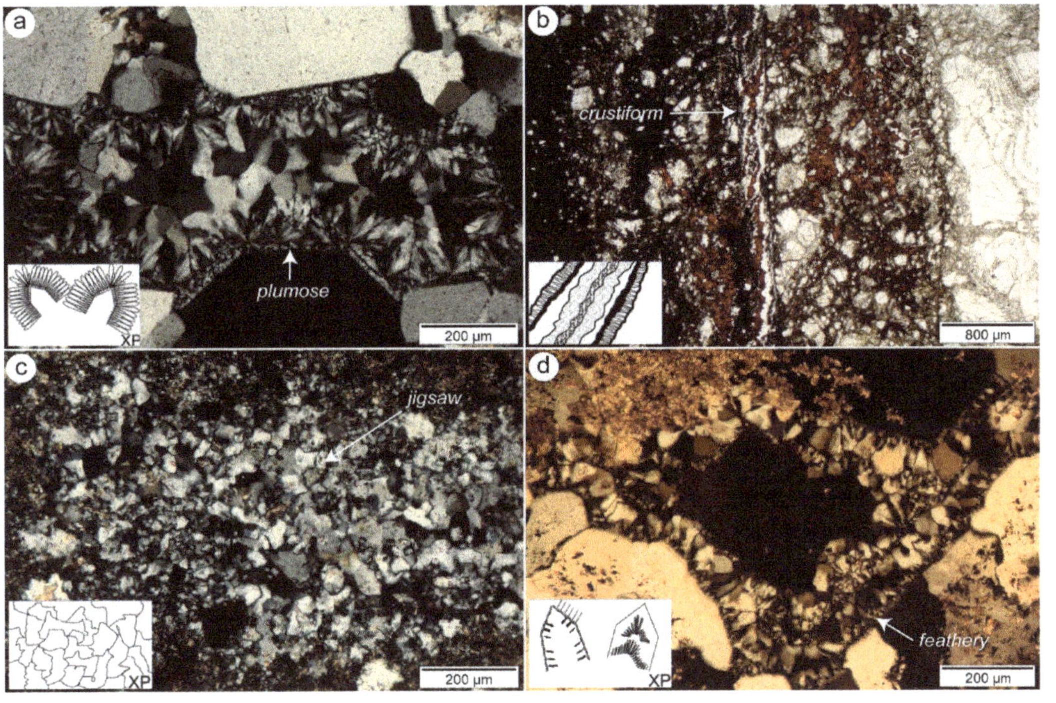

Figure 8. Photomicrographs (cross–polarized light) of the pre–ore quartz showing different textures: (**a**) plumose, (**b**) crustiform, (**c**) jigsaw, and (**d**) feathery. Sketches adapted and modified by Moncada et al. [55].

Late calcite occurs in veins (<1 cm) that cut gangue containing ore minerals or as crystals filling cavities. The cavity fillings consist of large, euhedral crystals with lesser quartz and dolomite.

4.5. Supergene Minerals

Secondary minerals formed by surface weathering include pararealgar, hörnesite [$Mg_3(AsO_4)_2{\cdot}8(H_2O)$], gypsum, and unidentified Fe–Mn–Ni oxides that contain minor amounts of As and Sb. Hörnesite in several ore samples was identified by XRD (Figure 21) and is characterized by globular, white crystals ≤100 μm. It is locally associated and intergrown with pristine realgar crystals and likely formed by the weathering of primary As–bearing sulfide minerals occurring as large masses of radiating crystals on mine waste. The Ni oxide is recognized by an aqua–green color and globular masses that are ≤1 cm across. In contrast, gypsum was identified as small, transparent, tabular crystals. These minerals dominantly occur as thin films on outcrops of gossan in metaophicalcites and metaserpentinites.

5. Discussion

The mineralization studied in this work is located along the Matra Fault, which displaced different tectonic units of the Schistes Lustrés Complex (Figure 1; [5]). These units include ophiolitic rocks and their carbonate sedimentary cover, which were both faulted and underwent hydrothermal mineralization processes [2,4].

5.1. Ore Textures and Geochemistry

From the meso– to the microscale, the mineralogy and tectonic setting of the Matra deposit indicates that this ore deposit formed in a dynamic system where the multi–stage activation of the Matra Fault drove fluid circulation and sulfide crystallization. Ore minerals occur in breccia clasts, veinlets, gangue, and along the fault plane as slickensides. Specifically, analyses of pyrite and stibnite were conducted to characterize their microtextures and geochemistry.

Pyrite is ubiquitous in ore deposits and can host various trace elements that reflect the physico–chemical conditions of formation (e.g., redox state, pressure, and temperature) and the nature of their sources. Therefore, pyrite is often used to study ore–forming processes and to develop genetic models for different types of ore deposits [56,57]. As stibnite is also an important ore constituent of the MD, it could provide additional information to develop a genetic model.

A proposed sequence of mineralization at the MD was developed relative to reactivation of the Matra Fault. Py–I has the largest size and often exhibits euhedral forms, suggesting it probably formed during an early stage of faulting (Figure 3). This is supported by the results of George et al. [58] that demonstrated how continuous, or renewed, fluid circulation and crystallization can produce euhedral pyrite crystals. Conditions in the Matra hydrothermal system likely changed with the crystallization of anhedral Py–II, which is widespread and occurs either as aggregates with Py–I or within gangue (Figure 3).

The evolution of the textures from Py–I to Py–II might be driven by the changing thermal condition that existed during formation of the MD, with temperature decreasing over time. As shown by relationships in other sulfide deposits (e.g., [59,60]), colloform pyrites occur in low–temperature (<300 °C) hydrothermal settings, forming the characteristic texture we observe in MD for Py–II (Figure 3). Yet the different and highly disjointed texture of Py–III suggests mechanical alteration of Py–II due to microfracturing during faulting. Continued or repeated movement along the Matra Fault produced strain partitioning, which resulted in the fracturing of Py–III to form the characteristic highly fractured and fine-grained crystals.

Although these differences in pyrite textures are proposed to record changes in the mineralizing system, there is little chemical variation from Py–I to Py–III (Figures 9 and 10). Due to this and documentation that mineralization was syndeformational, there is not compelling evidence to indicate that these represent different generations of pyrite.

Trace elements hosted by pyrite can exist in the mineral structure or occur as micro– to nanoscale particles (NPs; [61–63]). At MD, the elements that likely underwent isovalent substitutions with Fe^{2+} include Mn, Co, Ni, Cu, and Zn. The Ni enrichment observed in the pyrite of Matra ranges between 50 and 5000 ppm, with higher values for Py–III (Figure 5). A similar Ni concentration was identified by Picot and Johan [64] in Alpine Corsica close to Matra. These authors described Ni–bearing minerals in the ophiolitic rocks of the Santo Pietro di Tenda Unit. We therefore suggest that the Ni in Matra pyrite may derive from fluid interaction with mafic and ultramafic host rocks widely exposed in the area (Figure 1).

The Co/Ni ratios can provide information regarding the different settings in which pyrite forms [65]. Those determinated for Matra are very low (Co/Ni < 0.1) (Figure 10a) and are incompatible with sedimentary, hydrothermal, or magmatic/volcanological origins [66,67]. Although there is a linear relationship between these two elements, the ratio is influenced by a positive Ni anomaly due to the mafic nature of country rocks hosting mineralization.

The occurrence of a second group of elements that are abundant in Matra samples (e.g., As^{3+}, Sb^{3+}, and Tl^{+}) can be explained by As enrichment in pyrite. In fact, As can cause structural defects, such as heterovalent substitutions, and promote metal enrichment in the pyrite structure [68,69]. The substitution process involving $2Fe^{2+} \leftrightarrow Tl^{+} + Sb^{3+}$ or (Cu^{+} and Ag^{+}) was proposed by D'Orazio et al. [70] and George et al. [68]. This suggests that structural substitutions of Tl, Sb, and As may occur via isovalent and heterovalent processes or NPs. In addition, our data show that the LA–ICP–MS spectra follow a linear trend and lack any signs of peaks (Figure 11). This supports the absence of micro-inclusions of other minerals and is confirmed by the petrographic observations (Figure 3).

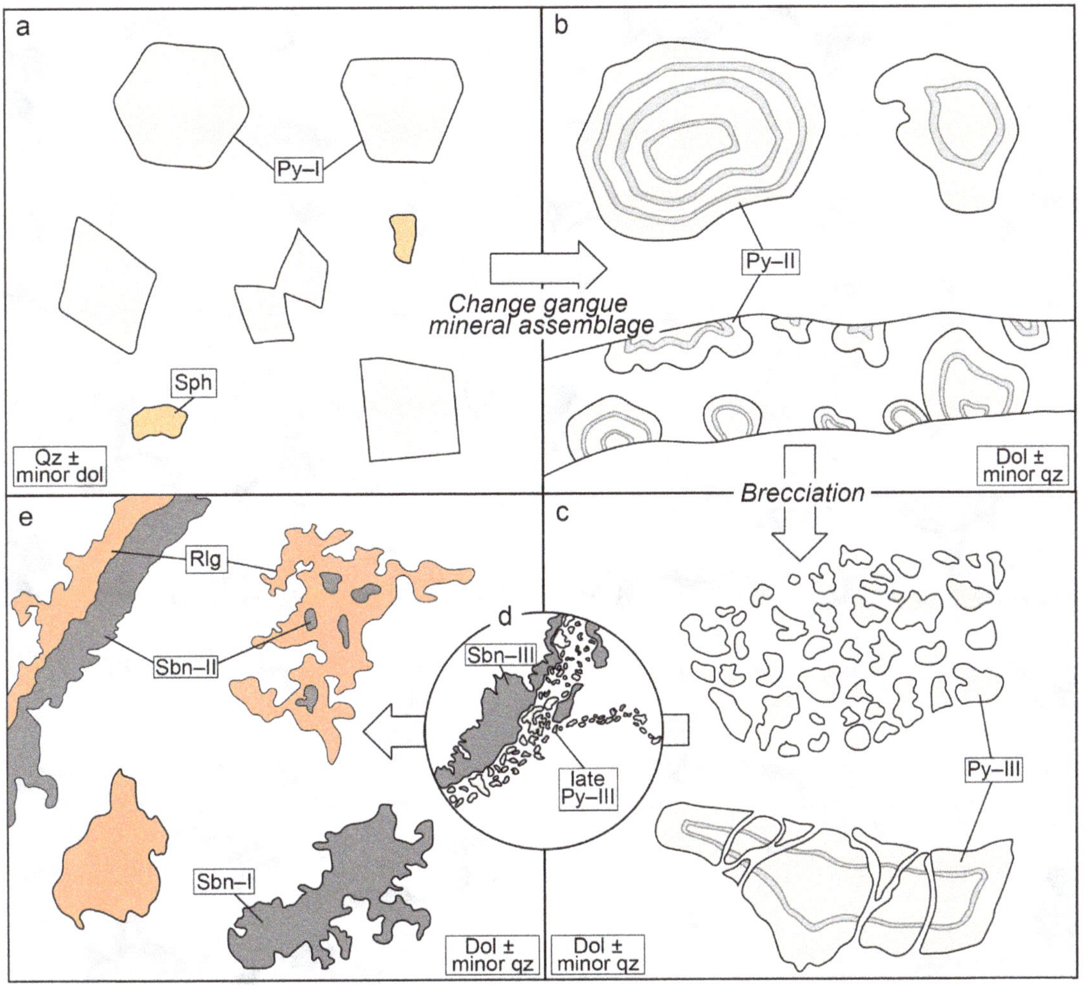

Figure 9. Schematic representation depicting the textural evolution of sulfides in the Matra Mine. (**a**) Pyrite (Py–I) and sphalerite in prevalent quartz gangue. (**b**) Pyrite (Py–II) in prevalent dolomite gangue. (**c**) Pyrite (Py–III) in prevalent dolomite gangue. (**d**) Pyrite (late Py–III) and stibnite (Sbn–III) in prevalent dolomite gangue. (**e**) Stibnite (Sbn–I and Sbn–II) and realgar in dolomite and quartz gangue.

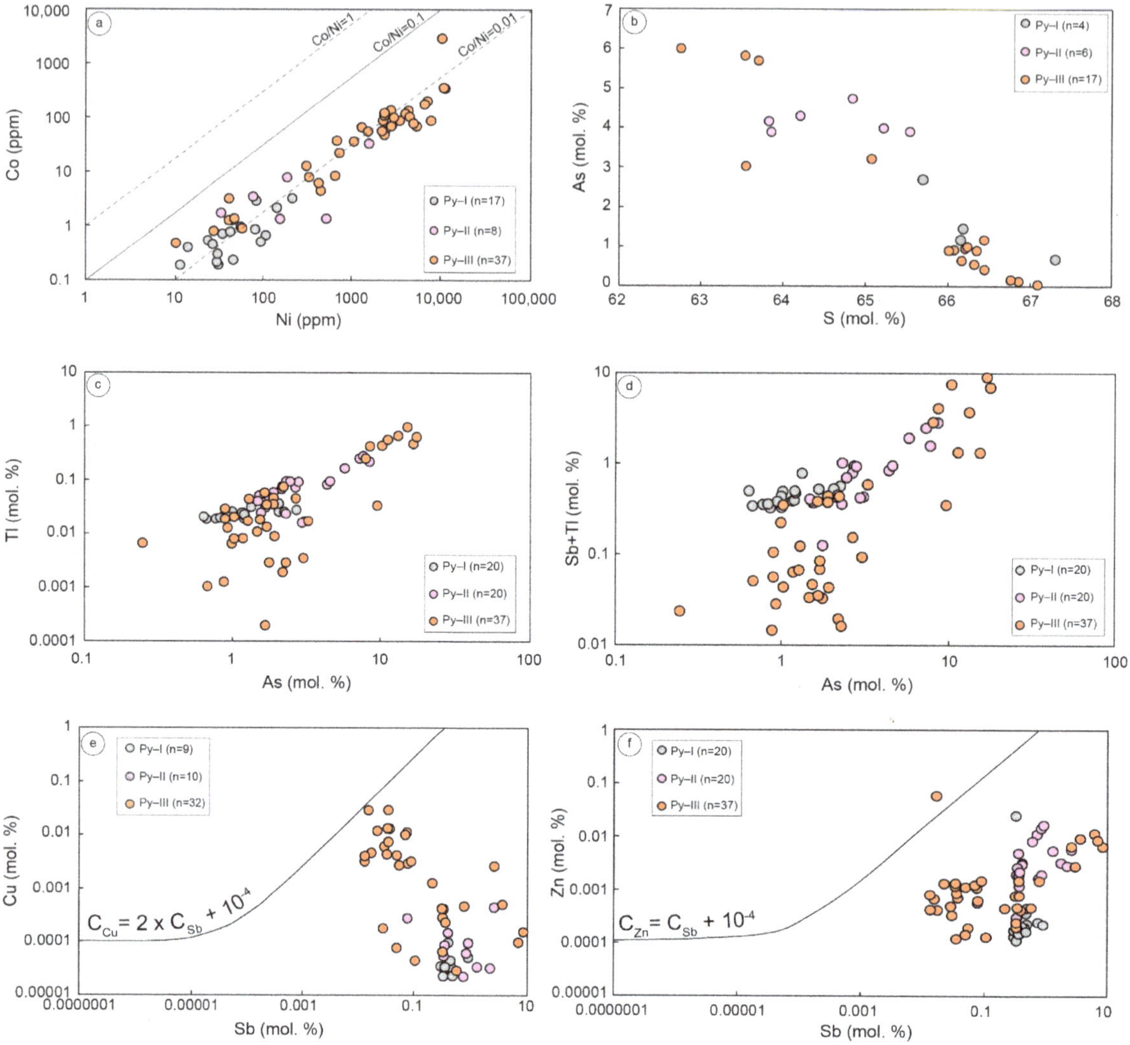

Figure 10. (**a**) Ni (ppm) versus Co (ppm) diagram showing data generated by LA–ICP–MS analysis of pyrite ($R^2 = 0.55$); (**b**) linear relationship between S (mol. %) and As (mol. %) based on data from EPMA point analysis ($R^2 = -0.91$); (**c**) relationship of As (mol. %) and Tl (mol. %) in pyrite samples shown by data from LA–ICP–MS analysis ($R^2 = 0.91$); (**d**) relationship of As (mol. %) and Sb+Tl (mol. %) in pyrite samples shown by data from LA–ICP–MS analysis ($R^2 = 0.84$); (**e**) Sb (mol. %) and Cu (mol. %) plot with a solubility line ($C_{Cu} = 2 \times C_{Sb} + 10^{-4}$, where C_{Cu} and C_{Sb} are the mol. % concentrations of Cu and Sb in pyrite), proposed by George et al. [58]; and (**f**) Sb (mol. %) and Zn (mol. %) plot with a solubility line ($C_{Zn} = C_{Sb} + 10^{-4}$, where C_{Zn} and C_{Sb} are the mol. % concentrations of Zn and Sb in pyrite), proposed by George et al. [58].

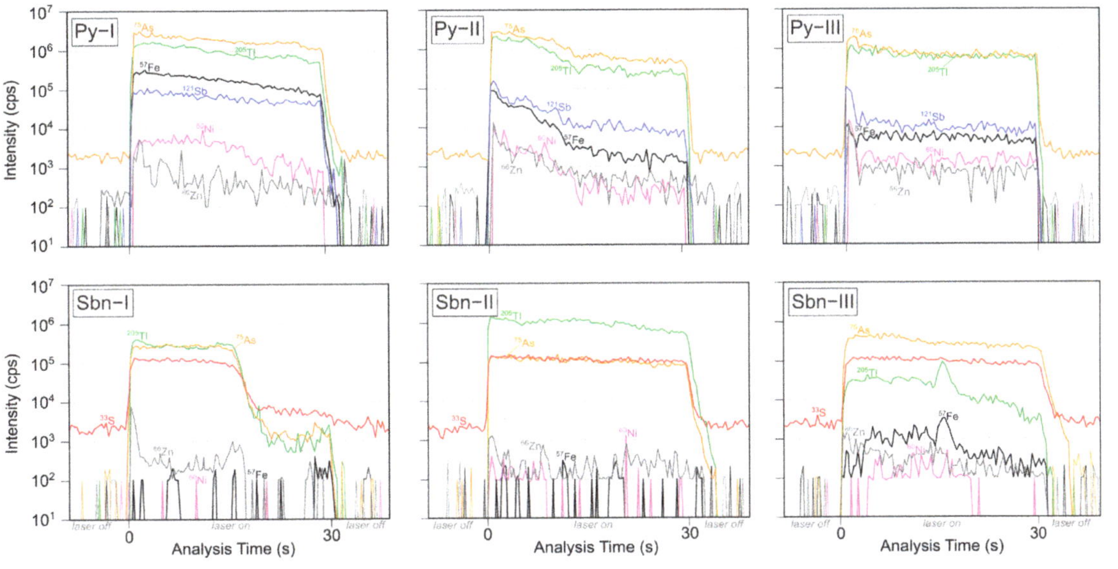

Figure 11. Representative LA–ICP–MS depth–concentration profiles of selected trace elements in pyrite (Py–I, Py–II, and Py–III) and stibnite (Sbn–I, Sbn–II, and Sbn–III). In most cases, the spectra appear flat and only rare peaks possibly related to the presence of NP are documented.

Arsenic in pyrite can have different oxidation states. This is represented by As^{1-} [71], As^{2+} [72], As^{3+} [73], or amorphous As–Fe–S nanoparticles (As^0; [74]). The negative correlation between As and S (R = −0.91; Figure 10b) indicates that As enters the structure as an anion [74]. This is confirmed by plotting data generated by EPMA analyses of Py–I, Py–II, and Py–III in a ternary diagram (Figure 12). Deditius et al. [73–75] stated that, as is evidenced in the MD, the majority of pyrite exhibits an oxidation state of 1+, which is the case for Matra pyrite (Figure 12). However, a secondary distribution typical of Matra pyrite, in which the As has a 3+ oxidation state, appears to occur (Figure 12). This suggests an incipient evolution of the system's oxidation state that led to the crystallization of realgar, orpiment, and hörnesite during the hydrothermal event.

The different types of pyrite in samples from the MD are also enriched in Sb and Tl. This is illustrated by the broad linear relationship (R = 0.91) between As and Tl concentrations (Figure 10c,d), as reported by D'Orazio et al. [70]. Nevertheless, the relationship between the As, Sb, and Tl (R = 0.82) is less clear, yet it suggests a close relation. What is evident is that data for Py–I and Py–II plot as two distinct clusters, whereas Py–III is easily distinguished due to the large range in compositions (Figure 10d).

George et al. [58] proposed that it is possible to verify whether Sb controls how elements (e.g., Cu and Zn) are concentrated in the pyrite structure. This was deduced by hypothesizing solubility lines and considering the following relationships: $C_{Zn} = C_{Sb} + 10^{-4}$ and $C_{Cu} = 2 \times C_{Sb} + 10^{-4}$ (where C_{Zn}, C_{Cu}, and C_{Sb} are the mol. % concentrations of Zn, Cu, and Sb in pyrite). If data plot below the line (e.g., a solubility curve), it may be expected that a solid solution occurs. Conversely, where data plot above the line, this likely indicates the presence of nanoparticles or larger inclusions. Data for pyrite from the MD are located below the line in a plot of Cu vs. Sb, suggesting that Cu occurs in solid solution with all the pyrite types (Figure 10e). Similarly, data in a Zn vs. Sb plot, excluding one point, are located below the line and suggest that Zn occurs in solid solution with pyrite (Figure 10f). The anomalous point likely indicates the presence of sphalerite nanoparticles.

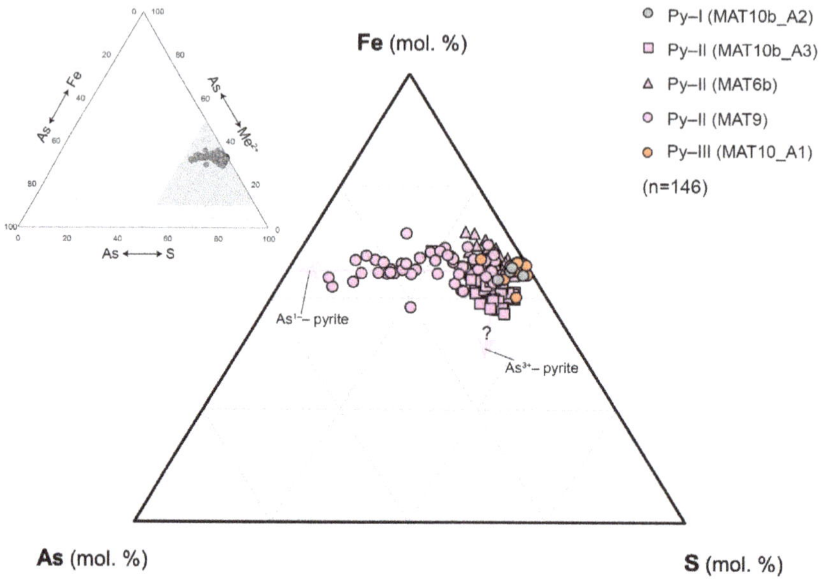

Figure 12. Ternary As–Fe–S diagram showing the data from EPMA points and micro-maps for pyrite samples from the Matra Mine. The main trends related to the As oxidation states are taken from Deditius et al. [75].

Stibnite mainly occurs as small euhedral crystals that are either disseminated in gangue or combine to form veins. Although the three types of stibnite are associated with different mineral assemblages, there is little variation in their geochemical composition (Figure 7). As with pyrite, there is no compelling evidence to indicate there are different generations of stibnite. Rather, they are probably the same generation, with slightly different textures.

Element substitutions in stibnite are not as well understood compared with pyrite [76–86]. The $Sb^{3+} \leftrightarrow As^{3+}$ replacement mechanism in stibnite is favored because of the similar ionic radii and chemical properties of these elements [87,88]. In fact, a linear relationship between As and Sb (R = –0.97; Figure 13a) is observed in MD stibnite, supporting this replacement mechanism. Linear relationships are also evident for Pb and Cu (R = 0.48; Figure 13b), as previously demonstrated [76]. Despite these relationships, it is unclear how the elements enter stibnite, and it is also unknown how high concentrations of Fe, Zn, and Tl can occur. Substitution mechanisms such as $Cu^{+} + Pb^{2+} + As^{3+} \leftrightarrow 2Sb^{3+} + \leftrightarrow$ and $3Sb^{3+} \leftrightarrow As^{3+} + 2Cu^{+} + Hg^{2+} + Pb^{2+}$ were proposed by Fu et al. [76] and Song et al. [78], respectively. However, these are not applicable to stibnite from the MD. When comparing trace element compositions for stibnite documented in the literature, we observe that for MD ore, As and Pb + Hg are at least one order of magnitude higher and lower, respectively (Figure 13c). The abundance in As is not surprising for the MD, and the small contents of Pb + Hg in stibnite are supported by a lack of Pb–Hg–bearing minerals [49]. Additionally, Hg was not measured during LA–ICP–MS acquisition, as Hg contents in bulk ore samples were below 31 ppb [89]. In contrast, Tl has average concentrations of 100–1000 ppm (Figure 13d). Unlike mineralizations in other hydrothermal systems where Cu + Ag + Tl contents in stibnite define different trends when compared with Fe, the influence of Cu and Ag is negligible and no linear relationship between Fe and Tl can be observed in data for the MD (Figure 13e).

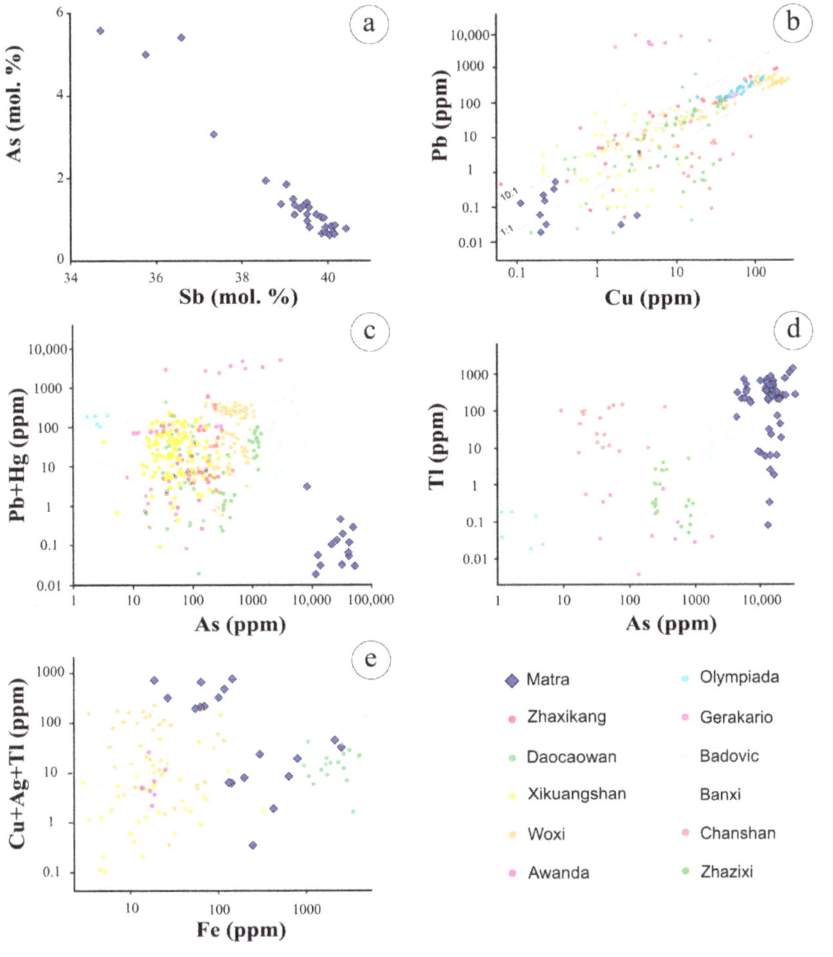

Figure 13. (**a**) Plot of Sb (mol. %) versus As (mol. %) in stibnite of the Matra deposit based on EPMA data ($R^2 = -0.97$). (**b–e**) Comparison of the stibnite from the Matra deposit with other published data [76–86]. (**b**) Cu (ppm) versus Pb (ppm). (**c**) As (ppm) versus Pb + Hg (ppm). (**d**) As (ppm) versus Tl (ppm). (**e**) Fe (ppm) versus Cu + Ag + Tl (ppm).

Although the mechanisms by which metals are incorporated into the pyrite structure are not entirely clear, Reich et al. [71] suggested that As–rich pyrites formed under nonequilibrium conditions may have precipitated rapidly at low temperatures (T < 250 °C), incorporating trace metals into the mineral structure. This incorporation is facilitated by the presence of extra vacancies and surface defects resulting from rapid growth and disequilibrium. Due to the limited reference data available, it is possible that during formation under nonequilibrium conditions, stibnite precipitates rapidly in low–temperature systems (T < 250 °C) to become enriched in trace metals.

5.2. Paragenetic Sequence

The mineral paragenesis developed for the MD records pre–ore and main–ore stages, which are subdivided into a first and second event, followed by a supergene stage (Figure 14).

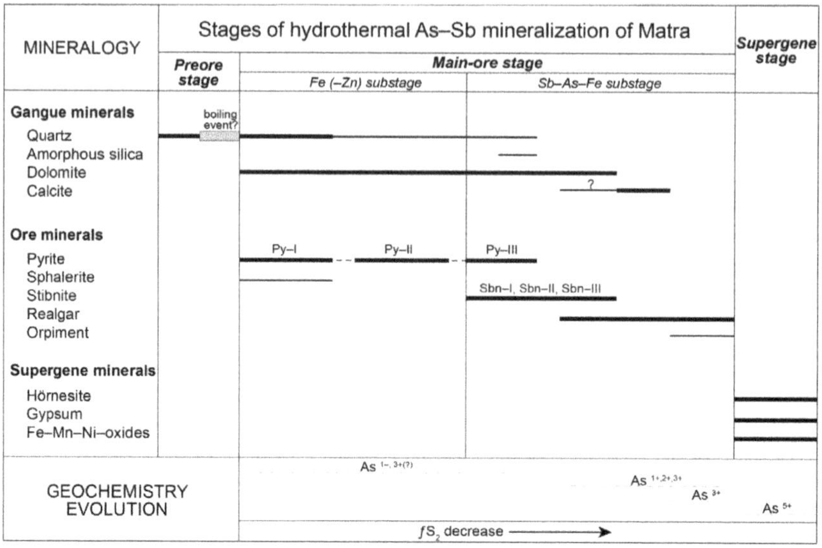

Figure 14. Paragenetic sequence for the Matra deposit showing pre–ore, main–ore, and supergene stages.

The pre–ore stage is represented by the formation of quartz devoid of sulfide minerals. Plumose, crustiform, jigsaw, and feathery textures in quartz (Figure 8) record a late, short–lived boiling event [55,90]. This was most likely due to tectonic decompression caused by the reactivation of the Matra Fault. Although the boiling event was likely brief, there were two important consequences. The first is that Hg partitioned into the vapor phase, which explains the low concentrations of Hg ($\leq$31 ppb) in ore samples [89], while Sb and As remained in the aqueous phase at low temperatures [91]. The second consequence is that a loss of CO_2 prompted the precipitation of ferroan dolomite.

The main–ore stage began with the Fe (–Zn) substage with the deposition of dolomite containing Py–I, Py–II, and minor sphalerite (Figure 14). Although the textures of pyrite changed over time (from euhedral to anhedral), the only potentially important geochemical change was a slight decrease in As content from Py–I to Py–II. There is no evidence that boiling occurred during the formation of gangue quartz and dolomite or associated sulfides.

Subsequently, the main–ore stage evolved into the Sb–As–Fe substage, which is represented by the deposition of As–Sb sulfides and Py–III (Figure 14). At this point, the hydrothermal system was open due to reactivation of the Matra Fault, which brecciated Py–III and permitted continuous fluid flow to allow mineralization, including stibnite veins (e.g., Sbn–I, Sbn–II, and Sbn–III). An important point is that the deposition of realgar indicates a change in the oxidation state of As relative to As in earlier pyrite. The formation of As–As bonds in realgar suggests that As was not completely oxidized by S, and the average oxidation state is considered to be As^{2+}. In realgar, three different scenarios may be envisaged for the bonding of As atoms: (1) one bond to three S atoms; (2) two bonds to two S atoms; and (3) one bond to two As atoms and one S atom (e.g., [92]). These scenarios result in oxidation states of 3+, 2+, and 1+, with the average oxidation state for realgar being considered 2+ [93]. The main–ore stage ended with crystallization of orpiment and hörnesite. These minerals contain As atoms in different oxidation states. Macroscopic observations of orpiment and realgar crystals deposited in the same vugs indicate that both minerals were stable based on the pristine nature of the crystals, even though As atoms have oxidation states of 3+ in orpiment and 2+ in realgar. A similar relationship is documented by the coexistence of hörnesite and realgar in unoxidized ore samples. Even though the oxidation states of As atoms are 5+ for hörnesite and 2+ in realgar, the realgar crystals exhibit a vitreous luster and appear pristine.

The main mechanism for the precipitation of sulfides is therefore considered to be changes in sulfur fugacity. These changes occurred because the reactivation of the Matra Fault allowed an influx of oxygenated meteoric water into the hydrothermal system that resulted in significant fluid mixing. After the deposition of pyrite at the MD, mineralogical changes in the main–ore stage document a transition from reducing to oxidizing conditions that caused changes in the oxidation state of As atoms (Figure 14). The distribution of data in a Fe–As–S ternary diagram (Figure 12), interpreted by following Deditius et al. [75], shows that As enters the structure of pyrite in a reduced state (As^{1-}) compared to an oxidized state (As^{2+}) for realgar. Due to the continued influx of meteoric water, the oxidation state of As changed from 3+ to 5+ during the deposition of orpiment and hörnesite, respectively. An important point is that the meteoric water did not introduce significant quantities of other elements to alter the composition of the hydrothermal fluids. Instead, cooling caused by fluid mixing decreased the solubility of As and Sb [91], and a concurrent decrease in fS_2 can explain the formation of stibnite and realgar in late–stage hydrothermal systems. A salient characteristic of the hydrothermal fluids that produced mineralization was that they were neutral or slightly reduced, as indicated by the widespread occurrence of dolomite gangue during the main–ore stage. Fluid mixing and interaction with the carbonate gangue likely caused changes in the fugacity of oxygen (fO_2) and carbon dioxide (fCO_2). This could explain how arsenic sulfides (As^{2+} and As^{3+}) and arsenates (As^{5+}) can coexist.

Another possible factor that could contribute to the precipitation of sulfides is a decrease in temperature. Stibnite could be most sensitive to temperature, as demonstrated by other authors [94–96]. To verify this hypothesis, particular attention was paid to the twinning of dolomite gangue, which has been shown to provide a simple constraint to discriminate <350 °C [97]. The results seem to be consistent with temperatures <200 °C, but it is not clear whether calcite forms at the same time as stibnite or only later. Therefore, the role of temperature in the precipitation of sulfides along the Matra Fault requires future investigation.

As noted above, the reactivation and movement of the Matra Fault allowed increasing amounts of oxidized meteoric water to enter the hydrothermal system. Cooling associated with fluid mixing is an efficient mechanism for the deposition of stibnite [97] and As–sulfide minerals [91]. When hydrothermal fluids ascended along the Matra Fault, changes in physico–chemical conditions (e.g., temperature and pH) caused a decrease in the solubility of metals and precipitation of As and Sb minerals. Even if Sb was transported as a hydroxide complex [e.g., $Sb(OH)_3$], significant precipitation would occur once temperatures were below 250–300 °C [98–100]. The sensitivity of Sb to temperature is further demonstrated by how the solubility of stibnite decreases with temperature [98]. The absence of base metals at the MD, except for minor sphalerite (Figure 14), supports the hypothesis that chloride ions were not complexing agents and that fluid temperatures were <300 °C [101,102]. Therefore, it is possible that several parameters could have changed the valence state of a metalloid, allowing the same element to crystallize with different valence states.

The supergene minerals are not related to the hydrothermal system and formed due to surface weathering. This is evident in the widespread occurrence of hörnesite in mine waste and outcrops of gossan containing As–bearing minerals. Gypsum likely formed due to the oxidation of sulfides. Other metals that were mobilized by weathering formed thin films consisting of Ni and Fe–Mn oxides.

5.3. Geodynamic Evolution of the Matra Deposit

A comparison of the characteristics of mineralization at the MD and Tuscany Carlin–style deposits is made in this section. Similarities and differences in their geochemistry are examined in the context of tectonic and magmatic evolution. Regarding ore mineralogy, many similarities between the MD and Carlin–style gold deposits in Tuscany can be observed, which could suggest a possible genetic link. A common sulfide assemblage at both localities consists of stibnite and pyrite ± realgar ± sphalerite. However, there are important differences in geochemical signatures, with Tuscany represented by the assemblage of Au–Hg–Tl–Ba [8,9] compared with the MD, which is characterized by Tl and the absence of the other elements. Significant differences also exist in the composition of hydrothermal fluids that formed quartz prior to the realgar–orpiment–stibnite–carbonate mineralization. In Tuscany, acidic ore fluids decalcified ferroan limestones that released the Fe necessary for sulfidation and Au deposition in arsenian pyrite hosted by jasperoids [10,103]. At the MD, there is no evidence of pre–existing ferroan carbonates, acidic ore fluids, or decalcification of the host rock. As a consequence, jasperoids did not form at the MD. Another important feature of the MD is the pre–ore quartz, which displays textures that are indicative of boiling, which has never been reported in Tuscany mineralization. The two–phase liquid–rich fluid inclusions in the Tuscany deposits have T_h values of 123°–245 °C and salinities <7 wt.% NaCl equiv. [8]. All these factors support a significant meteoric water component in the hydrothermal system, as we have proposed for the MD. Currently, there are no fluid inclusion or stable isotope data for the MD to allow direct comparison.

Structures also had a prominent role in mineralization in both the MD and Tuscany deposits. Filimon et al. [5] established that mineralization at the MD was structurally controlled by the Matra Fault, which developed during a post–Alpine event linked to the opening of the Tyrrhenian Sea [104]. At a regional scale, the larger N–S trending fault network is well–preserved in Alpine Corsica [6] and the northern Tyrrhenian Sea [105]. It formed due to slab rollback and the consequent northeastward migration of the Apennine frontal thrust [45]. This extension also triggered the post–orogenic magmatism of the Tuscany Magmatic Province, from eastern Corsica to southern Tuscany.

Magmatic evolution is the last point to be examined. The only magmatic activity in Corsica is represented by the small lamproite outcrops of Sisco (~14 Ma; [46]). Due to the absence of large–scale and prolonged magmatic activity in eastern Corsica, there was no differentiation in a deeper magma to supply a metal–rich (e.g., Au) vapor phase. Instead, extensional tectonism provided pathways for fluid circulation and the leaching of metals from mafic–ultramafic country rocks by hydrothermal solutions of meteoric origin.

After sporadic manifestations of igneous activity in the Tyrrhenian Sea (e.g., Capraia: ~7 Ma, Elba: 8.4–5.9 Ma, Montecristo: 7 Ma, and Giglio: 5 Ma), magmatism migrated eastward. This is represented by the emplacement of intrusions; the formation of volcanic complexes (e.g., San Vincenzo and Campiglia Marittima: 5.7–4.3 Ma, Gavorrano: 4.4 Ma, Roccastrada: 2.5–2.2 Ma, Amiata: 0.3–0.2 Ma, and Latera: 0.6–0.2 Ma); and the active geothermal fields of Larderello, Monte Amiata, and Latera ([10,106]; Figure 1a). This shows how the intensity of igneous activity increased with time in the Tuscany Magmatic Province and elevated its role in the formation of the Tuscany Carlin–style gold deposits [10]. The greater areal extent of hydrothermal alteration and the existence of large volumes of more–evolved igneous rocks are factors that support the enrichment of ore metals that generated Carlin–style gold deposits proximal to the geothermal fields at Larderello, Monte Amiata, and Latera [10].

The above comparison highlights how mineralization in the MD and Tuscany ore deposits formed in different ways and at different times. However, given their common

origin related to Tyrrhenian Sea extension from the Miocene onward, further constraints related to the metallogenesis in the peri–Tyrrhenian area are needed.

6. Conclusions

The As–Sb mineralization at the MD formed in a hydrothermal system partly controlled by reactivation of the Matra Fault, which was linked to the opening of the Tyrrhenian back–arc basin. This normal fault juxtaposes mafic–ultramafic rocks, with associated carbonate sedimentary covers. The structurally controlled mineralization consists of pre–ore quartz that transitioned to main–ore–stage dolomite after a short–lived boiling event. Sulfide mineralization in the Fe (–Zn) substage of the main–ore stage is characterized by three types of pyrite that include idiomorphic (Py–I) and anhedral (Py–II) forms. The disjointed/fractured texture of (Py–III) pyrite records activation of the Matra Fault prior to the Sb–As–Fe substage of the main–ore event. This event coincided with the influx of meteoric water into the hydrothermal system. Supporting evidence is a change from reducing (high–fS_2) to oxidizing (low–fS_2) conditions indicated by oxidation numbers of $As^{1-,3+}$ (pyrite), As^{2+} (realgar), As^{3+} (orpiment), and As^{5+} (hörnesite). Three types of stibnite were also identified by distinct textures, but there is little variation in geochemistry. The trace elements with the highest concentrations in pyrite and stibnite are As–Sb–Tl–Ni and As–Fe–Tl, respectively. These elements entered the lattice of sulfide minerals under a nonequilibrium state at a low temperature (<250 °C). The combination of these conditions with rapid mineral precipitation can promote the scavenging of metals and allow metals to be incorporated into sulfides. Analogous pyrite + stibnite ± realgar ± sphalerite mineralization is found in the Tuscany Magmatic Province, developed east of the Tyrrhenian Sea and during the migration of the Apennine orogenic system. Although there is an indisputable tectonic link between the MD and Carlin–style gold mineralization in Tuscany, the MD has unique features that distinguish it from all the other peri–Tyrrhenian age deposits.

Supplementary Materials: The following supporting information can be downloaded at https://www.mdpi.com/article/10.3390/min15080814/s1: Table S1: X-ray fluorescence (XRF) analyses of major (Wt%) and trace (ppm) element composition of samples from the Matra mine; Table S2: Trace elements concentration (ppm) in pyrites from the Matra mine; Table S3: Trace elements concentration (ppm) in stibnite from Matra mine; Table S4: Major elements (wt. %) data of sulfides and dolomite from Matra mine; Table S5: XRD spectra and data of orpiment and hörnesite from Matra mine (* https://rruff.info).

Author Contributions: Conceptualization, D.I.F. and M.D.R.; methodology, J.A.G. and E.S.; software, D.I.F.; validation, J.A.G. and E.S.; formal analysis, D.I.F. and E.S.; investigation, J.A.G.; resources, M.D.R.; data curation, D.I.F. and E.S.; writing—original draft preparation, D.I.F.; writing—review and editing, E.S.; visualization, J.A.G.; supervision, J.A.G. and E.S.; project administration, M.D.R.; funding acquisition, E.S. and M.D.R. All authors have read and agreed to the published version of the manuscript.

Funding: This research was funded by "PRA project, grant number PRA2022–Zarone" by the University of Pisa and "FAR-2024–Saccani" by the University of Ferrara.

Data Availability Statement: The original contributions presented in this study are included in the article/Supplementary Materials. Further inquiries can be directed to the corresponding author.

Acknowledgments: The authors would like to thank two anonymous reviewers for their constructive reviews and the editor for managing this paper. Technical support and discussions with A. Risplendente, B. Ledérsert, C. Biagioni, S. Pagnotta, and M. Masotta also improved the manuscript. Financial support was provided by V. Zarone (PRA2022 project, Università di Pisa) and the University of Ferrara (FAR–2024–Saccani).

Conflicts of Interest: The authors declare no conflicts of interest.

References

1. Mori, C.; Orsini, A.; Migon, C. Impact of arsenic and antimony contamination on benthic invertebrates in a minor Corsican river. *Hydrobiologia* **1999**, *392*, 73–80. [CrossRef]
2. Orcel, J. The realgar deposit of Matra (Corsica). *Bull. Société Française Minéralogie Cristallogr.* **1921**, *44*, 98–104. (In French) [CrossRef]
3. Orcel, J. Mineralogical and petrographic notes on Corsica. *Bull. Société Des Sci. Hist. Naturelles Corse* **1924**, *44*, 65–127. (In French)
4. Feraud, J. Les Gisements De Sulfures D'arsenic Du Sud-Est De La France Mineralization Liées-Aux-Strates Et Gites Filoniens. Ph.D. Thesis, Universitè de Paris VI, Paris, France, 1974. (In French).
5. Filimon, D.I.; Vannucci, G.; La Rosa, A.; Zarone, V.; Di Rosa, M. The geological setting of Matra ore deposit (France): Insights from lithostratigraphy and polyphase deformation history. *J. Maps* **2024**, *20*, 2394502. [CrossRef]
6. Gueydan, F.; Brun, J.-P.; Phillippon, M.; Noury, M. Sequential extension as a record of Corsica Rotation during Apennines slab roll-back. *Tectonophysics* **2017**, *710*, 149–161. [CrossRef]
7. Chamot-Rooke, N.; Gaulier, J.M.; Jestin, F. Constraints on Moho depth and crustal thickness in the Liguro-Provençal basin from a 3D gravity inversion: Geodynamic implications. *Geol. Soc. Lond. Spec. Publ.* **1999**, *156*, 37–61. [CrossRef]
8. Lattanzi, P. Epithermal precious metal deposits of Italy–an overview. *Miner. Depos.* **1999**, *34*, 630–638. [CrossRef]
9. Brogi, A.; Fulignati, P. Tectonic control on hydrothermal circulation and fluid evolution in the Pietratonda–Poggio Peloso (southern Tuscany, Italy) carbonate-hosted Sb-mineralization. *Ore Geol. Rev.* **2012**, *44*, 158–171. [CrossRef]
10. Sillitoe, R.H.; Brogi, A. Geothermal systems in the northern Apennines, Italy: Modern analogues of Carlin-style gold deposits. *Econ. Geol.* **2021**, *116*, 1491–1501. [CrossRef]
11. European Commission. *Directorate-General for Internal Market, Industry, Entrepreneurship and SMEs, Study on the Critical Raw Materials for the EU (2023)–Final Report*; Publications Office of the European Union: Luxembourg, Luxembourg, 2023.
12. Durand-Delga, M. Main features of Alpine Corsica and correlation with the Ligurian Alps. *Mem. Della Soc. Geol. Ital.* **1984**, *28*, 285–329. (In French)
13. Rossi, P.; Oggiano, G.; Cocherie, A. A restored section of the "southern Variscan realm" across the Corsica–Sardinia microcontinent. *Comptes Rendus. Géoscience* **2009**, *341*, 224–238. [CrossRef]
14. Malavieille, J.; Chemenda, A.; Larroque, C. Evolutionary model for the Alpine Corsica: Mechanism for ophiolite emplacement and exhumation of high-pressure rocks. *Terra Nova* **1998**, *10*, 317–322. [CrossRef]
15. Marroni, M.; Pandolfi, L. The architecture of an incipient oceanic basin: A tentative reconstruction of the Jurassic Liguria-Piemonte basin along the Northern Apennines-Alpine Corsica transect. *Int. J. Earth Sci.* **2007**, *96*, 1059–1078. [CrossRef]
16. Vitale Brovarone, A.; Beyssac, O.; Malavieille, J.; Molli, G.; Beltrando, M.; Compagnoni, R. Stacking and metamorphism of continuous segments of subducted lithosphere in a high-pressure wedge: The example of Alpine Corsica (France). *Earth Sci. Rev.* **2013**, *116*, 35–56. [CrossRef]
17. Di Rosa, M.; Frassi, C.; Malasoma, A.; Marroni, M.; Meneghini, F.; Pandolfi, L. Syn-exhumation coupling of oceanic and continental units along the western edge of the Alpine Corsica: A review. *Ofioliti* **2020**, *45*, 71–102.
18. Saccani, E.; Principi, G.; Garfagnoli, F.; Menna, F. Corsica ophiolites: Geochemistry and petrogenesis of basaltic and metabasaltic rocks. *Ofioliti* **2008**, *33*, 187–207.
19. Saccani, E. A new method of discriminating different types of post-Archean ophiolitic basalts and their tectonic significance using Th–Nb and Ce–Dy–Yb systematics. *Geosci. Front.* **2015**, *6*, 481–501. [CrossRef]
20. Ribes, C.; Manatschal, G.; Ghienne, J.F.; Karner, G.D.; Johnson, C.A.; Figueredo, P.H.; Incerpi, N.; Epin, M.A. The syn-rift stratigraphic record across a fossil hyper-extended rifted margin: The example of the northwestern Adriatic margin exposed in the Central Alps. *Int. J. Earth Sci.* **2019**, *108*, 2071–2095. [CrossRef]
21. Cavazza, W.; Zattin, M.; Ventura, B.; Zuffa, G.G. Apatite fission track analysis of Neogene exhumation in northern Corsica (France). *Terra Nova* **2001**, *13*, 51–57. [CrossRef]
22. Marroni, M.; Pandolfi, L. Deformation history of the ophiolite sequence from the Balagne Nappe, northern Corsica: Insights in the tectonic evolution of the Alpine Corsica. *Geol. J.* **2003**, *38*, 67–83. [CrossRef]
23. Durand-Delga, M.; Peybernès, B.; Rossi, P. Arguments en faveur de la position, au Jurassique, des ophiolites de Balagne (Haute-Corse, France) au voisinage de la marge continentale européenne. *Comptes Rendus De. Académié De. Sci. Paris.* **1997**, *325*, 973–981. [CrossRef]
24. Beccaluva, L.; Deriu, M.; Macciotta, G.; Savelli, C.; Venturelli, G. Geochronology and Magmatic Character of the Pliocene-Pleistocene Volcanism in Sardinia (Italy). *Bull. Volcanol.* **1977**, *40*, 153–168. [CrossRef]
25. Venturelli, G.; Capedri, S.; Thorpe, R.S.; Potts, P.J. Rare-earth and other element distribution in some ophiolitic metabasalts of Corsica, Western Mediterranean. *Chem. Geol.* **1979**, *24*, 39–353. [CrossRef]
26. Di Rosa, M.; Meneghini, F.; Marroni, M.; Frassi, C.; Pandolfi, L. The coupling of high-pressure oceanic and continental units in Alpine Corsica: Evidence for syn-exhumation tectonic erosion at the roof of the plate interface. *Lithos* **2020**, *354*, 105328. [CrossRef]

27. Venturelli, G.; Thorpe, R.S.; Potts, P.J. Rare earth and trace element characteristics of ophiolitic metabasalts from the Alpine-Apennine belt. *Earth Planet. Sci. Lett.* **1981**, *53*, 109–123. [CrossRef]
28. Guelfi, R.; Maremmani, A.; Di Rosa, M.; Meneghini, F.; Pandolfi, L.; Marroni, M. Adding pieces to the puzzle of the Western Tethys Oceanic Basin structure: The architecture of the Inzecca Unit in the Noceta-Vezzani area (Alpine Corsica, France). *Episodes* **2025**, *48*, 163–179. [CrossRef] [PubMed]
29. De Cesari, F.; Di Rosa, M.; Sanità, E.; Pandolfi, L.; Marroni, M. Long-lived sedimentation in the Western Tethys oceanic basin: Revisiting the metasediments of Bagliacone-Riventosa Formation (Corsica, France). *Int. Geol. Rev.* **2024**, *67*, 906–932. [CrossRef]
30. Ohnenstetter, M. La serie ophiolitifere de Rospigliani (Corse) est- elle un temoin des phenomenes tectoniques, sedimentaires et magmatiques lies au fonctionnement des zones transformantes? *Comptes Rendus De. L'academie Des. Sci. De. Paris.* **1979**, *289*, 1199–1202. (In French)
31. Lahondére, D. Les schistes blues et les éclogites à lawsonite des unités continentals et océanique de la Corse alpine: Nouvelles donnée pétrologique et structurales (Corse). Ph.D. Thesis, Université Montpellier 2 Sciences et Techniques, Montpellier, France, 1996.
32. Padoa, E. Les ophiolites du massif d'Inzecca (Corse alpine): Litostratigraphie, structure géologique et évolution géodynamique. *Géologie De. Fr.* **1999**, *3*, 37–48.
33. Elter, P.; Pertusati, P.C. Considerations on the Alps–Apennines boundary and its relationship with the arc of the Western Alps. *Mem. Della Soc. Geol. Ital.* **1973**, *12*, 359–375. (In Italian)
34. Lagabrielle, Y.; Polino, R. A structural diagram of the domain of lustrous ophiolitic shales to the northwest of the Mont Viso massif (Southwestern Alps) and its implications. *Comptes Rendus De. L'académie Des. sciences. Série 2 Mécanique Phys. Chim. Sci. De. L'univers Sci. De. La. Terre* **1988**, *306*, 921–928. (In French)
35. Marroni, M.; Meneghini, F.; Pandolfi, L. A revised subduction inception model to explain the Late Cretaceous, double-vergent orogen in the precollisional western Tethys: Evidence from the Northern Apennines. *Tectonics* **2017**, *36*, 2227–2249. [CrossRef]
36. Di Rosa, M.; Sanità, E.; Frassi, C.; Lardeaux, J.M.; Corsini, M.; Marroni, M.; Pandolfi, L. A journey of the continental crust to and from mantle depth: The P–T–t–d path of Venaco unit, Alpine Corsica (France). *Geol. J.* **2023**, *59*, 422–440. [CrossRef]
37. Zarki-Jakni, B.; Van Der Beek, P.; Poupeau, G.; Sosson, M.; Labrin, E.; Rossi, P.; Ferrandini, J. Cenozoic denudation of Corsica in response to Ligurian and Tyrrhenian extension: Results from apatite fission track thermochronology. *Tectonics* **2004**, *23*, TC1003. [CrossRef]
38. Lahondére, D.; Guerrot, C. Datation Sm-Nd du métamorphisme éclogitique en Corse Alpine: Un argument pour l'existence au Crétacé Supérieur d'une zone de subduction active localisée sous le bloc corso-sarde. *Géologie De. Fr.* **1997**, *3*, 3–11. (In French)
39. Gueguen, E.; Doglioni, C.; Fernandez, M. On the post-25 Ma geodynamic evolution of the western Mediterranean. *Tectonophysics* **1998**, *298*, 259–269. [CrossRef]
40. Fellin, M.G.; Picotti, V.; Zattin, M. Neogene to Quaternary rifting and inversion in Corsica: Retreat and collision in the western Mediterranean. *Tectonics* **2005**, *24*, TC1011. [CrossRef]
41. Jolivet, L.; Daniel, J.M.; Fournier, M. Geometry and Kinematics of ductile extension in Alpine Corsica. *Earth Planet. Sci. Lett.* **1991**, *104*, 278–291. [CrossRef]
42. Brunet, C.; Monié, P.; Jolivet, L.; Cadet, J.P. Migration of compression and extension in the Tyrrhenian Sea, insights from $^{40}Ar/^{39}Ar$ ages on micas along a transect from Corsica to Tuscany. *Tectonophysics* **2000**, *321*, 127–155. [CrossRef]
43. Malinverno, A.; Ryan, W.B.F. Extension in the Tyrrhenian Sea and shortening in the Apennines as a result of arc migration driven by the sinking of the lithosphere. *Tectonics* **1986**, *5*, 227–245. [CrossRef]
44. Kastens, K.A.; Mascle, J.; Auroux, C. *Proc. ODP, Init. Repts. 107 College Station, TX;* (Ocean Drilling Program): Brazos County, TX, USA, 1987. [CrossRef]
45. Rosenbaum, G.; Lister, G.S.; Duboz, C. Relative motion of Africa, Iberia and Europe during Alpine orogeny. *Tectonophysics* **2002**, *359*, 117–129. [CrossRef]
46. Serri, G.; Innocenti, F.; Manetti, P. Geochemical and petrological evidence of the subduction of delaminated Adriatic continental lithosphere in the genesis of the Neogene-Quaternary magmatism of central Italy. *Tectonophysics* **1993**, *223*, 117–147. [CrossRef]
47. Peccerillo, A.; Poli, G.; Donati, C. The Plio-Quaternary magmatism of Southern Tuscany and Northern Latium: Compositional characteristics, genesis and geodynamic significance. *Ofioliti* **2001**, *26*, 229–238.
48. Caron, J.M.; Delcey, R. Lithostratigraphy of the Corsican Schistes Lustrés: Diversity of post-ophiolitic sequences. *Comptes Rendus De. Académié De. Sci. Paris.* **1979**, *288*, 1525–1528. (In French)
49. Caron, J.M.; Loye-Pilot, M.D.; Conchon, O.; Scius, H. Carte géologique de France, feuille Pietra-di-Verde (1115). *BRGM*, 1990; scale 1: 50000, 1 sheet. (In French)
50. Warr, L.N. IMA–CNMNC approved mineral symbols. *Mineral. Mag.* **2021**, *85*, 291–320. [CrossRef]
51. Lanari, P.; Vidal, O.; De Andrade, V.; Dubacq, B.; Lewin, E.; Grosch, E.G.; Schwartz, S. XMapTools: A MATLAB©-based program for electron microprobe X-ray image processing and geothermobarometry. *Comput. Geosci.* **2014**, *62*, 227–240. [CrossRef]

52. Paton, C.; Hellstrom, J.; Paul, B.; Woodhead, J.; Hergt, J. Iolite: Freeware for the visualisation and processing of mass spectrometric data. *J. Anal. At. Spectrom.* **2011**, *26*, 2508–2518. [CrossRef]
53. Belousov, I.; Danyushevsky, L.; Goemann, K.; Gilbert, S.; Olin, P.; Thompson, J.; Lounejecaa, E.; Garbe-Shonberg, D. STDGL3, a Reference Material for Analysis of Sulfide Minerals by Laser Ablation ICP-MS: An Assessment of Matrix Effects and the Impact of Laser Wavelengths and Pulse Widths. *Geostand. Geoanalytical Res.* **2023**, *47*, 493–508. [CrossRef]
54. Kampman, N.; Bertier, P.; Busch, A.; Snippe, J.; Harrington, J.; Pipich, V.; Maskell, A.; Bickle, M. Validating reactive transport models of CO_2-brine-rock reactions in caprocks using observations from a natural CO_2 reservoir. *Energy Procedia* **2017**, *114*, 4902–4916. [CrossRef]
55. Moncada, D.; Mutchler, S.; Nieto, A.; Reynolds, T.J.; Rimstidt, J.D.; Bodnar, R.J. Mineral textures and fluid inclusion petrography of the epithermal Ag–Au deposits at Guanajuato, Mexico: Application to exploration. *J. Geochem. Explor.* **2012**, *114*, 20–35. [CrossRef]
56. Large, R.R.; Maslennikov, V.V.; Robert, F.; Danyushevsky, L.V.; Chang, Z. Multistage sedimentary and metamorphic origin of pyrite and gold in the giant Sukhoi Log Deposit, Lena Gold Province. *Russia. Econ. Geol.* **2007**, *102*, 1233–1267. [CrossRef]
57. Reich, M.; Deditius, A.; Chryssoulis, S.; Li, J.; Ma, C.; Parada, M.A.; Barra, F.; Mittermayr, F. Pyrite as a record of hydrothermal fluid evolution in a porphyry copper system: A SIMS/EMPA trace element study. *Geochem. Cosmochim. Acta* **2013**, *104*, 42–62. [CrossRef]
58. George, L.L.; Biagioni, C.; D'Orazio, M.; Cook, N.J. Textural and trace element evolution of pyrite during greenschist facies metamorphic recrystallization in the southern Apuan Alps (Tuscany, Italy): Influence on the formation of Tl-rich sulfosalt melt. *Ore Geol. Rev.* **2018**, *102*, 59–105. [CrossRef]
59. Liu, K.; Huang, F.; Gao, S.; Zhang, Z.; Ren, Y.; An, B. Morphology of framboidal pyrite and its textural evolution: Evidence from the Logatchev area, Mid-Atlantic Ridge. *Ore Geol. Rev.* **2022**, *141*, 104630. [CrossRef]
60. Fan, L.; Wang, G.; Holzheid, A.; Zoheir, B.; Shi, X.; Lei, Q. Systematic variation in trace element composition of pyrites from the 26°S hydrothermal field, Mid-Atlantic Ridge. *Ore Geol. Rev.* **2022**, *148*, 105006. [CrossRef]
61. Palenik, C.S.; Utsunomiya, S.; Reich, M.; Kesler, S.E.; Wang, L.; Ewing, R.C. "Invisible" gold revealed: Direct imaging of gold nanoparticles in a Carlin-type deposit. *Amer Min.* **2004**, *89*, 1359–1366. [CrossRef]
62. Deditius, A.P.; Reich, M. Constraints on the solid solubility of Hg, Tl, and Cd in arsenian pyrite. *Amer Min.* **2016**, *101*, 1451–1459. [CrossRef]
63. Zhao, H.X.; Frimmel, H.E.; Jiang, S.Y.; Dai, B.Z. LA-ICP-MS trace element analysis of pyrite from the Xiaoqinling gold district, China: Implications for ore genesis. *Ore Geol. Rev.* **2011**, *43*, 142–153. [CrossRef]
64. Picot, P.; Johan, Z. *Atlas des Minéraux Métalliques*; BRGM Elsevier: Amsterdam, The Netherlands, 1977.
65. Loftus-Hills, G.; Solomon, M. Cobalt, nickel and selenium in sulphides as indicators of ore genesis. *Miner. Depos.* **1967**, *2*, 228–242. [CrossRef]
66. Brill, B.A. Trace-element contents and partitioning of elements in ore minerals from the CSA Cu–Pb–Zn deposit, Australia, and implications for ore genesis. *Canad Miner.* **1989**, *27*, 263–274.
67. Reich, M.; Simon, A.C.; Deditius, A.; Barra, F.; Chryssoulis, S.; Lagas, G.; Tardani, D.; Knipping, J.; Bilenker, L.; Sánchez-Alfaro, P.; et al. Trace element signature of pyrite from the Los Colorados iron oxide-apatite (IOA) deposit, Chile: A missing link between Andean IOA and iron oxide–copper–gold systems? *Econ. Geol.* **2016**, *111*, 743–761. [CrossRef]
68. George, L.L.; Biagioni, C.; Lepore, G.O.; Lacalamita, M.; Agrosì, G.; Capitani, G.C.; Bonaccorsi, E.; d'Acapito, F. The speciation of thallium in (Tl, Sb, As)-rich pyrite. *Ore Geol. Rev.* **2019**, *107*, 364–380. [CrossRef]
69. Mederski, S.; Pršek, J.; Majzlan, J.; Kiefer, S.; Dimitrova, D.; Milovský, R.; Bender Koch, C.; Kozień, D. Geochemistry and textural evolution of As–Tl–Sb–Hg-rich pyrite from a sediment-hosted As–Sb–Tl–Pb±Hg±Au mineralization in Janjevo, Kosovo. *Ore Geol. Rev.* **2022**, *151*, 105221. [CrossRef]
70. D'Orazio, M.; Biagioni, C.; Dini, A.; Vezzoni, S. Thallium-rich pyrite ores from the Apuan Alps, Tuscany, Italy: Constraints for their origin and environmental concerns. *Miner. Depos.* **2017**, *52*, 687–707. [CrossRef]
71. Reich, M.; Kesler, S.E.; Utsunomiya, S.; Palenik, C.S.; Chryssoulis, S.L.; Ewing, R.C. Solubility of gold in arsenian pyrite. *Geochem. Cosmochim. Acta* **2005**, *69*, 2781–2796. [CrossRef]
72. Qian, G.; Brugger, J.; Testamale, D.; Skiner, W.; Pring, A. Formation of As(II)-pyrite during experimental replacement of magnetite under hydrothermal conditions. *Geochim. Cosmochim. Acta* **2013**, *100*, 1–10. [CrossRef]
73. Deditius, A.P.; Utsunomiya, S.; Renock, D.; Ewing, R.C.; Ramana, C.V.; Becker, U.; Kesler, S.E. A proposed new type of arsenian pyrite: Composition, nanostructure and geological significance. *Geochim. Cosmochim. Acta* **2008**, *72*, 2919–2933. [CrossRef]
74. Deditius, A.P.; Utsunomiya, S.; Ewing, R.C.; Kesler, S.E. Nanoscale "liquid" inclusions of As–Fe–S in arsenian pyrite. *Amer Min.* **2009**, *94*, 391–394. [CrossRef]
75. Deditius, A.P.; Reich, M.; Kesler, S.E.; Utsunomiya, S.; Chryssoulis, S.L.; Walshe, J.; Ewing, R.C. The coupled geochemistry of Au and As in pyrite from hydrothermal ore deposits. *Geochim. Cosmochim. Acta* **2014**, *140*, 644–670. [CrossRef]

76. Fu, S.; Hu, R.; Bi, X.; Sullivan, N.A.; Yan, J. Trace element composition of stibnite: Substitution mechanism and implications for the genesis of Sb deposits in southern China. *Appl. Geochem.* **2020**, *118*, 104637. [CrossRef]
77. Fu, S.; Wang, T.; Yan, J.; Pan, L.; Wei, L.; Lan, Q.; Fu, S. Formation of the Banxi Sb deposit in Eastern Yangtze Block: Evidence from individual fluid inclusion analyses, trace element chemistry, and He-Ar-S isotopes. *Ore Geol. Rev.* **2022**, *146*, 104949. [CrossRef]
78. Song, X.; Lai, J.; Xu, J.; Liu, X.; Li, B.; He, H.; Wang, Y.; Shi, J.; Wang, C.; Wen, C. Material Source and Genesis of the Daocaowan Sb Deposit in the Xikuangshan Ore Field: LA–ICP–MS Trace Elements and Sulfur Isotope Evidence from Stibnite. *Minerals* **2022**, *12*, 1407. [CrossRef]
79. Silyanov, S.A.; Sazonov, A.M.; Naumov, E.A.; Lobastov, B.M.; Zvyagina, Y.A.; Artemyev, D.A.; Nekrasova, N.A.; Pirajno, F. Mineral Paragenesis, formation stages and trace elements in sulfides of the Olympiada gold deposit (Yenisei Ridge, Russia). *Ore Geol. Rev.* **2022**, *143*, 104750. [CrossRef]
80. Stergiou, C.L.; Melfos, V.; Voudouris, P.; Papadopoulou, L.; Spry, P.G.; Peytcheva, I.; Dimitrova, D.; Stefanova, E. A fluid inclusion and critical/rare metal study of epithermal quartz-stibnite veins associated with the Gerakario Porphyry Deposit, Northern Greece. *Appl. Sci.* **2022**, *12*, 909. [CrossRef]
81. Mederski, S.; Pršek, J.; Dimitrova, D. Distribution of In, Sn, Ga, Ge, and other critical metals in sulfide ores from epithermal listvenite-associated Badovc Pb–Zn–Sb–Ni deposit (Kosovo): Insights from mineralogy and geochemistry. *Ore Geol. Rev.* **2024**, *164*, 105824. [CrossRef]
82. Wang, T.; Fu, S.; Sullivan, N.A.; Lan, J.; Wei, L. Deciphering the ore-forming process of Sb-W deposits through scheelite and stibnite trace element geochemistry. *J. Geochem. Explor.* **2024**, *257*, 107367. [CrossRef]
83. Qui, Z.; Wu, J.; Voudouris, P.; Tombros, S.; Liu, J.; Zhai, D. LA-ICP-MS Trace Element Characteristics and Geological Significance of Stibnite in the Zhaxikang Pb–Zn–Ag–Sb Deposit, Southern Tibet, SW China. *Minerals* **2024**, *14*, 1294. [CrossRef]
84. Ma, G.; Zhao, X.; Mao, Q.; Xue, C.; Wan, J.; Li, J. Gold and antimony metallogenic relationship of the Awanda Au (Sb) deposit, South Tianshan, NW China: Constraints from textures, in-situ trace elements and s-pb isotope systematics. *Ore Geol. Rev.* **2024**, *176*, 106384. [CrossRef]
85. Zhang, J.; Yang, Y.; Wang, Q.; Zhang, H.; Zhu, F. Trace element composition and significance of quartz and stibnite in the Woxi Au-Sb-W deposit, Hunan. *Ore Energy Resour. Geol.* **2025**, *18*, 100082. [CrossRef]
86. Mederski, S.; Pršek, J.; Kluza, K.; Dimitrova, D. LA-ICP-MS Trace Element Composition of Stibnite from the Kizhnica-Hajvalia Badovc Ore Field, Kosovo. In Proceedings of the Mineral Resources in a Changing World 17th Biennial SGA Meeting, Zurich, Switzerland; 2023; pp. 296–299.
87. Rytuba, J.J. Mercury from mineral deposits and potential environmental impact. *Environ. Geol.* **2003**, *43*, 326–338. [CrossRef]
88. Wilson, S.C.; Lockwood, P.V.; Ashley, P.M.; Tighe, M. The chemistry and behaviour of antimony in the soil environment with comparisons to arsenic: A critical review. *Environ. Pollut.* **2010**, *158*, 1169–1181. [CrossRef] [PubMed]
89. Filimon, D.I. Metallogenesis of As-Sb sulfides Ore Deposit of Matra (Alpine Corsica, France). Master's Thesis, University of Pisa, Pisa, Italy, 2023.
90. André-Mayer, A.S.; Leroy, J.; Bailly, L.; Chauvet, A.; Marcoux, E.; Grancea, L.; Llosa, F.; Rosas, J. Boiling and vertical mineralization zoning. A case study from the Apacheta low-sulfidation epithermal gold-silver deposit, south Peru. *Miner. Depos.* **2002**, *37*, 452–464. [CrossRef]
91. Spycher, N.F.; Reed, M.H. Evolution of a broadlands-type epithermal ore fluid along alternative P–T paths; implications for the transport and deposition of base, precious, and volatile metals. *Econ. Geol.* **1989**, *84*, 328–359. [CrossRef]
92. Vaughan, D.J.; Craig, J.R. Mineral Chemistry of Metal Sulfides. Cambridge University Press: Cambridge, UK, 1978.
93. Jovanovski, G.; Makreski, P. Intriguing minerals: Photoinduced solid-state transition of realgar to pararealgar—Direct atomic scale observation and visualization. *ChemTexts* **2020**, *6*, 5. [CrossRef]
94. Krupp, R.E. Solubility of stibnite in hydrogen sulfide solution, speciation, and equilibrium constants, from 25 to 350 °C. *Geochim. Cosmochim. Acta* **1988**, *52*, 3005–3015. [CrossRef]
95. Wilson, N.; Webster-Brown, J.; Brown, K. Controls on stibnite precipitation at two New Zealand geothermal power stations. *Geothermics* **2007**, *36*, 330–347. [CrossRef]
96. Yu, H.C.; Qiu, K.F.; Simon, A.C.; Wang, D.; Mathur, R.; Wan, R.Q.; Jiang, X.-Y.; Deng, J. Telescoped boiling and cooling mechanisms triggered hydrothermal stibnite precipitation: Insights from the world's largest antimony deposit in Xikuangshan China. *Amer Min.* **2023**, *108*, 1213–1223. [CrossRef]
97. Ferrill, D.A.; Morris, A.P.; Evans, M.A.; Burkhard, M.; Groshong, R.H., Jr.; Onasch, C.M. Calcite twin morphology: A low-temperature deformation geothermometer. *J. Struct. Geol.* **2004**, *26*, 1521–1529. [CrossRef]
98. Hagemann, S.; Lüders, V. P-T-X conditions of hydrothermal fluids and precipitation mechanism of stibnite-gold mineralization at the Wiluna lode-gold deposits, Western Australia: Conventional and infrared microthermometric constraints. *Miner. Depos.* **2003**, *38*, 936–952. [CrossRef]
99. Dickson, F.W.; Radke, A.S.; Weissberg, B.G.; Heropoulos, C. Solid solutions of antimony, arsenic, and gold in stibnite (Sb_2S_3) Orpiment (As_2S_3), and realgar (As_2S_2). *Econ. Geol.* **1975**, *70*, 591–594. [CrossRef]

100. Williams-Jones, A.E.; Normand, C. Controls of mineral parageneses in the system Fe–Sb–S–O. *Econ. Geol.* **1997**, *92*, 308–324. [CrossRef]
101. Seward, T.M. The formation of lead (II) chloride complexes to 300 °C: A spectrophotometric study. *Geochim. Cosmochim. Acta* **1984**, *48*, 121–134. [CrossRef]
102. Zhong, R.C.; Brugger, J.; Chen, Y.J.; Li, W.B. Contrasting regimes of Cu, Zn and Pb transport in ore-forming hydrothermal fluids. *Chem. Geol.* **2015**, *395*, 154–164. [CrossRef]
103. Gliozzo, E.; Brogi, A.; Ruggieri, G.; Langone, A. Sb–Au-bearing chalcedonies in hot geothermal systems: Insights from the jasperoids of Poggio Peloso (southern Tuscany, Italy). *Geol. Mag.* **2023**, *160*, 712–731. [CrossRef]
104. Di Rosa, M.; Filimon, D.I.; Groff, J.A.; Marroni, M. U–Pb dating of carbonate gangue with associated As–Sb mineralization in the Matra Fault (Alpine Corsica, France): Constraints for the rifting stage in the Tyrrhenian Sea. *Journal of Geodynamics* **2025**, *165*, 102110. [CrossRef]
105. Moeller, S.; Grevemeyer, I.; Ranero, C.R.; Berndt, C.; Kleschen, D.; Sallares, V.; Zitellini, N.; de Franco, R. Early-stage rifting of the northern Tyrrhenian Sea Basin: Results from a combined wide-angle and multichannel seismic study. *Geochem. Geophys. Geosystatistics* **2013**, *14*, 3032–3052. [CrossRef]
106. Dini, A.; Gianelli, G.; Puxeddu, M.; Ruggieri, G. Origin and evolution of Pliocene–Pleistocene granites from the Larderello geothermal field (Tuscan Magmatic Province, Italy). *Lithos* **2005**, *81*, 1–31. [CrossRef]

minerals

MDPI

Article

Genetic Mineralogical Characteristics of Pyrite and Quartz from the Qiubudong Silver Deposit, Central North China Craton: Implications for Ore Genesis and Exploration

Wenyan Sun [1], Jianling Xue [2,*], Zhiqiang Tong [3,*], Xueyi Zhang [4], Jun Wang [5], Shengrong Li [6] and Min Wang [1]

1 Chinese Academy of Geological Sciences, Beijing 100037, China; swy_hx@163.com (W.S.); mwang@foxmail.com (M.W.)
2 Development and Research Center, China Geological Survey, Beijing 100037, China
3 Mudanjiang Natural Resources Survey Center, China Geological Survey, Changchun 130102, China
4 China Metallurgical Geology Bureau, Beijing 100025, China; zhangxueyi_252@foxmail.com
5 Xinjiang Geological Survey Institute, China Chemical Geology and Mine Bureau, Urumqi 830000, China; xuenang@foxmail.com
6 School of Earth Sciences and Resources, China University of Geosciences, Beijing 100083, China; lisr@cugb.edu.cn
* Correspondence: xjianling@mail.cgs.gov.cn (J.X.); tongzhiqiang@mail.cgs.gov.cn (Z.T.)

Abstract: The Qiubudong silver deposit on the western margin of the Fuping ore cluster in the central North China Craton is a representative breccia-type deposit characterized by relatively high-grade ores, thick mineralized zones, and extensive alteration, indicating considerable potential for economic resource development and further exploration. Previous studies on this deposit have not addressed its genetic mineralogical characteristics. This study focuses on pyrite and quartz to investigate their typomorphic features, such as crystal morphology, trace element composition, thermoelectric properties, and luminescence characteristics, and their implications for ore-forming processes. Pyrite crystals are predominantly cubic in early stages, while pentagonal dodecahedral and cubic–dodecahedral combinations peak during the main mineralization stage. The pyrite is sulfur-deficient and iron-rich, enriched in Au, and relatively high in Ag, Cu, Pb, and Bi contents during the main ore-forming stage. Rare earth element (REE) concentrations are low, with weak LREE-HREE fractionation and a strong negative Eu anomaly. The thermoelectric coefficient of pyrite ranges from −328.9 to +335.6 μV/°C, with a mean of +197.63 μV/°C; P-type conduction dominates, with an occurrence rate of 58%–100% and an average of 88.78%. A weak–low temperature and a strong–high temperature peak characterize quartz thermoluminescence during the main mineralization stage. Fluid inclusions in quartz include liquid-rich, vapor-rich, and two-phase types, with salinities ranging from 10.11% to 12.62% NaCl equiv. (average 11.16%) and densities from 0.91 to 0.95 g/cm^3 (average 0.90 g/cm^3). The ore-forming fluids are interpreted as F-rich, low-salinity, low-density hydrothermal fluids of volcanic origin at medium–low temperatures. The abundance of pentagonal dodecahedral pyrite, low Co/Ni ratios, high Cu contents, and complex quartz thermoluminescence signatures are key mineralogical indicators for deep prospecting. Combined with thermoelectric data and morphological analysis, the depth interval around 800 m between drill holes ZK3204 and ZK3201 has high mineralization potential. This study fills a research gap on the genetic mineralogy of the Qiubudong deposit and provides a scientific basis for deep exploration.

Keywords: pyrite; quartz; genetic mineralogy; thermoelectric; thermoluminescence properties; ore-forming fluid evolution; deep prospecting

1. Introduction

Genetic mineralogy is a discipline that investigates the origins of minerals and their applications. As a branch of mineralogy, it focuses on the study of minerals as products of their natural environments. Minerals preserve critical information about their formation processes and exploration potential, and unraveling this information constitutes the central objective of genetic mineralogy. Genetic mineralogy was initiated by Soviet geologists in the 20th century, and was introduced to China in the 1950s by Guangyuan Chen [1,2]. Since then, significant progress has been made, especially in the genetic study of minerals such as amphibole, biotite, quartz, and pyrite.

The establishment of the mineral typomorphism theory represents one of the most significant achievements in 20th-century mineralogical development. Mineral typomorphism refers to mineralogical phenomena that reflect specific formation conditions, encompassing three aspects: typomorphic mineral assemblages, typomorphic mineral species, and typomorphic mineral characteristics [2]. This study primarily focuses on typomorphic mineral characteristics, which denote the distinguishable variations in properties exhibited by the same mineral species formed under different geological periods, geological processes, and geological bodies. These variations serve as diagnostic indicators for determining their formation conditions. Typomorphic characteristics include morphological typomorphism, compositional typomorphism, structural typomorphism, physical property typomorphism, and spectroscopic typomorphism. These characteristics collectively provide critical genetic information for reconstructing mineral-forming environments and guiding exploration strategies [3–6].

In recent years, the development of in situ microanalytical techniques (e.g., EPMA, LA-ICP-MS, SIMS) has further propelled genetic mineralogy into a new phase, allowing detailed analysis of trace elements and revealing the physicochemical evolution of ore-forming fluids [7]. Among ore minerals, pyrite is abundant in hydrothermal systems and highly sensitive to variations in ore fluid conditions. Its typomorphic features, such as crystal morphology, trace element signatures, Co/Ni ratios, thermoelectric properties, and rare earth element (REE) patterns, can provide critical insights into ore-forming processes and metal enrichment mechanisms [8]. Notably, arsenic enrichment in pyrite is often linked to its gold-bearing potential [9–13]. Quartz, the dominant gangue mineral in most hydrothermal deposits, can also record thermal and fluid evolution through its thermoluminescence response and fluid inclusion characteristics [14–16].

The Fuping ore cluster, located in the central segment of the Taihang Mountains, is one of the most significant metallogenic belts in the central North China Craton (Figure 1). It hosts dozens of gold, silver, molybdenum, and lead–zinc deposits and occurrences. The Qiubudong silver deposit, situated on the western margin of the cluster, is a breccia-type deposit with an estimated silver resource of approximately 800 tonnes, indicating medium to large-scale potential [17]. Previous investigations at Qiubudong have addressed multiple aspects of its geological framework: the characteristics of ore-bearing breccias [18], the division of mineralization stages [19], fluid inclusion and isotopic geochemistry [19,20], geochronology [20], and the geophysical signatures of ore-hosting rocks [21]. However, despite these efforts, no systematic study has been conducted on the genetic mineralogical characteristics of pyrite and quartz, the two most critical minerals associated with mineralization at the deposit.

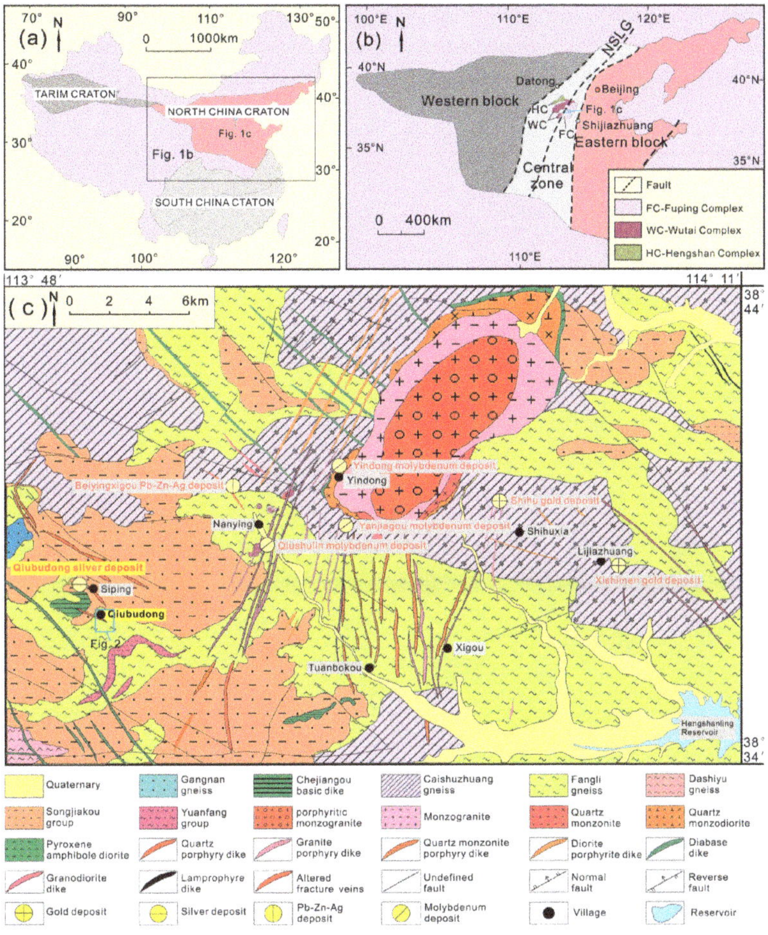

Figure 1. (**a**) Regional geological map of the North China Craton and the location of the Qiubudong deposit; (**b**) simplified tectonic framework of the central North China Craton, showing the Fuping (FC), Wutai (WC), and Hengshan (HC) complexes; (**c**) local geological map of the Qiubudong deposit area (modified after [20]).

The lack of genetic mineralogical data from the Qiubudong deposit has limited the understanding of its ore-forming mechanisms and the ability to establish mineralogical indicators for deep prospecting. In this study, we investigate the genetic mineralogical characteristics of pyrite and quartz based on their crystal morphology, trace and rare earth element compositions, thermoelectric properties, thermoluminescence behavior, and fluid inclusion features. The integrated mineralogical and geochemical data aim to elucidate the evolution of ore-forming fluids and redox conditions and identify reliable typomorphic indicators that can guide deep exploration. By filling this knowledge gap, the study contributes to a better genetic understanding of breccia-type Ag deposits in the Taihang region and provides a framework for applying mineralogical indicators in similar metallogenic settings.

2. Geological Setting

The Qiubudong silver deposit is located in the central segment of the Taihang Mountains. The exposed strata primarily belong to the Archean Songjiakou Group of the Fuping

Complex, comprising potassium feldspar granitic gneiss, plagioclase amphibolite, and syenogranitic rocks (Figure 2). Quaternary unconsolidated sediments are limited in distribution, mainly along valleys and streambeds. Intrusive rocks are categorized into two main groups: early-stage metamorphosed plutonic rocks, mainly amphibole-biotite plagioclase gneiss, and late-stage Yanshanian intermediate-acidic dikes, including quartz porphyry, aplite, quartz vein rocks, and pegmatite. These dikes intrude the basement gneisses and overlying units [17,18].

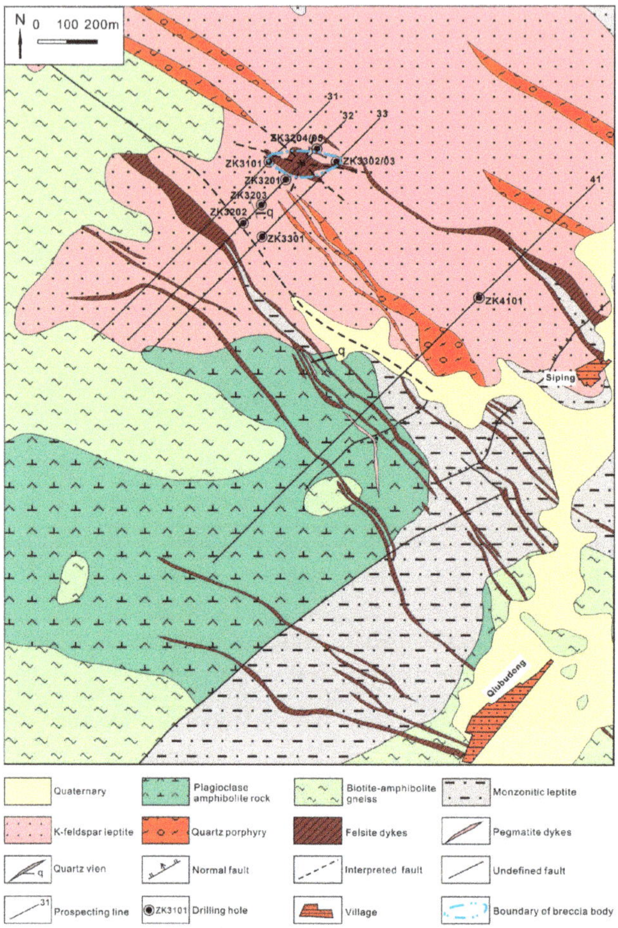

Figure 2. Local geological map of the Qiubudong silver deposit area, showing the distribution of rock types, alteration zones, and orebody outlines (modified after [20]).

Structurally, the deposit lies on the northern limb of the Junchagou–Shangpansong dome and forms part of the Xifujian–Chenbugou composite anticline. Two major fault systems intersect near the deposit: NW-trending faults, which dominate in orientation and control the spatial distribution of ore bodies, and NE-trending faults, which locally offset or terminate the NW-striking structures. The interaction between these fault sets creates a structural dilation environment that facilitated the migration and emplacement of hydrothermal fluids, thereby enhancing ore deposition at their intersections. These faults were later filled by post-mineral quartz porphyry and pegmatite dikes [17]. The structural framework is consistent with the NE–SW compressional regime of the Taihang orogenic belt, a major metallogenic zone in the central North China Craton that hosts numerous polymetallic deposits [22,23].

Two types of orebodies are recognized: breccia-type (dominant) (Figure 3) and altered-rock type (Figure 4). The breccia-type ores display semi-euhedral granular and brecciated textures with angular clasts and a matrix cemented by fine-grained hydrothermal minerals. Ore minerals such as pyrite, native silver, electrum, argentite, and stromeyerite are mainly distributed within the matrix and along open spaces between clasts, indicating mineralization postdates brecciation. Additionally, localized mineral grains are observed within the clasts themselves, suggesting multiphase fluid activity. In contrast, the altered-rock type ores show porphyritic and cryptocrystalline groundmass textures, characterized by fine vein-disseminated and cataclastic structures. Ore minerals in this type are dispersed along microfractures and replacement zones. Associated sulfides include sphalerite, chalcopyrite, galena, and limonite. Gangue minerals are dominated by quartz and feldspar, with subordinate kaolinite, sericite, chlorite, and calcite [17]. These mineralogical and textural relationships collectively imply a complex metallogenic history involving repeated brecciation, fluid percolation, and metal precipitation.

Figure 3. Representative field and core photographs of different breccia types from the Qiubudong silver deposit. (**a**) Tectonic breccia showing angular clasts and limited cementation; (**b**) explosive breccia exposed in outcrop, characterized by chaotic clast distribution; (**c**) Porphyry explosive breccia, with angular porphyritic fragments cemented by hydrothermal matrix; (**d**) Leptite explosive breccia, dominated by fine-grained felsic volcanic clasts; (**e**) compound component explosive breccia with mixed clast types and complex cement; (**f**) Felsite explosive breccia, containing felsitic fragments in altered matrix.

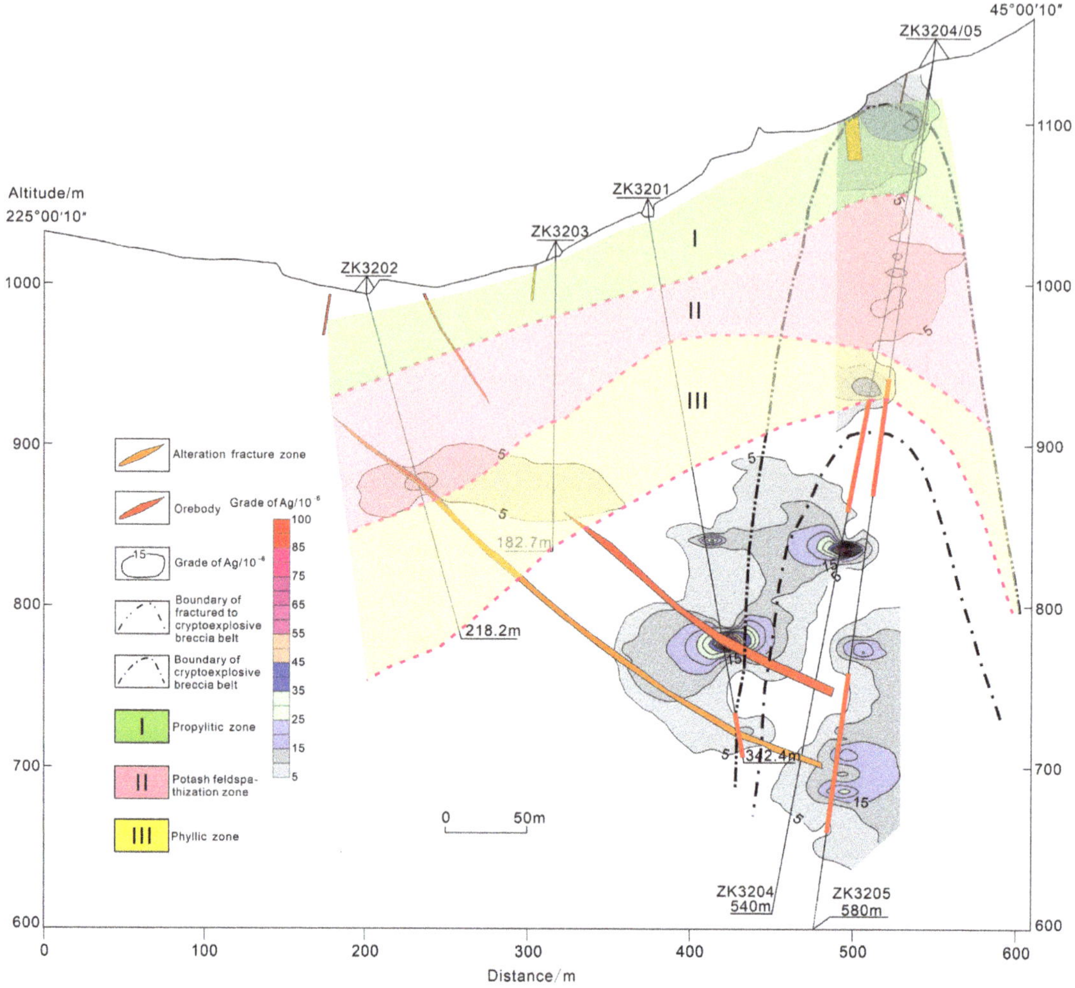

Figure 4. Alteration zoning profile along exploration line No. 32, highlighting zonation of silicification, sericitization, chloritization, and pyritization relative to mineralized zones.

The mineralization process is divided into four stages (Figure 5): (1) sericitic alteration stage (Ser–W), (2) quartz–pyrite stage (Q–Py), (3) polymetallic sulfide stage (Sul), and (4) quartz–carbonate stage (Cc). Stages (2) and (3) are the principal ore-forming episodes. These hydrothermal stages are characterized by variable fluid temperatures and redox states, resulting in successive sulfide and carbonate mineral precipitation.

A paragenetic sequence based on petrographic observations reveals the relative timing and association of gangue and ore minerals:

In Stage 1 (Ser–W), sericitization and weak pyritization occur, typically with disseminated fine-grained pyrite and sericite in altered wall rocks.

Stage 2 (Q–Py) is marked by the deposition of early euhedral quartz and massive to euhedral pyrite, often with pentagonal dodecahedral morphology.

In Stage 3 (Sul), intensive mineralization occurs with abundant galena, sphalerite, chalcopyrite, and late-stage coarse pyrite, intergrown with quartz.

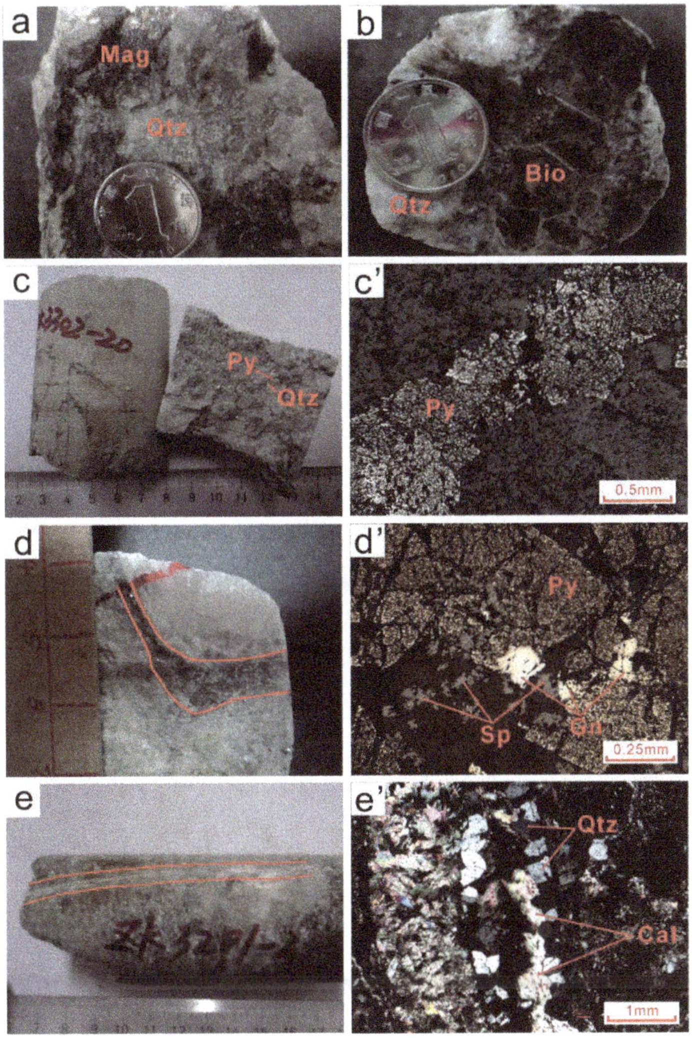

Figure 5. Schematic diagram illustrating the paragenetic sequence of the Qiubudong silver deposit. The diagram delineates the temporal relationships among alteration stages, ore and gangue mineral precipitation, and associated fluid evolution (after [20]). (**a**) Sericitic alteration stage (Ser–W) (I). (**b**) Quartz–pyrite stage (Q–Py) (II). (**c**,**c′**) Polymetallic sulfide stage (Sul) (III). (**d**,**d′**) Quartz–carbonate stage (Cc) (IV). (**c**–**e**): Hand specimen. (**c′**–**e′**): Photomicrographs. Mag—magnetite, Py—pyrite, Sp—sphalerite, Gn—galena; Qtz—quartz, Bio—biotite, Cal—calcite are shown with relative timing of formation and alteration overprint.

Stage 4 (Cc) is characterized by crosscutting quartz–calcite veinlets with minor sulfide remnants and local supergene alteration.

The wall rocks in the Qiubudong deposit primarily consist of Mesoproterozoic metamorphic rocks, including gneisses and schists, which have undergone extensive hydrothermal alteration. The alteration assemblages comprise K-feldspar alteration, silicification, epidotization, sericitization, phyllic alteration, chloritization, pyritization, and kaolinization. These alteration zones occur as halos surrounding ore bodies and vary in intensity and width depending on the stage of mineralization. Their spatial distribution provides valuable indicators for ore targeting and delineating mineralized structures.

3. Occurrence Characteristics of Pyrite and Quartz

Pyrite is the most critical gold-bearing mineral in the Qiubudong silver deposit and occurs throughout all stages of mineralization. As hydrothermal activity evolved, the crystal morphology, color, grain size, and occurrence of pyrite changed systematically, reflecting distinct paragenetic stages. Specifically, in the sericitic alteration stage (Stage 1), pyrite is typically fine-grained and disseminated within altered wall rocks. During the quartz–pyrite stage (Stage 2), it occurs as euhedral to subhedral crystals with pentagonal dodecahedral forms, often coexisting with early quartz. In the polymetallic sulfide stage (Stage 3), pyrite becomes coarser and intergrows with sphalerite, galena, and chalcopyrite, indicating peak ore precipitation. Finally, in the quartz–carbonate stage (Stage 4), only minor residual pyrite is found, often reworked or rimmed by carbonates. These textural and compositional variations are consistent with the mineralization zoning shown in Figure 4 and support the stage-wise evolution of the hydrothermal system.

In the early sericitic alteration stage, pyrite is mainly hosted in shallow intrusive rocks, quartz porphyry, and aplite. It appears as euhedral to subhedral cubic grains, with a light brassy color and grain size of 0.5–2 mm, typically occurring as disseminations or local aggregates. During the quartz–pyrite stage, pyrite becomes more abundant (50%–95%), appearing as bright yellow subhedral to euhedral grains (0.05–1 mm), and occurs mainly in fine veinlets and stockwork zones. This stage marks the main ore precipitation phase and is characterized by intimate pyrite–quartz associations.

In the polymetallic sulfide stage, pyrite shows subhedral to anhedral forms with pale brassy color and finer grain sizes (0.03–0.25 mm). It may enclose sphalerite or be partially replaced by galena, often acting as cement in breccia zones composed of quartz porphyry or shallow granitic fragments. Similar textures have been documented in other hydrothermal systems ([9,11]). In the final quartz–carbonate stage, pyrite becomes irregular in shape (0.01–3 mm), with a light coppery color, and occurs in net-like veinlets (10%–40% abundance), commonly intergrown with calcite and quartz [24].

Quartz, the dominant gangue mineral, occurs throughout mineralization and records key information about thermal and fluid evolution. In the sericitic alteration stage, quartz is gray-white, medium to coarse-grained, and massive in texture, often associated with sericite, indicating intense silicification. In the quartz–pyrite stage, quartz appears gray to smoky gray and occurs mainly as fine veinlets (20%–65%), coexisting with pyrite, marking the peak of ore-forming activity. In the polymetallic sulfide stage, quartz becomes dark gray and is commonly intergrown with pyrite, sphalerite, galena, and chalcopyrite in veins or breccia cements. In the quartz–carbonate stage, quartz is typically distributed along the margins of carbonate veins and often displays comb-like textures, reflecting waning hydrothermal activity and decreasing fluid temperatures [15].

4. Samples and Analytical Methods

Pyrite and quartz samples used in this study were collected from multiple mineralization stages intersected by drill holes ZK3101, ZK3201, ZK3202, ZK3204, and ZK3302 at the Qiubudong silver deposit. The sampling strategy was designed to represent the complete paragenetic sequence of the deposit.

4.1. Major and Trace Element Analysis of Pyrite

Major and trace elements of single pyrite grains were analyzed at the Electron Microprobe Laboratory, Institute of Geology, Chinese Academy of Geological Sciences. Analyses were performed using a JEOL JXA-8230 electron probe microanalyzer (EPMA) (JEOL, Akishima, Tokyo, Japan) at an accelerating voltage of 20 kV, beam current of 2×10^{-8} A, and a beam diameter of 5 μm. Counting time was set to 20 s for peak positions and 10 s for background. Standards used included pyrite (Fe, S), cobalt metal (Co), and nickel metal (Ni). Data were corrected using the ZAF method and calibrated against SPI reference materials [25]). The relative analytical uncertainty was within 2%–5% for major elements and within 5%–10% for trace elements above detection limits.

Rare earth elements (REEs) in pyrite were analyzed at the Beijing Research Institute of Uranium Geology using an Element I inductively coupled plasma mass spectrometer (ICP-MS) (Finnigan MAT, Bremen, Germany). The instrument was operated at a resolution of 300 and a radio frequency (RF) power of 1.25 kW. The analysis was performed under laboratory conditions at 20 °C and 30% relative humidity, following standard ICP-MS protocols [26,27].

4.2. Thermoelectric Measurements of Pyrite

Thermoelectric properties of pyrite were measured at the Genetic Mineralogy Laboratory, China University of Geosciences (Beijing), using a BHTE-06 thermoelectric coefficient analyzer (Beihang University, Beijing, China). The active temperature difference (ΔT) was maintained at 60 ± 3 °C. Sample preparation included the following steps: (1) crushing and sieving the sample to a particle size of 40–60 mesh; (2) hand-picking pure pyrite grains; (3) cleaning surfaces by ultrasonic washing in ethanol; and (4) conducting thermoelectric coefficient measurements. Each sample was measured three times to ensure reproducibility, and the average value was used. The instrument resolution was 0.1 μV/°C, and tests were conducted at a relative humidity (RH) below 85%, with ambient temperature maintained at 25 ± 1 °C. The polarity and magnitude of the thermoelectric coefficient (Q value) were used to distinguish between p-type (positive conduction) and n-type (negative conduction) conductivity of pyrite, which reflects the redox state of ore-forming fluids [8].

4.3. Thermoluminescence Measurements of Quartz

Quartz thermoluminescence was also measured at the Genetic Mineralogy Laboratory, China University of Geosciences (Beijing), using an FJ427-A1 computer-controlled thermoluminescence dosimeter (Beijing Nuclear Instrument Factory, Beijing, China). Quartz grains with >95% purity were selected after crushing and sieving to 40–60 mesh. Approximately 50 mg of sample was weighed with a precision of 0.0001 g and evenly spread on the heating tray. The heating program included: an initial temperature of 40 °C, maximum heating temperature of 420 °C, a ramp rate of 2 °C/s, preheating time of 5 s, total measurement duration of 210 s, and an annealing time of 20 s [14,15].

4.4. Microthermometry of Fluid Inclusions in Quartz

Microthermometric measurements of fluid inclusions in quartz were performed using a Linkam THMSG600 heating–cooling stage (Renishaw, London, UK) at the Genetic Mineralogy Laboratory, China University of Geosciences (Beijing). The operating temperature range was −196 to 600 °C. The precision was ±1 °C from 0 to 600 °C and ±0.1 °C from −196 to 0 °C. The reproducibility error for homogenization temperature was <2 °C and <0.2 °C for freezing point measurements. Heating rates during observation were 5–10 °C/min, and were reduced to 1 °C/min near phase transition temperatures to ensure accurate determination of phase changes [28,29].

5. Results

Pyrite exhibits distinct crystal morphology variations across the mineralization stages at the Qiubudong silver deposit. Eight hundred ninety-eight pyrite grains were observed and classified (Table 1; Figure 6). The dominant crystal forms are cubic (49.22%) and pentagonal dodecahedral (19.71%), followed by cubic + pentagonal dodecahedral intergrowths (12.92%) and cubic + octahedral intergrowths (10.58%). Other morphologies, such as octahedral + pentagonal dodecahedral (1.45%) and combined cubic + octahedral + pentagonal dodecahedral forms (3.23%), are less common (Figure 7). The abundance of cubic pyrite decreases progressively from the sericitic alteration stage to the quartz–carbonate stage, with the most significant drop occurring during the quartz–pyrite stage. In contrast, the proportion of pentagonal dodecahedra and their intergrowths with cubic forms sharply increases during the quartz–pyrite and polymetallic sulfide stages, suggesting more dynamic fluid conditions during these stages. The cubic + octahedral type becomes more prominent in the late quartz–carbonate stage. These systematic morphological variations (Figure 8) imply a strong link between pyrite crystallization environments and hydrothermal evolution [7,24].

Table 1. Characteristics of pyrite crystal forms at different metallogenic stages.

Stage		Ser-W	Q-Py	Sul	Cc	Total
Grains		133	396	303	66	898
Single crystal	a	110 (82.71%)	193 (48.74%)	119 (39.27%)	20 (30.30%)	442 (49.22%)
	e	2 (1.50%)	72 (18.18%)	102 (33.66%)	1 (1.52%)	177 (19.71%)
	o	—	1 (0.25%)	—	—	1 (0.11%)
Combination	a + e	18 (13.53%)	41 (10.35%)	57 (18.81%)	—	116 (12.92%)
	a + o	2 (1.50%)	45 (11.36%)	10 (3.30%)	38 (57.58%)	95 (10.58%)
	o + e	—	9 (2.27%)	1 (0.33%)	3 (4.55%)	13 (1.45%)
	a + o + e	1 (0.75%)	10 (2.53%)	14 (4.62%)	4 (6.06%)	29 (3.23%)
Habitus	a	98.50%	72.98%	66.01%	93.94%	75.95%
	e	15.79%	33.33%	57.43%	12.12%	37.31%
	o	2.26%	16.41%	8.25%	68.18%	15.37%
Single crystal		84.21%	67.17%	72.94%	31.82%	69.04%
Combination		15.79%	32.83%	27.06%	68.18%	30.96%

Note: "a" = cube, "e" = pentagonal dodecahedron, "o" = octahedron; combined codes (e.g., "a + e") indicate combinate crystal form of pyrite observed in different domains of the same sample.

Electron microprobe and ICP-MS analyses (Table 2) indicate that the pyrite from all stages is compositionally characterized by Fe enrichment and S deficiency relative to stoichiometric FeS_2 (ideal Fe = 46.55 wt%, S = 53.45 wt%). In this study, Fe ranges from 46.74 to 48.21 wt% (avg. 47.42 wt%) and S from 50.76 to 52.94 wt% (avg. 51.94 wt%), indicating a sulfur-deficient type of pyrite, consistent with previous observations in other hydrothermal systems [30]. Gold contents are relatively high (103.35–458.31 ppm), suggesting that pyrite is a major gold carrier. Ag, Cu, Pb, and Bi concentrations are elevated in the polymetallic sulfide and quartz–carbonate stages. Zn and Ni show relative enrichment during the quartz–pyrite and polymetallic sulfide stages. Arsenic is widespread (72.22–10,528.66 ppm), consistent with its known affinity for gold in arsenian pyrite [9]. Total REE contents are low (0.83–131.12 ppm), with LREE generally enriched over HREE and weak overall fractionation [(La/Yb)N = 0.71–26.35]. Chondrite-normalized REE patterns show slightly right-leaning to flat curves, with pronounced negative Eu anomalies (δEu = 0.11–1.80) and weak Ce anomalies (δCe $\approx$ 0.94–1.18), reflecting moderate fluid evolution and possible feldspar interaction [31] (Table 3, Figure 9).

A total of 1210 pyrite grains from 29 samples were analyzed for thermoelectric properties (Table 4, Figure 10). The thermoelectric coefficient (TEC) ranges from −328.9 to +335.6 μV/°C, with a mean of +197.63 μV/°C. P-type conductivity dominates across all stages, with an average frequency of 88.78%. In the sericitic alteration and quartz–pyrite stages, P-type pyrite occurs at 98% and 97.31%, respectively, with only rare N-type occurrences. During the polymetallic sulfide stage, P-type frequency decreases to 81.84%, and N-type pyrite increases (18.16%), suggesting fluid evolution and redox fluctuations. In the quartz–carbonate stage, P- and N-type pyrite appear nearly balanced (P: 58%, N: 42%), indicative of a more heterogeneous fluid regime in the late mineralization phase. These changes in thermoelectric polarity and magnitude are consistent with variations in crystal morphology and trace element composition, and reflect progressive fluid evolution and redox dynamics during pyrite formation [8,11].

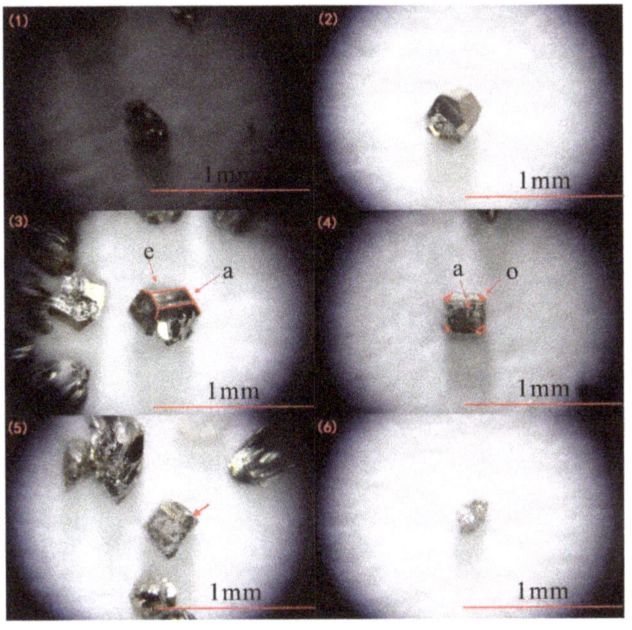

Figure 6. SEM images showing typical pyrite crystal morphologies in the Qiubudong deposit. Observed forms include cubic, pentagonal dodecahedral, and composite crystal intergrowths, indicative of different growth environments. (**1**) Cubic crystal form of pyrite; (**2**) pentagonal dodecahedral crystal form of pyrite; (**3**) combinate crystal form of pyrite (cube + pentagonal dodecahedron); (**4**) combinate crystal form of pyrite (cube + octahedron); (**5**) overgrowth phenomenon of pyrite; (**6**) intergrowth phenomenon of pyrite; a: cube; e: pentagonal dodecahedron; o: octahedron.

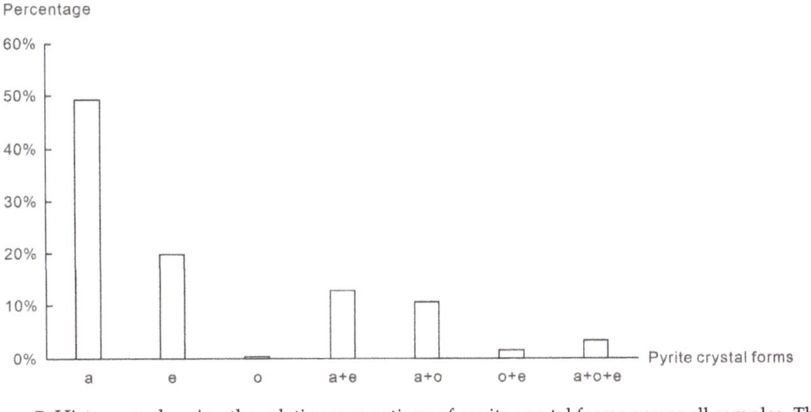

Figure 7. Histogram showing the relative proportions of pyrite crystal forms across all samples. The distribution highlights the dominance of cubic and pentagonal dodecahedral forms.

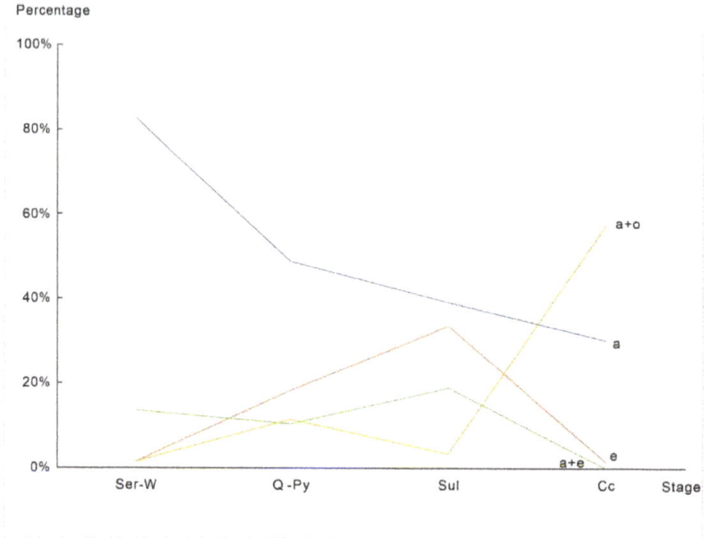

Figure 8. Variation in the occurrence rate of pyrite crystal forms across different mineralization stages. The main ore-forming stage shows elevated levels of pentagonal dodecahedral and intergrowth types.

Table 2. Results of electron microprobe of pyrite (S and Fe are expressed as percentages. The units of other elements are $\times 10^{-6}$).

Stage		**Q-Py**			**Sul**			**Cc**	
Sample No.	**ZK3201-10-3**	**ZK3204-4-7**	**ZK3302-20-3**	**ZK3101-8-5**	**ZK3204-5-4**	**ZK3204-20-5**	**ZK3101-8-2-2**	**ZK3101-8-1-2**	**ZK3201-2-3**
Co	508.72	874.65	539.11	660.03	669.71	566.08	499.15	1861.56	548.96
Ni	29.03	219.3	18.98	227.53	546.71	40.38	43.68	185.69	15.96
Au	260.62	344.05	203.76	424.19	368.56	458.31	236.84	236.82	103.35
Ag	0	20.98	54.37	27	126.17	88.56	101.11	21.81	49.69
Cu	9.8	0	0	23.09	38.91	40.06	24.29	86.97	0
Pb	940.64	2372.56	322.17	1159.91	1660.71	1368.15	1429.96	2108.27	1261.65
Zn	132.79	1225.57	196.54	230.47	193.49	3.87	62.59	110	41.56
Fe	0.473	0.4685	0.4755	0.4674	0.4743	0.4716	0.4781	0.4821	0.4775
S	0.5141	0.523	0.5213	0.5294	0.5213	0.52	0.5184	0.5076	0.5198
As	10,528.66	2895.88	1389.79	289.43	444.23	5513.47	72.22	5165.47	468.31
Sb	61.28	83.73	84.49	67.86	88.12	103.83	87.45	501.85	99.11
Se	61.28	5.69	0	0	60.92	15.4	212.42	4.93	41.51
Te	0	133.59	123.4	23.03	29.24	42.35	173.33	7.34	69.57
Bi	302.52	287.2	258.64	15.08	204.05	115.99	568.13	53.86	0
S/Fe	1.893	1.944	1.909	1.973	1.914	1.92	1.889	1.834	1.896
Chemical formula	$Fe_{1.057}S_2$	$Fe_{1.029}S_2$	$Fe_{1.048}S_2$	$Fe_{1.014}S_2$	$Fe_{1.045}S_2$	$Fe_{1.041}S_2$	$Fe_{1.059}S_2$	$Fe_{1.091}S_2$	$Fe_{1.055}S_2$
Co/Ni	17.526	3.988	28.398	2.901	1.225	14.018	11.427	10.025	34.387
Au/Ag	-	16.402	3.748	15.711	2.921	5.175	2.342	10.86	2.08
Se/Te	-	0.043	0	0	2.083	0.364	1.226	0.671	0.597
As/Sb	171.818	34.587	16.449	4.265	5.041	53.101	0.826	10.293	4.725
S/Se	8390.228	91,923.243	-	-	8557.814	33,766.379	2440.608	103,011.513	12,522.46
Fe + S	0.9872	0.9915	0.9968	0.9969	0.9956	0.9916	0.9965	0.9897	0.9973
Co + Ni	537.74	1093.95	558.09	887.57	1216.41	606.46	542.84	2047.25	564.93
Au + Ag	260.62	365.03	258.13	451.18	494.73	546.86	337.95	258.63	153.05
Cu + Pb + Zn	1083.23	3598.13	518.71	1413.47	1893.11	1412.08	1516.84	2305.24	1303.21
Se + Te	61.28	139.28	123.4	23.03	90.16	57.75	385.74	12.27	111.08
As + Sb + bi	10,892.45	3266.81	1732.92	372.37	736.39	5733.29	727.81	5721.18	567.43

Table 3. REE data and characteristics of pyrite from the Qiubudong silver deposit.

Sample No.	**3204-20**	**3101-12**	**3201-10**	**3302-20**	**3101-6**	**3101-7**	**3201-20**	**3204-23**	**3101-8**
Stage		**Q-Py**				**Sul**			**Cc**
La	0.09	3.90	3.38	1.96	3.26	31.20	0.56	0.72	74.2
Ce	0.24	7.65	6.35	4.25	6.54	60.00	1.33	1.50	135.00

Table 3. *Cont.*

Sample No.	3204-20	3101-12	3201-10	3302-20	3101-6	3101-7	3201-20	3204-23	3101-8
Stage	Q-Py				Sul				Cc
Pr	0.03	0.84	0.67	0.05	0.778	6.66	0.20	0.15	14.90
Nd	0.12	3.18	2.54	2.24	3.38	22.10	1.04	0.65	52.10
Sm	0.05	0.54	0.38	0.60	0.83	3.53	0.38	0.12	8.03
Eu	0.02	0.06	0.08	0.02	0.07	0.14	0.03	0.04	0.27
Gd	0.02	0.46	0.32	0.38	0.60	2.66	0.41	0.10	6.11
Tb	0.10	0.09	0.01	0.05	0.14	0.37	0.04	0.02	0.89
Dy	0.05	0.49	0.19	0.26	0.84	1.9	0.68	0.13	4.40
Ho	0.01	0.11	0.02	0.04	0.18	0.31	0.14	0.03	0.73
Er	0.05	0.41	0.07	0.13	0.60	0.83	0.51	0.09	2.37
Tm	0.01	0.08	0.01	0.03	0.10	0.13	0.09	0.02	0.29
Yb	0.04	0.52	0.11	0.17	0.70	1.09	0.56	0.12	2.02
Lu	0.01	0.08	0.01	0.03	0.11	0.16	0.07	0.02	0.25
Y	0.34	2.92	0.68	1.18	5.42	8.73	4.29	0.76	21.70
ΣREE	0.83	18.39	14.13	10.20	18.13	131.12	6.04	3.70	301.55
LREE	0.54	16.17	13.40	9.13	14.85	123.63	3.53	3.18	284.50
HREE	0.29	2.22	0.73	1.07	3.27	7.49	2.51	0.53	17.05
LREE/HREE	1.85	7.28	18.38	8.53	4.54	16.50	4.41	6.05	16.69
La_N/Yb_N	1.74	5.38	22.45	8.47	3.34	20.53	0.71	4.35	26.35
δEu	1.80	0.37	0.67	0.14	0.29	0.13	0.20	1.01	0.11
δCe	1.18	0.99	0.97	1.57	0.97	0.97	0.97	1.06	0.94

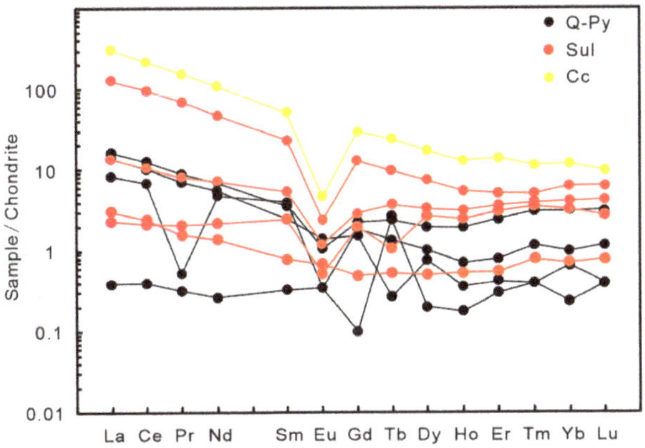

Figure 9. Chondrite-normalized REE patterns of pyrite from different ore stages. The plots show low total REE contents, weak LREE-HREE fractionation, and consistent negative Eu anomalies.

Table 4. Characteristics of pyroelectricity parameters of pyrite at different stages.

Stage	Grains	N-type (μV/°C)				P-type (μV/°C)				Averages /Mean Value of α (μV/°C)
		Minimum	Maximum	Mean	Occurrence Rate	Minimum	Maximum	Mean	Occurrence Rate	
Ser	100	−69	−6.7	−37.85	2%	5.1	333.3	258.5	98%	251.7
Q-Py	520	−168.1	−8.3	−59.8	2.69%	1.7	335.6	257.1	97.31%	248.6
Sul	490	−328.9	−6.7	−87.4	18.16%	6.7	335	200.6	81.84%	148.3
Cc	100	−165.3	−11.6	−90.2	42%	6.8	328.4	209.1	58%	83.4
	1210	−328.9	−6.7	−84.90	11.22%	1.7	335.6	215.37	88.78%	197.63

Quartz thermoluminescence (TL) properties vary significantly across different mineralization stages in the Qiubudong silver deposit (Table 5, Figure 11). In general, TL glow curves exhibit a characteristic bimodal shape, with the high-temperature peak consistently more intense than the low-temperature peak. In the sericitic alteration stage, TL analyses of two samples show complex bimodal patterns with sharp peaks: low-temperature peaks appear at 185–190 °C, high-temperature peaks at 263–285 °C. The full width at half maximum (FWHM) ranges from 60 to 95 °C, peak intensities from 550 to 1080 cps, and total luminescence from 46,667 to 102,001 cps. During the quartz–pyrite stage, five samples were analyzed, showing both unimodal and bimodal curves. Low-temperature peaks are blunt and occur at 183–199 °C, while high-temperature peaks remain sharp at 263–275 °C. The peak intensity spans 360–2900 cps and total TL emission ranges from 31,200 to 167,558 cps, indicating enhanced hydrothermal activity. One sample from the polymetallic sulfide stage displays a strong bimodal pattern with high-temperature peaks dominating, yielding the highest total TL intensity (up to 222,226 cps). In the quartz–carbonate stage, both unimodal and bimodal curves occur; the single peak is centered at 203 °C with relatively low intensity (~77 cps), while bimodal curves present peaks at 208 °C and 275 °C. Overall, TL curve evolution follows a systematic sequence: bimodal → mixed → bimodal → mixed, with the main ore-forming stage showing the strongest and sharpest TL responses, reflecting high thermal gradients and robust fluid activity [14,15].

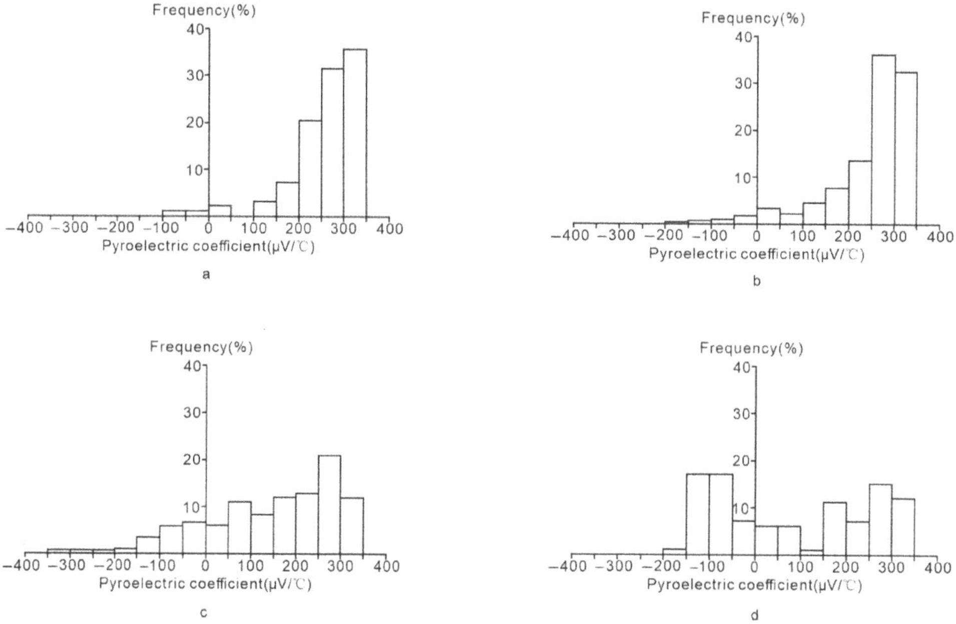

Figure 10. Distribution histogram of pyrite thermoelectric coefficients by different stages. P-type conductivity dominates, with limited N-type samples observed in later mineralization. (**a**)—sericitic alteration stage, (**b**)—Quartz-pyrite stage, (**c**)—Polymetallic sulfide stage, (**d**)—Quartz-carbonate stage.

Fluid inclusions (FIs) in quartz exhibit distinct physical and thermal behaviors that correspond to different mineralization stages (Table 6, Figures 12 and 13). Three main types of FIs were observed based on phase states at room temperature and their microthermometric behavior: liquid-rich, vapor-rich, and two-phase (liquid + vapor) inclusions. During freezing and heating experiments, two-phase inclusions typically showed vapor bubble disappearance upon heating (indicating homogenization to the liquid phase), whereas vapor-rich inclusions homogenized to vapor, and liquid-rich inclusions rarely developed vapor bubbles until close to homogenization. These phase behaviors were used to distinguish the fluid inclusion types during thermal analysis.

Table 5. Thermo-luminescence characteristic parameters of quartz from Qiubudong silver deposit.

Sample No.	Stage	Peak Type	Peak Temperature/°C	Half Peak Temperature/°C	Peak Strength/cps	Total Intensity/cps
ZK4101-34	Ser-W	Bimodal	190	60	700	46,667
			285	85	1080	102,001
ZK4101-46	Ser-W	Bimodal	185	85	550	51,945
			263	95	770	81,279
ZK3204-4	Q-Py	Unimodal	195	70	750	58,334
ZK3205-29	Q-Py	Bimodal	183	78	360	31,200
			275	120	860	114,668
ZK3101-12	Q-Py	Bimodal	199	75	1250	104,168
			275	52	2900	167,558
ZK3201-10	Q-Py	Bimodal	185	80	520	46,223
			263	115	410	52,389
ZK3204-23	Q-Py	Unimodal	208	60	1150	76,667
ZK3101-8	Sul	Bimodal	189	73	1650	133,835
			263	50	4000	222,226
ZK3201-2	Cc	Bimodal	208	70	425	33,056
			275	95	525	55,417
ZK3202-9	Cc	Unimodal	203	56	77	4791

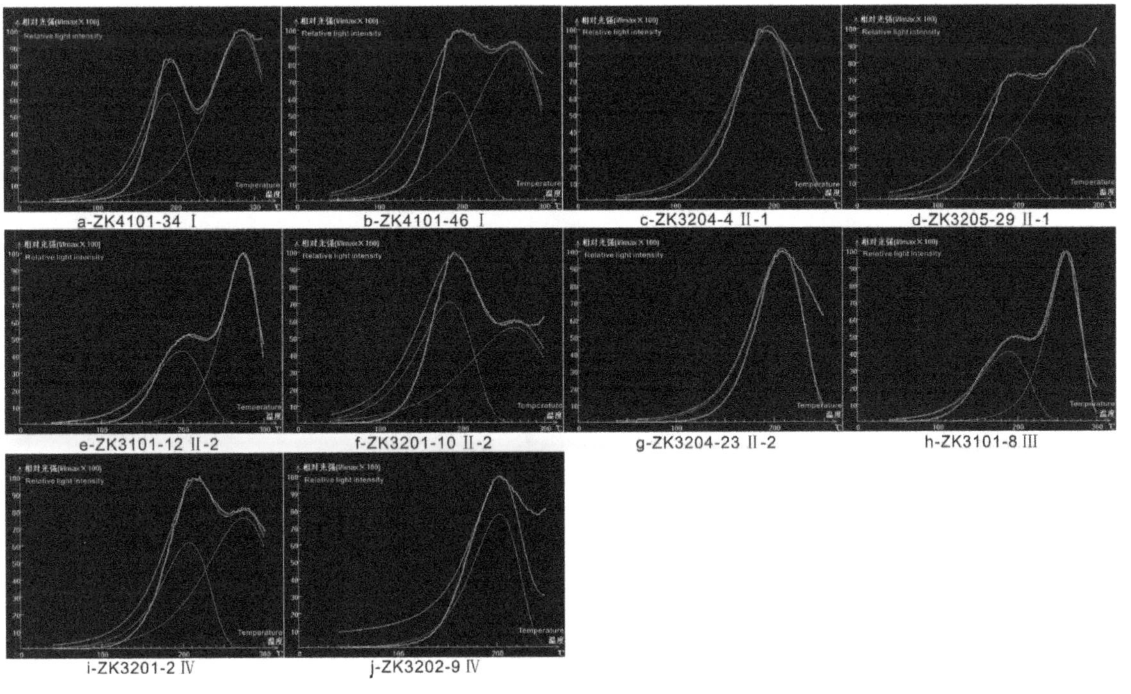

Figure 11. Thermoluminescence glow curves of quartz. The main ore-forming stage is characterized by strong bimodal emission, with a broad low-temperature and sharp high-temperature peak. I—phyllic stage, II—quartz-pyrite stage, III—polymetallic sulfide stage, IV—quartz-carbonate stage.

Table 6. Temperature (uncorrected for pressure), salinity, and density results of fluid inclusions in different metallogenic stages of the Qubudong silver deposit.

Stage	Homogenization Temperature/°C		Freezing Temperature/°C	Salinity (wt%NaCl)		Density (g/cm^3)	
	Range	Mean		Range	Mean	Range	Mean
Q-Py	170.2~359.8	254.5	−13~−1.6	2.74~16.89	10.22	0.73~0.98	0.86
Sul	240.3~330.6	271.2	−11.3~−4.9	7.73~15.27	11.8	0.80~0.94	0.87
Cc	146.7~258.4	204.7	−9.6~−6.5	9.86~13.51	11.47	0.92~0.99	0.95

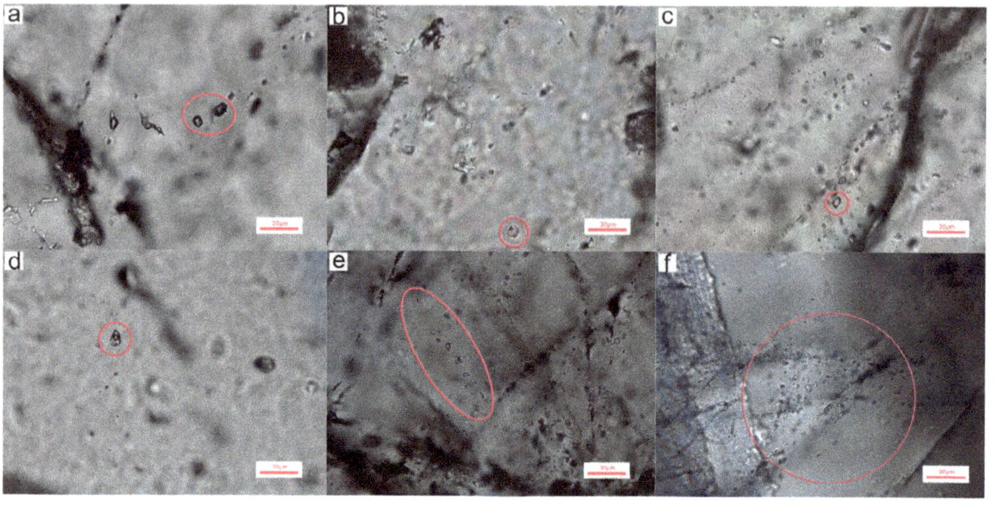

Figure 12. Characteristics of fluid inclusions in different metallogenic stages of the Qubudong silver deposit. (**a**)—quartz-pyrite stage, vapor-phase inclusion; (**b**)—quartz-pyrite stage, regular-shaped gas-liquid inclusion; (**c**)—quartz-pyrite stage, liquid-phase inclusion; (**d**)—polymetallic sulfide stage, gas-liquid inclusion; (**e**)—Polymetallic sulfide stage, gas-liquid inclusions showing linear distribution; (**f**)—quartz-carbonate stage, inclusions occurring in clusters. Variation in homogenization temperature and salinity across mineralization stages illustrates fluid cooling and dilution trends. Stage-wise clustering of inclusion types also suggests episodic confinement and possible meteoric mixing during late-stage mineralization.

In the quartz–pyrite stage (FIA-1), inclusions range from 5 to 25 μm in diameter and are dominated by two-phase types (85%), with homogenization temperatures (Th) ranging from 170.2 to 359.8 °C (average 254.5 °C), and freezing points from −13 to −1.6 °C, indicating low to moderate salinity. In the polymetallic sulfide stage (FIA-2), inclusions are 4–18 μm in size, again dominated by two-phase types (90%), with Th values of 240.3–330.6 °C (average 271.2 °C), and freezing points from −11.3 to −4.9 °C. In the quartz–carbonate stage (FIA-3), smaller inclusions (2–9 μm) are observed, with Th values of 146.7–258.4 °C (average 204.7 °C), and freezing points from −9.6 to −6.5 °C.

Overall, the FI data indicate that ore formation at Qiubudong occurred under fluctuating thermal and chemical regimes. The transition from moderate- to high-temperature, low-salinity fluids to lower-temperature, denser, confined systems is consistent with waning magmatic input and increased fluid–rock interaction. This interpretation is further supported by quartz thermoluminescence patterns and pyrite typomorphism across mineralization stages.

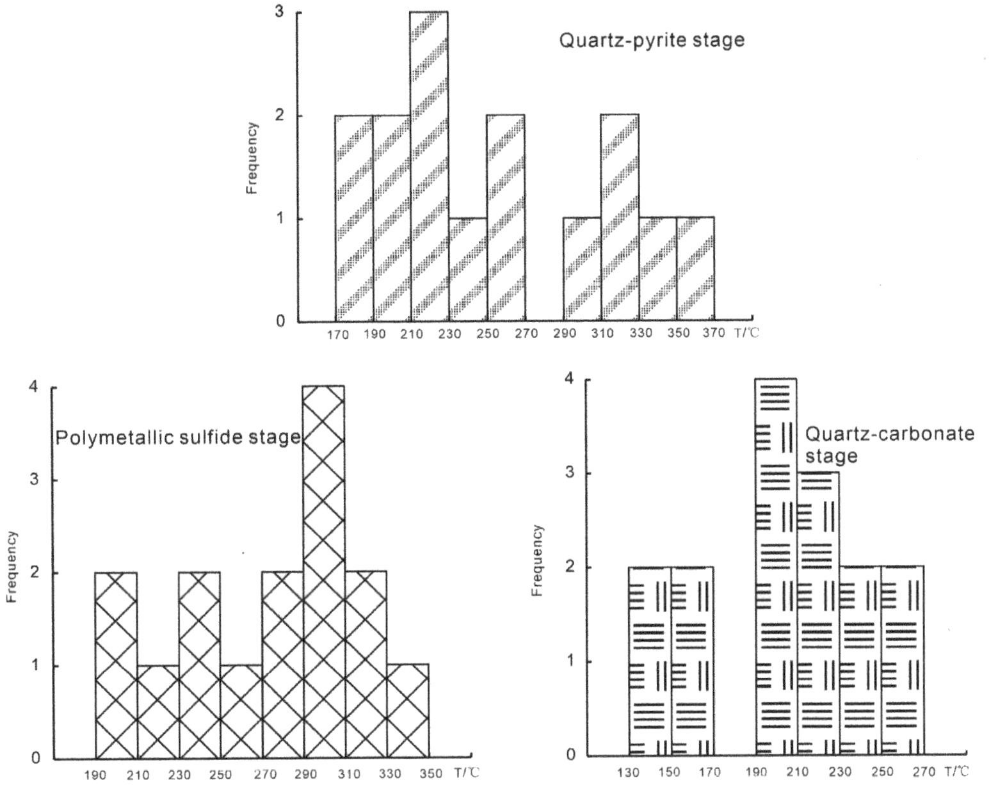

Figure 13. Histogram of homogenization temperature of fluid inclusions in different metallogenic stages of the Qubudong silver deposit.

Salinities of ore-forming fluids were calculated based on the relationship between freezing point depression and NaCl equivalent [32], ranging from 2.74 to 16.89 wt.% NaCl eq., with most values clustering between 10.11 and 12.62 wt.% (average 11.16 wt.%). Specifically, salinities are 10.22 wt.% (quartz–pyrite stage), 11.80 wt.% (polymetallic sulfide stage), and 11.47 wt.% (quartz–carbonate stage), all corresponding to low-salinity fluids (typically <26.3 wt.%). Fluid densities range from 0.73 to 0.99 g/cm^3, mainly between 0.91 and 0.95 g/cm^3 (average 0.90 g/cm^3). From early to late stages, the average fluid density increases from 0.86 to 0.95 g/cm^3, coupled with a decline in homogenization temperatures from ~254.5 °C to ~204.7 °C. This trend reflects a gradual cooling of the hydrothermal system and increasing fluid confinement during the waning stages of mineralization. The minor increase in salinity from early to middle stages, followed by stabilization in late fluids, may indicate limited mixing with residual evolved fluids or restricted input from external low-salinity sources such as meteoric water. These thermal and compositional evolutions, in conjunction with complex thermoluminescence responses and decreasing inclusion size and abundance, collectively suggest progressive cooling, minor fluid mixing, and partial confinement in the late mineralization stage.

Such fluid behavior is consistent with volcanic–hydrothermal systems where ore precipitation is governed by episodic fluid injection, structural compartmentalization, and cooling-driven solubility changes [29,33,34].

6. Discussion

6.1. Characteristics of Ore-Forming Hydrothermal Fluids

The mineralization at the Qiubudong silver deposit is primarily controlled by a medium- to low-temperature hydrothermal system dominated by volcanic-derived fluids, with minor meteoric water input. Pyrite geochemistry shows high Co (499.15–1861.56 ppm) and low Ni (15.96–546.71 ppm), with Co/Ni ratios mostly greater than 5, indicating a volcanic origin. These results are consistent with Co–Ni classification diagrams (Figure 14 [35–40]). Sulfur isotope data ($\delta^{34}S$ = 0.5–3.4‰, avg. 1.6‰), lead isotopes ($^{206}Pb/^{204}Pb$ = 15.799–16.049), and quartz $\delta^{18}O$ (8.6–12‰) and δD (−95.2 to −83.2‰) further support a dominantly magmatic fluid source with limited meteoric mixing [20].

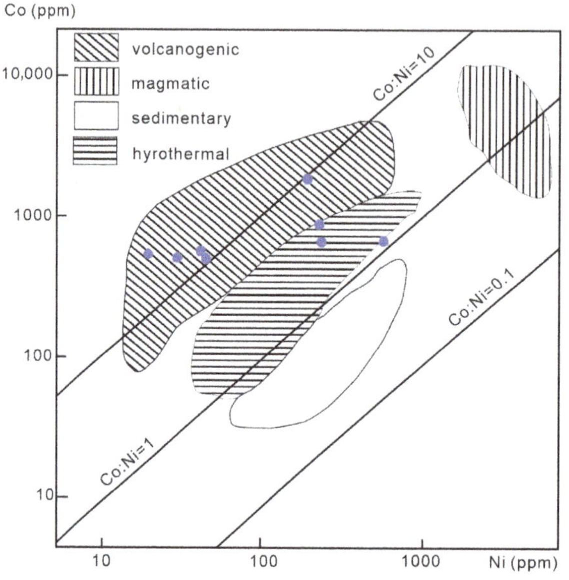

Figure 14. Binary plot of Co vs. Ni content in pyrite (Base map according to [41,42]). Most samples fall within the volcanic and hydrothermal genesis fields, supporting a mixed fluid origin.

Volatile and trace elements also provide insights into the hydrothermal environment. Pyrite Se/Te ratios (0–2.083, avg. 0.668) and S/Se ratios (2440.6–10,3011.5) indicate a low- to mid-temperature hydrothermal system. Pyrite thermoelectric coefficients (Figure 15) suggest crystallization temperatures of 110.3–369.0 °C, aligning with quartz fluid inclusion homogenization temperatures: 170.2–359.8 °C in the quartz–pyrite stage, 240.3–330.6 °C in the polymetallic sulfide stage, and 146.7–258.4 °C in the quartz–carbonate stage. The evolution of pyrite REE patterns from near-parallel to weakly LREE-enriched curves suggests increasing F content in late-stage fluids, with minor Cl enrichment [43,44]. Fluid inclusion salinities (10.11–12.62 wt.% NaCl eq.) and densities (0.91–0.95 g/cm^3) indicate that the ore-forming fluids were low-salinity and low-density throughout mineralization.

Comparison with other breccia-type Ag deposits in the Taihang orogenic belt (e.g., Zhaobao) reveals common F-enriched, low-salinity hydrothermal signatures. However, the widespread development of pentagonal dodecahedral pyrite and its relation to thermoelectric polarity in Qiubudong offers a novel perspective on fluid evolution and ore-forming mechanisms.

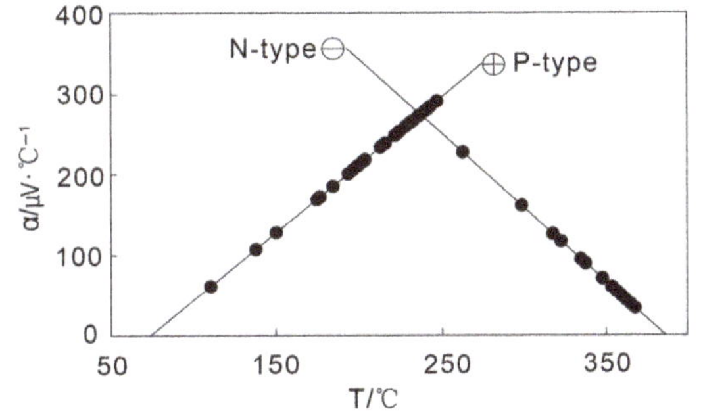

Figure 15. Diagram of pyrite thermoelectricity vs. temperature in the Qubudong silver deposit (base map after [45]), P-Type: positive conduction, N-Type: negative conduction.

6.2. Ore Prospecting Indicators

Distinct mineralogical and geochemical signatures provide effective criteria for ore prospecting. The predominance of cubic and pentagonal dodecahedral pyrite morphologies, especially the high abundance of pentagonal dodecahedral and their intergrowth forms during the main mineralization stage, correlates with high ore-forming potential. Additionally, pyrite trace element characteristics, such as strong Au–Cu positive correlation and Au–Co/Ni negative correlation (Figure 16), support their use in mineralization targeting. These element associations suggest that high Cu content and low Co/Ni ratios can be used as geochemical pathfinders for silver–gold mineralization.

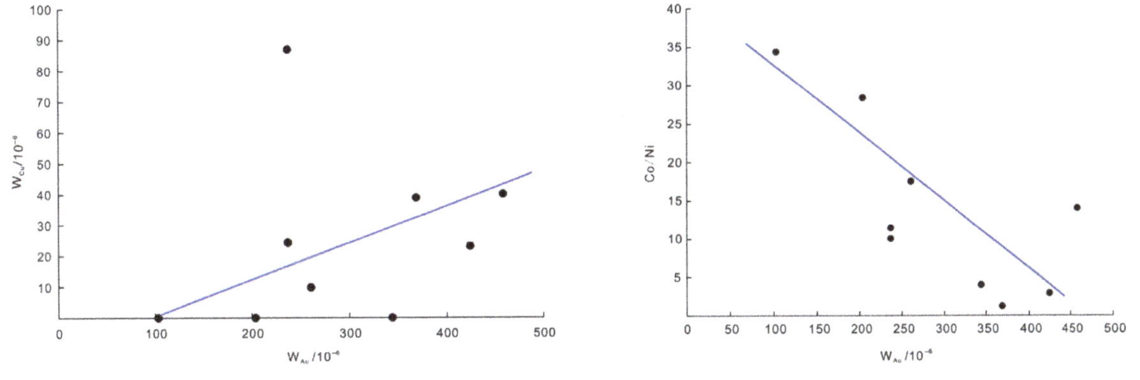

Figure 16. Plot showing correlation between Au and Cu concentrations, and inverse correlation between Au and Co/Ni ratio in pyrite. These element associations support their use as ore exploration indicators.

Thermoluminescence (TL) characteristics of quartz further serve as useful indicators. The main ore-forming stage is characterized by strong, complex bimodal TL glow curves with sharp high-temperature peaks and broad low-temperature peaks. These features are consistent with robust hydrothermal conditions and can aid in identifying mineralization stages during exploration.

6.3. Prediction of Mineralization Potential

The prediction of deep mineralization potential is based on the integrated analysis of pyrite typomorphism and thermoelectric properties [46]. Pentagonal dodecahedral pyrite, typically associated with gold enrichment, is considered a reliable typomorphic marker of mineralized zones. In deeper parts of the orebody, pyrite shows higher Co, Ni, and Se contents and displays N-type thermoelectric behavior, which is commonly linked to higher temperature, reducing ore-forming environments where magmatic–hydrothermal fluids predominate [8,11]. Conversely, pyrite from shallow and marginal zones is often enriched in As and S and exhibits P-type thermoelectricity, which may reflect lower temperature, oxidizing conditions influenced by meteoric water ingress or supergene modification [7]. These patterns provide useful indicators for evaluating vertical zoning and preservation state of mineralization.

Exploration modeling along section line 32 (Figure 17) shows that below the 800 m elevation level, particularly between drill holes ZK3204 and ZK3201, pyrite aggregates and pentagonal dodecahedral forms become dominant. Between 850 and 900 m elevation, these morphologies form the core of the mineralized zone. P-type thermoelectricity dominates overall, reflecting a stable mineralizing environment. However, in ZK3201 below 800 m, a marked increase in N-type pyrite suggests changes in physico-chemical conditions, possibly related to magmatic fluid input. These observations point to significant concealed ore potential in the deep zone between ZK3204 and ZK3201, meriting further exploration.

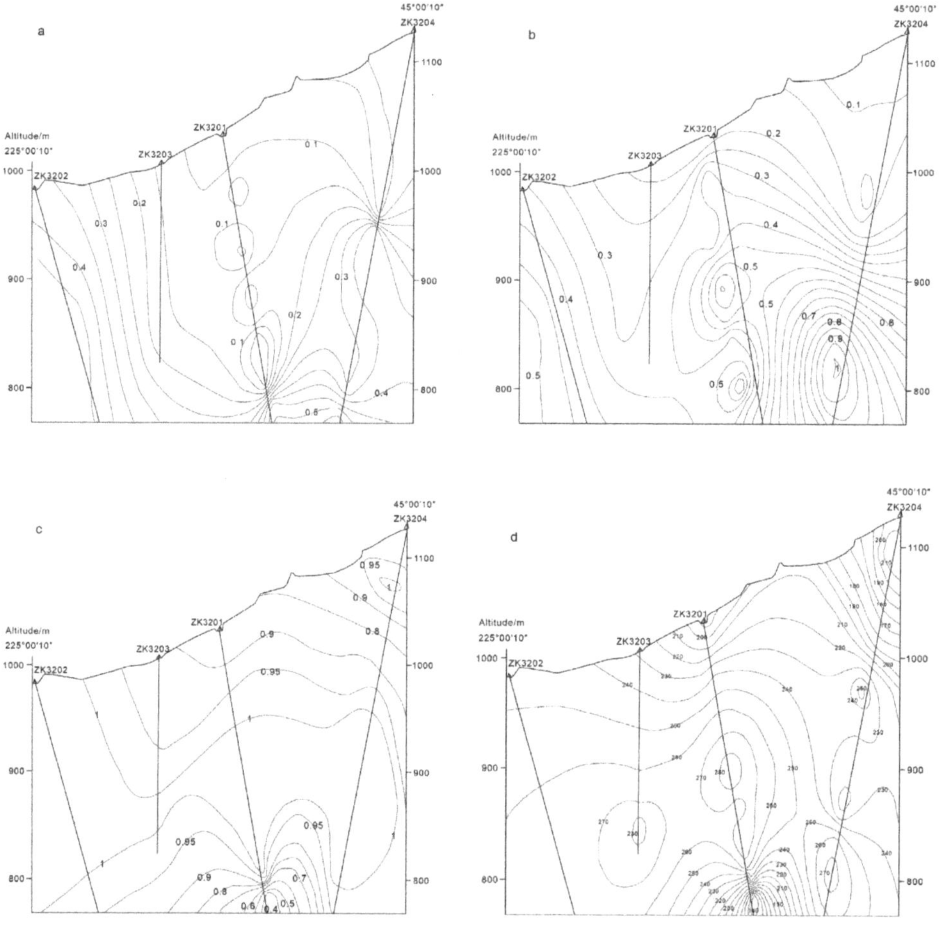

Figure 17. Exploration model along Section 32. Areas between drill holes ZK3201 and ZK3204 below 800 m elevation show concentration of pentagonal dodecahedral pyrite and P-type thermoelectricity, suggesting concealed mineralization potential. (**a**)—The occurrence rate of combined crystal forms of pyrite; (**b**)—the occurrence rate of pentagonal dodecahedral habit of pyrite; (**c**)—the occurrence rate of P-type pyrite; (**d**)—thermoelectric coefficients of P-type pyrite.

In this study, we adopted an indicator-based visualization approach to illustrate the spatial distribution of individual proxies (pyrite morphology, thermoelectric properties, trace elements, and TL characteristics) along section line 32. While a composite predictive model, such as a weighted sum of proxies in spatial grids, could further enhance prospectivity mapping, its implementation was limited by the current data density and irregular drillhole spacing. The present drill grid does not yet support uniform 10 m × 10 m block modeling with consistent proxy coverage across all intervals. However, such an integrated modeling approach will be prioritized in future work, as additional data from ongoing exploration campaigns become available.

7. Conclusions

1. Pyrite from the Qiubudong silver deposit exhibits systematic variations in crystal morphology (cubic, pentagonal dodecahedral, and intergrowths), chemical composition (Fe-enrichment, S-deficiency, Au-rich), and thermoelectric behavior (predominantly P-type with stage-specific transitions). These features collectively reflect dynamic hydrothermal evolution under a low-salinity, F-rich, structurally controlled environment at medium–low temperatures. The pentagonal dodecahedral habit, in particular, is strongly linked to main-stage mineralization and Au enrichment, offering a robust mineralogical indicator.
2. Quartz thermoluminescence and fluid inclusion analyses reveal a progression from hotter, two-phase F-rich fluids in early stages to cooler, more confined low-density systems in late stages. The TL glow curves show characteristic bimodal patterns in ore-forming stages, while fluid inclusion data demonstrate consistent low salinities (10–12 wt.% NaCl eq.) and increasing fluid density over time. These thermal and compositional trends support a magmatic-hydrothermal system undergoing gradual cooling and partial fluid confinement.
3. Integrated geochemical and isotopic data, including Co/Ni and Se/Te ratios, sulfur–lead–hydrogen–oxygen isotopes, and trace element associations in pyrite, indicate that the ore-forming fluids were dominantly sourced from volcanic–magmatic systems, with limited meteoric water involvement. The geochemical proxies support a scenario of staged fluid evolution with increasing fluid–rock interaction at depth.
4. Several mineralogical and geochemical indicators, such as the abundance of pentagonal dodecahedral pyrite, strong Au–Cu and inverse Au–Co/Ni correlations, complex quartz TL responses, and stage-specific thermoelectric polarity, can be applied as effective exploration tools. The integration of these attributes in 3D modeling has delineated a concealed ore potential zone below 800 m between drill holes ZK3201 and ZK3204. These results provide a practical framework for applying genetic mineralogy in deep prospecting of breccia-type Ag systems.

Author Contributions: Conceptualization, W.S., J.X., Z.T. and S.L.; methodology, W.S.; software, W.S.; validation, J.W.; formal analysis, W.S.; investigation, W.S., J.X. and S.L.; resources, W.S. and S.L.; data curation, W.S. and S.L.; writing—original draft preparation, W.S.; writing—review and editing, X.Z. and W.S.; visualization, W.S.; supervision, W.S. and J.X.; project administration, M.W.; funding acquisition, W.S. and M.W. All authors have read and agreed to the published version of the manuscript.

Funding: This study was supported by the Project of China Geological Survey (No. DD20230100504), Project of the National Forestry and Grassland Administration (No. QT03), National Key Research and Development Program (No. 2016YF600106).

Data Availability Statement: The original contributions presented in this study are included in the article. Further inquiries can be directed to the corresponding author.

Acknowledgments: The authors thank Wentuo Cao of the Sixth Geological Team of Hebei Bureau of Geology and Mineral Resources Exploration for the guidance on the field work.

Conflicts of Interest: The authors declare no conflicts of interest.

References

1. Chen, G.Y.; Sun, D.S.; Yin, H.A. *Genetic Mineralogy and Prospecting Mineralogy*; Chongqing Publishing House: Chongqing, China, 1988. (In Chinese)
2. Li, S.R. *Crystallography and Mineralogy*; Geological Publishing House: Beijing, China, 2008. (In Chinese)
3. Gerasimov, B.; Beryozkin, V.; Kravchenko, A. Typomorphic Features of Placer Gold from the Billyakh Tectonic Melange Zone of the Anabar Shield and Its Potential Ore Sources (Northeastern Siberian Platform). *Minerals* **2020**, *10*, 281. [CrossRef]
4. Gerasimov, B. The Use of Typomorphic Features of Placer Gold of the Anabar Region for Determining Its Sources. *Minerals* **2023**, *13*, 480. [CrossRef]
5. Lalomov, A.; Grigorieva, A.; Kotov, A.; Ivanova, L. Typomorphic Features and Source of Native Gold from the Sykhoi Log Area Placer Deposits, Bodaibo Gold-Bearing District, Siberia, Russia. *Minerals* **2023**, *13*, 707. [CrossRef]
6. Kondratieva, L.A.; Anisimova, G.S.; Kardashevskaia, V.N. Ore Mineralogy and Typomorphism of Native Gold of the Spokoininsky Cluster of the Aldan–Stanovoy Gold Province. *Minerals* **2023**, *13*, 543. [CrossRef]
7. Cook, N.J.; Ciobanu, C.L.; Pring, A. The mineralogy of pyrite and its significance in gold deposits. *Minerals* **2016**, *6*, 50.
8. Yan, Y.T.; Li, S.R. Application of thermoelectricity in genetic mineralogy of pyrite. *Acta Petrol. Mineral.* **2011**, *30*, 403–414, (In Chinese with English Abstract).
9. Reich, M.; Kesler, S.E.; Utsunomiya, S.; Palenik, C.S.; Chryssoulis, S.L.; Ewing, R.C. Solubility of gold in arsenian pyrite. *Geochim. Cosmochim. Acta* **2005**, *69*, 2781–2796. [CrossRef]
10. Large, R.R.; Bull, S.W.; Maslennikov, V.V.; Thomas, H. A carbonaceous sediment-hosted gold deposit at Sukhoi Log, Russia: A large silica–pyrite–gold system of diagenetic origin. *Miner. Depos.* **2007**, *42*, 299–328.
11. Large, R.R.; Maslennikov, V.; Robert, F.; Danyushevsky, L.; Chang, Z. Multistage sedimentary and metamorphic origin of pyrite and gold in the giant Sukhoi Log deposit, Russia. *Econ. Geol.* **2007**, *102*, 1233–1267. [CrossRef]
12. Keith, M.; Smith, D.J.; Jenkin, G.R.; Holwell, D.A.; Dye, M.D. A review of Te and Se systematics in hydrothermal pyrite from precious metal deposits: Insights into ore-forming processes. *Ore Geol. Rev.* **2018**, *96*, 269–282. [CrossRef]
13. Salje, E.K.H. Ferroelastic Twinning in Minerals: A Source of Trace Elements, Conductivity, and Unexpected Piezoelectricity. *Minerals* **2021**, *11*, 478. [CrossRef]
14. Gorobets, B.S.; Rogovaya, M.L. Thermoluminescence of minerals and its application in ore geology. *Geol. Ore Depos.* **1992**, *34*, 203–212.
15. Müller, W.F.; Grobety, B. Thermoluminescence and cathodoluminescence in quartz: Application to ore genesis. *Mineral. Mag.* **2002**, *66*, 859–874.
16. Rusk, B. *Cathodoluminescent Textures and Trace Elements in Hydrothermal Quartz*; Springer: Berlin/Heidelberg, Germany, 2012. [CrossRef]
17. Cao, W.T.; Hou, J.R. Genesis of the Qiubudong silver deposit in Pingshan County, Hebei Province. *Geol. Explor.* **2018**, *54*, 250–259, (In Chinese with English Abstract).
18. Sun, W.Y.; Wang, C.Y.; Zhang, T. Structural framework and deep exploration potential of the silver-molybdenum metallogenic system in central Taihang Mountains. *Acta Geol. Sin.* **2021**, *95*, 4014–4032, (In Chinese with English Abstract).
19. Zhang, T. Study on The Geological Characteristics and Origin of QiubudongSilver Deposit in Pingshan County, Hebei Province. Master's Thesis, Jilin University, Changchun, China, 2018. (In Chinese with English Abstract).
20. Sun, W.-Y.; Li, S.-R.; Santosh, M.; Wang, X.; Zhang, L.-J. Isotopic and geochemical constraints on the genesis of the Qiubudong silver deposit, central North China Craton: Implications for ore genesis and lithospheric dynamics. *Ore Geol. Rev.* **2014**, *57*, 229–242. [CrossRef]
21. Kong, Z.Z. The Geoelectric Structure Characteristics and Mineral Prospecting Significance of the Sigou-Rock-Mass in Qiubodongsilver Deposit at Hebei. Master's Thesis, Jilin University, Changchun, China, 2016. (In Chinese with English Abstract).
22. Zhao, G.; Cawood, P.A.; Wilde, S.A.; Lu, L. Archean blocks and their boundaries in the North China Craton: Lithological, geochemical, structural and P–T metamorphic evidence. *Precambrian Res.* **2001**, *107*, 45–73. [CrossRef]
23. Santosh, M.; Maruyama, S.; Zhao, D. The making and breaking of supercontinents: Some speculations based on superplume events. *Gondwana Res.* **2012**, *21*, 7–34. [CrossRef]
24. Ciobanu, C.L.; Cook, N.J.; Catchpole, H.J. Typomorphic features of ore minerals: A review of genetic significance and analytical approaches. *Ore Geol. Rev.* **2021**, *135*, 104244. [CrossRef]
25. Goldstein, J.I.; Newbury, D.E.; Michael, J.R.; Ritchie, N.W.M.; Scott, J.H.J.; Joy, D.C. *Scanning Electron Microscopy and X-Ray Microanalysis*, 4th ed.; Springer: New York, NY, USA, 2018. [CrossRef]

26. Jarvis, K.E. Inductively coupled plasma mass spectrometry: A new technique for the rapid or ultra-trace level determination of REEs in geological materials. *Chem. Geol.* **1992**, *95*, 47–60. [CrossRef]
27. Eggins, S.M.; Woodhead, J.D.; Kinsley, L.P.J.; Mortimer, G.E.; Sylvester, P.; McCulloch, M.T.; Hergt, J.M.; Handler, M.R. A simple method for the precise determination of ≥40 trace elements in geological samples by ICPMS using enriched isotope internal standardisation. *Chem. Geol.* **1997**, *134*, 311–326. [CrossRef]
28. Roedder, E. Fluid inclusions. Mineralogical Society of America. *Rev. Mineral.* **1984**, *12*, 644.
29. Shepherd, T.J.; Rankin, A.H.; Alderton, D.H.M. *A Practical Guide to Fluid Inclusion Studies*; Blackie: Glasgow, UK, 1985.
30. Zhou, X.W.; Shao, J.L.; Bian, Q.J. Study on typomorphic characteristics of pyriyte from Dongbeizhai gold deposit, Sichuan Province. *Earth Sci.—J. China Univ. Geosci.* **1994**, *19*, 52–59, (In Chinese with English Abstract).
31. Bi, X.W.; Hu, R.Z.; Peng, J.T.; Wu, K.X. REE and HFSE geochemical characteristics of pyrites in Yao'an gold deposit: Tracing ore-forming fluid signatures. *Bullutin Mineral. Petrol. Geochem.* **2004**, *23*, 1–4, (In Chinese with English Abstract).
32. Hall, D.L.; Sterner, S.M.; Bodnar, R.J. Freezing point depression of NaCl–KCl–H_2O solutions. *Geochim. Cosmochim. Acta* **1988**, *52*, 495–504.
33. Heinrich, C.A. The physical and chemical evolution of low-salinity magmatic fluids at the porphyry to epithermal transition: A thermodynamic study. *Miner. Depos.* **2005**, *39*, 864–889. [CrossRef]
34. Goldstein, R.H.; Reynolds, T.J. *Systematics of Fluid Inclusions in Diagenetic Minerals*; SEPM Short Course; Geo Science World: McLean, VA, USA, 1994; Volume 31.
35. Hegemann, E. The geochemistry of cobalt and nickel in pyrite. *Econ. Geol.* **1943**, *38*, 527–547.
36. Davidson, D.T. Geochemical behavior of nickel and cobalt. *Geochim. Cosmochim. Acta* **1962**, *26*, 1–18.
37. Bralia, A.; Sabatini, G.; Troja, F. Nickel and cobalt distribution in pyrite as a guide to ore genesis. *Mineral. Mag.* **1979**, *43*, 869–873.
38. Wang, Y.; Xiong, L.; Zhong, Z.; Ren, S.; Zhang, G.; Wang, J.; Zhang, Y.; Song, C. Typomorphic Characteristics of Gold-Bearing Pyrite and Its Genetic Implications for the Fang'an Gold Deposit, the Bengbu Uplift, Eastern China. *Minerals* **2022**, *12*, 1196. [CrossRef]
39. Gul, M.A.; Zhang, H.; Li, Y.; Yang, X.; Sun, C.; Zhao, X.; Ren, G.; Kanwal, A.; Hafeez, M.; Yang, Y.; et al. Machine learning-driven classification of Pb Zn ore deposits using pyrite trace elements and isotopic signatures: A case study of the Gunga deposit. *J. Geochem. Explor.* **2025**, *272*, 107693. [CrossRef]
40. Shi, D.; Fan, S.; Li, G.; Zhu, Y.; Yan, Q.; Jia, M.; Faisal, M. Genesis of Yongping copper deposit in the Qin-Hang Metallogenic Belt, SE China: Insights from sulfide geochemistry and sulfur isotopic data. *Ore Geol. Rev.* **2024**, *173*, 15. [CrossRef]
41. Bralia, A.; Sabatini, G.; Troja, F. A Revaluation of the Co/Ni Ratio in Pyrite as Geochemical Tool in Ore Genesis Problems. *Miner. Depos.* **1979**, *14*, 353–374. [CrossRef]
42. Brill, B.A. Trace-element contents and partitioning of elements in ore minerals from the CSA Cu-Pb-Zn deposit, Australia. *Can. Mineral.* **1989**, *27*, 263–274. [CrossRef]
43. Oreskes, N.; Einaudi, M.T. Origin of rare earth element-enriched hematite breccias at the Olympic Dam Cu-U-Au-Ag deposit, Roxby Downs, South Australia. *Econ. Geol.* **1990**, *85*, 1–28. [CrossRef]
44. Keppler, H. Constraints from partitioning experiments on the composition of subduction-zone fluids. *Nature* **1996**, *380*, 237–240. [CrossRef]
45. Андреев, Б.С. *Пирит золоторудных месторождений*; Изд.Наука: Москва, Russia, 1992; pp. 51–63.
46. Wang, J.; Liu, Z.; Wang, K.; Zeng, X.; Liu, J.; Zhang, F. Typomorphic Characteristics of Pyrites from the Shuangwang Gold Deposit, Shaanxi, China: Index to Deep Ore Exploration. *Minerals* **2019**, *9*, 383. [CrossRef]

minerals

MDPI

Article

Chronological and Geochemical Characteristics of a Newly Discovered Biotite Granite Porphyry in the Zhuxi W-Cu Polymetallic Deposit, Jiangxi Province, South China: Implications for Cu Mineralization

Yongpeng Ouyang [1,2,3,*], Qi Chen [2,3], Runling Zeng [2,3] and Tongfei Li [3,4]

1 School of Earth Sciences and Resources, China University of Geosciences, Beijing 100083, China
2 The Tenth Geological Brigade of Jiangxi Geological Bureau, Jiangxi Geological Bureau, Yingtan 335001, China; chenqi202310@163.com (Q.C.); zengrl_jx@163.com (R.Z.)
3 Yingtan Key Laboratory of Exploration and Research of Scarce and Advantage Minerals, The Tenth Geological Brigade of Jiangxi Geological Bureau, Yingtan 335001, China; 23000032@wxc.edu.cn
4 College of Environment and Tourism, West Anhui University, Lu'an 237012, China
* Correspondence: yongpeng0524@163.com

Abstract: Multiple occurrences of adakitic rocks, with crystallization ages clustering around ~160 Ma, have been documented in the Zhuxi district, northeast Jiangxi Province, South China. This research identifies a new adakitic biotite granite porphyry within the Zhuxi W-Cu polymetallic deposit. Zircon U-Pb geochronology of this porphyry yields a crystallization age of 161.6 ± 2.1 Ma. Integrated with previously published data, the adakitic rocks in the study area—comprising diorite porphyrite, biotite quartz monzonite porphyry, and the newly identified biotite granite porphyry—are predominantly calc-alkaline and peraluminous. They exhibit enrichment in light rare-earth elements (LREEs) and depletion in heavy rare-earth elements (HREEs), with slight negative Eu anomalies. The trace element patterns are characterized by enrichment in Ba, U, K, Pb, and Sr, alongside negative Nb, Ta, P, and Ti anomalies, indicative of arc-like magmatic signatures. Comparative analysis of geological and geochemical characteristics suggests that these three rock types are not comagmatic. Petrogenesis of the Zhuxi adakitic suite is linked to a dynamic tectonic regime involving Mesozoic crustal thickening, subsequent delamination, and lithospheric extension. Asthenospheric upwelling likely triggered partial melting of the overlying metasomatized lithospheric mantle, generating primary mantle-derived magmas. Underplating and advection of heat by these magmas induced partial melting of the thickened lower crust, forming the biotite granite porphyry. Partial melting of delaminated lower crustal material, interacting with the asthenosphere or asthenosphere-derived melts, likely generated the diorite porphyrite. The biotite quartz monzonite porphyry is interpreted to have formed from mantle-derived magmas that underwent assimilation of, or mixing with, silicic crustal melts during ascent. The ~160 Ma crystallization ages of these adakitic rocks are broadly contemporaneous with W-Mo mineralization in the Taqian mining area of the Zhuxi district. Furthermore, their geochemical signatures imply a prospective metallogenic setting for Cu-Mo mineralization around this period in the Taqian area.

Keywords: adakitic rock; zircon U-Pb dating; geochemical characteristics; copper deposit; Zhuxi district

1. Introduction

Adakites were initially defined as calc-alkaline dacitic and andesitic volcanic rocks characterized by distinctive geochemical signatures: high Al_2O_3 (>15 wt%), elevated Na_2O (typically 3.5–7.5 wt%), high Sr concentrations (>400 ppm), low Yb (<1.9 ppm) and Y (<18 ppm) contents, and the absence of significant negative Eu anomalies relative to typical island-arc calc-alkaline magmas [1,2]. Subsequent research has demonstrated that adakitic rocks can also form in non-plate-subduction tectonic settings, for instance, through partial melting of low-potassium tholeiitic basalt that has undergone granulite-facies metamorphism under high confining pressures [3]. In recent years, numerous intermediate-to-acidic volcanic and intrusive rocks exhibiting adakite-like geochemical characteristics have been identified across China [4–8]. These are frequently associated with significant porphyry-type Cu-Mo-Au deposits [9–17], including prominent examples such as the Dexing porphyry Cu deposit in Jiangxi Province and the Gangdese porphyry Cu belt in Tibet [18–21].

The Zhuxi district in northeast Jiangxi Province represents one of South China's most significant W-Cu polymetallic metallogenic provinces. Key deposits within this district include the Zhuxi W-Cu polymetallic deposit, the Taqian porphyry W-Mo deposit, and the Yuexing vein-type Cu-Pb-Zn polymetallic deposit. The Zhuxi W-Cu polymetallic deposit alone hosts estimated WO_3 resources of approximately 3.634 million tons (at an average grade of 0.537 wt%) and ~361,000 tons of Cu metal, ranking it as the largest known W-polymetallic deposit globally [22]. While extensive research has elucidated many geological features, petrogenetic aspects, and metallogenic controls of this deposit [23–36], the "W-Cu coexistence" phenomenon observed here remains inadequately explained. Tungsten mineralization is conventionally linked to crustally derived magmas [37,38], whereas copper mineralization is typically associated with mantle-derived magmas and processes related to oceanic crust subduction [39–41]. Given these disparate magmatic origins, W and Cu mineralizations are generally not expected to coexist within the same mineral system. However, drilling at the Zhuxi W-Cu polymetallic deposit has revealed widespread spatial association and paragenetic complexity of W and Cu ore bodies.

Recent investigations in the Zhuxi district have identified multiple intermediate-to-acidic intrusions with adakitic affinities that appear closely related to Cu-Mo mineralization, with crystallization ages ranging from 157 to 160 Ma [23–25]. Notably, a diorite porphyrite displaying adakitic geochemical signatures, with a crystallization age of 162.7 ± 1.4 Ma, was discovered within altered granite from borehole ZK4212 in the Zhuxi mining area [42]. Furthermore, this study reports a newly discovered biotite granite porphyry in borehole ZK2105 within the same mining area. Field geological observations suggest that this intrusive body may have been tectonically emplaced to its current position by regional thrusting. Considering the recognized metallogenic potential of adakitic magmas, particularly for porphyry Cu systems, this study aims to (1) determine the precise emplacement age of the newly identified biotite granite porphyry, (2) analyze its detailed geochemical characteristics, (3) investigate its petrogenesis and explore its genetic relationship with other adakitic rocks in the vicinity, and (4) elucidate its implications for W-Cu mineralization. These objectives are pursued to refine petrogenetic and metallogenic models for the Zhuxi district, and to provide robust geological constraints for future mineral exploration endeavors in the region.

2. Regional Geology and Sample Description

2.1. Geological Background

The Zhuxi district is located at the eastern margin of the Pingle Depression, within the northern segment of the Jiangnan Orogen, marking a transition zone between the Yangtze Block to the northwest and the Cathaysia Block to the southeast [43,44] (Figure 1a). The district lies northwest of the Northeast Jiangxi Deep Fault system and is geographically adjacent to the major Dexing Cu-Au-Ag and Jiurui Cu-Au mining areas. Since the Neoproterozoic (Jingningian period), the region has experienced multiple protracted tectonothermal cycles involving sedimentation, deformation, metamorphism, and magmatism. This complex geological evolution generated conditions highly favorable for polymetallic mineralization, leading to the formation of significant W, Cu, Mo, and Au deposits, among which the Zhuxi W-Cu polymetallic deposit is the largest discovered to date.

The regional stratigraphy comprises a metamorphic basement unconformably overlain by a sedimentary cover sequence (Figure 1b). The most extensively exposed basement units belong to the Neoproterozoic Wannian and Shuangqiaoshan groups, predominantly consisting of low-grade metavolcanic and metasedimentary rocks. These are overlain by Carboniferous, Permian, Triassic, Jurassic, Cretaceous, Neogene, and Quaternary strata [23].

Recurrent magmatism accompanied the region's prolonged and complex tectonic evolution. Magmatic activity, ranging from mafic–ultramafic to felsic (granitic), occurred episodically from the Neoproterozoic (Jingningian) to the Cenozoic (Himalayan). Mafic-to-intermediate rocks primarily comprise lamprophyre, diorite porphyrite, and quartz diorite porphyrite, which typically occur as dikes emplaced at relatively shallow crustal levels within the mining district. Intermediate-to-acidic intrusions include granodiorite porphyry, biotite granite, altered granite, two-mica granite, muscovite granite, and granite porphyry. Among these, the acidic granites generally form dike swarms or larger plutons at deeper levels, whereas granodiorite porphyries occur predominantly as small stocks or dykes at shallower levels. Metallogenesis in the district is closely linked to the emplacement of Yanshanian (Jurassic–Cretaceous) intermediate-to-acidic intrusions, particularly granites and the biotite granite porphyry relevant to this study [45].

The basement rocks are interpreted as metamorphosed marine volcano-sedimentary sequences, largely derived from Neoproterozoic volcanic-arc and back-arc basin systems. Structurally, the study area lies within a zone of intense deformation resulting from the Paleozoic collision and amalgamation between the Yangtze and Cathaysia blocks, followed by significant Mesozoic intracontinental deformation. Key tectonic events include (1) strong Neoproterozoic (Jingningian) folding and metamorphism; (2) Caledonian folding and thrusting; (3) Variscan–Indosinian intracontinental deformation, including northeast-trending ductile shearing and orogeny; and (4) Yanshanian–Himalayan intracontinental reactivation, characterized by further orogeny, basin formation, and significant lithospheric extension. Consequently, regional structures are complex, featuring prominent folds, extensive fault systems (including major thrusts and strike-slip faults), and localized ductile shear zones.

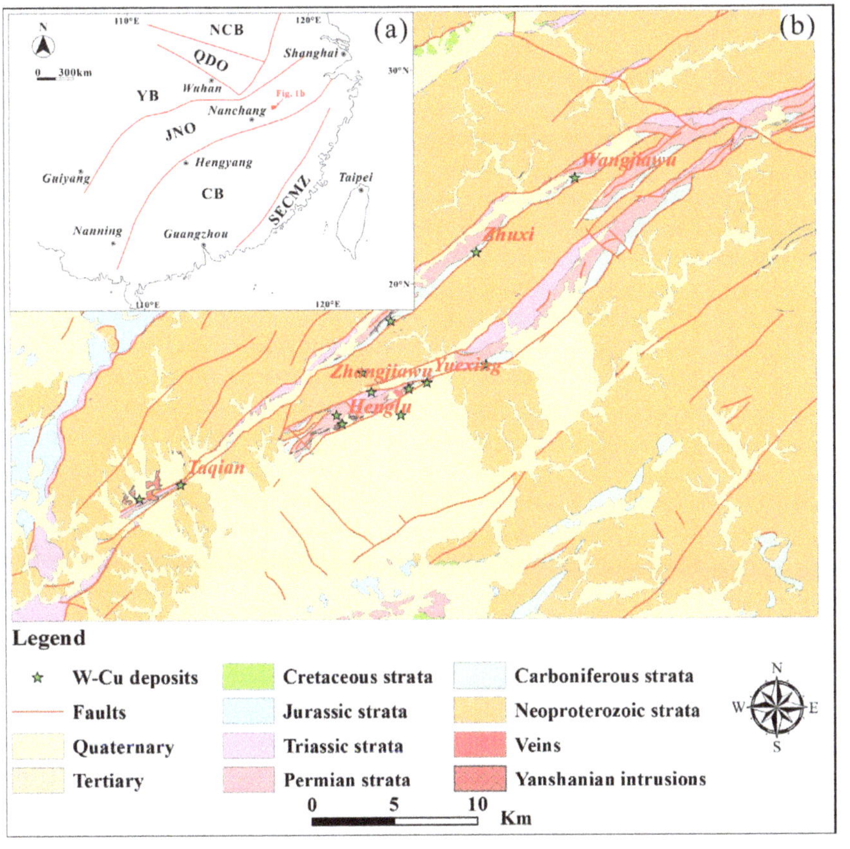

Figure 1. Geological sketched map of the study area: (**a**) Tectonic sketched map of the study area (NCB denotes the North China Block, QDO denotes the Qinling–Dabie Orogen, YB denotes the Yangtze Block, JNO denotes the Jiangnan Orogen, CB denotes the Cathaysian Block, and SECMZ denotes the Southeast China Magmatic Zone) (modified from [22]). (**b**) Simplified geological map of the Zhuxi district (modified from [46]).

The study area hosts abundant and diverse mineral resources, encompassing metallic, nonmetallic, and energy minerals. Metallic deposits, including W-Cu polymetallic, W-Mo polymetallic, and Cu polymetallic systems, are concentrated within the Taqian–Fuchun metallogenic belt. These are primarily hosted within late Paleozoic carbonate strata and at or near the contact zones with Yanshanian intermediate-to-acidic intrusions [47]. Non-metallic mineral occurrences are largely stratabound or associated with dikes. For instance, limestone deposits are developed in the Carboniferous Huanglong Formation (northeastern area) and the Permian Qixia and Changxing formations. Dolomite is exploited from the dolomite member of the Carboniferous Huanglong Formation in the northeastern region, while clay minerals are found in shale units within the Permian Xiaojiangbian Formation (northeastern area). Coal deposits are hosted by the Permian Leping Formation and the Triassic Anyuan Formation.

Based on ore genesis and style of mineralization, three primary deposit types are recognized in the district (Figure 1b). Skarn-type W-Cu polymetallic deposits, including major occurrences such as the Zhuxi, Taqian, Zhangjiawu, and Tanling deposits, are predominantly developed within late Paleozoic carbonate strata, typically concentrated at the contacts between Yanshanian intermediate-to-acidic intrusions and the carbonate host rocks. Intrusion-hosted disseminated/stockwork W-Cu-Mo-Pb-Zn deposits, exemplified by the Yangcaojian and Xiyuanwu deposits, encompass mineralization found primarily within Yanshanian intermediate-to-acidic intrusions (e.g., biotite granite, altered granite, granodiorite, and granite porphyry) or their immediate endo-/exo-contact zones; this category is sometimes referred to locally as "altered granite (porphyry) type". Additionally, vein-type (fracture-controlled) Cu-W-Pb-Zn polymetallic deposits, represented by the Yuexing deposits, are predominantly structurally controlled, localized within faults, fracture zones, and detachment structures, and can occur in a variety of country rock types [22].

2.2. Sample Description

This study investigates several intrusive rocks from the Zhuxi W-Cu polymetallic deposit. While incorporating previously published data for a biotite quartz monzonite porphyry from the Zaolin ore block (Figure 2a) and a diorite porphyrite hosted within altered granite [42], the primary focus involves new analyses of a biotite granite porphyry intersected between −1285.2 m and −1317.4 m depth in drill hole ZK2105 (indicated in Figure 2b).

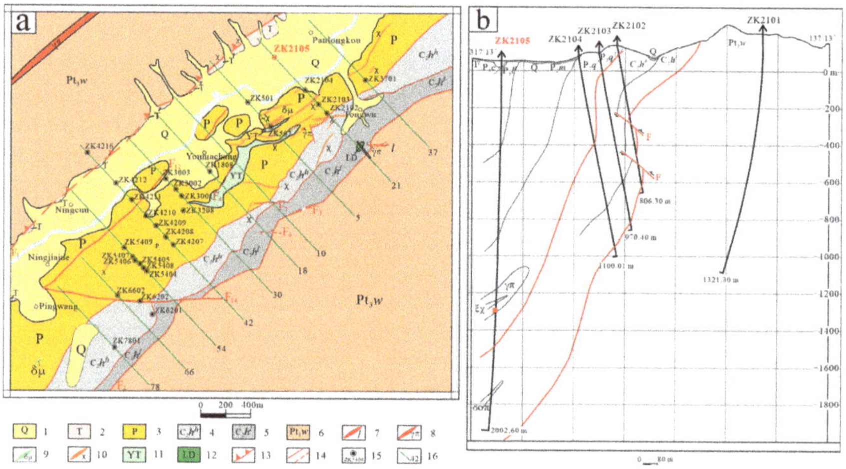

Figure 2. Simplified map of the deployment of the exploration project of the Zhuxi W-Cu polymetallic deposit: (**a**) Simplified geological map of the Zhuxi mining area (1—Quaternary strata; 2—Triassic strata; 3—Permian strata; 4—Upper Carboniferous Series Huanglong Formation; 5—Upper Carboniferous Series Huanglong Formation; 6—Upper Proterozoic Wannian Group; 7—aplite; 8—granite porphyry; 9—diorite porphyrite; 10—lamprophyre; 11—tremolite actinization zone; 12—green alteration zone; 13—thrust nappe fault; 14—measured (presumed) fault; 15—locations and the number of drill holes; 16—exploration line and the number of exploration lines) (revised from [29]). (**b**) Profile map of the 21st exploration line in the Zaolin ore section in the Zhuxi mining area.

Samples of the biotite granite porphyry collected for this study are grayish-white and exhibit a blocky structure in hand specimen, with a distinct porphyritic texture (Figure 3a). Petrographic analysis reveals phenocrysts set in a finer-grained matrix. Phenocrysts constitute approximately 35% of the rock volume and primarily consist of feldspar (~20%) and biotite (~15%). Biotite occurs as subhedral-to-euhedral flakes or laths, ranging from 0.1 to 1.8 mm in diameter, typically exhibiting well-defined cleavage; some grains show partial alteration to chlorite (Figure 3b). Feldspar phenocrysts are subhedral to anhedral, tabular to prismatic in shape, with grain sizes ranging from 0.2 to 4 mm. Some larger feldspar crystals show evidence of partial alteration (e.g., sericitization), resulting in slightly clouded surfaces (Figure 3b). The matrix (~65% volume) is microcrystalline and composed predominantly of quartz, feldspar, and mica.

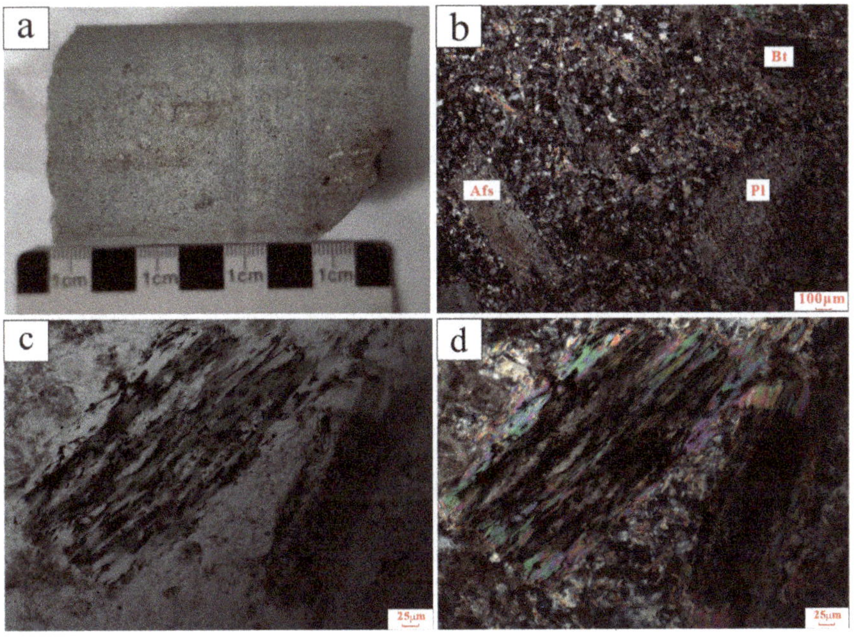

Figure 3. Hand samples and microscopic characteristics of the biotite granite porphyry in the Zhuxi district: (**a**) hand samples; (**b**) porphyritic texture of the sample (Pl—plagioclase; Afs—alkaline feldspar; Bt—biotite); (**c**) chloritized biotite (under a single polarizer); (**d**) chloritized biotite (under an orthogonal polarizer).

3. Methodology

3.1. LA-ICP-MS Zircon U-Pb Dating

The selection of biotite granite porphyry zircons for this study was accomplished via Guangzhou Tuoyan Testing Technology Co., Ltd. (Guangzhou, China) The selection criteria were euhedral zircons with a good crystal shapes, no cracks, and transparency and cleanliness. After selection, they were polished to the zircon centers and secured with epoxy glue. Subsequently, micrographs were obtained using transmitted light, reflected light, and cathodoluminescene (CL) techniques with a scanner with an analytical electron microscope (JSM-IT300) from JEOL (Tokyo, Japan) linked to a Delmic Sparc system. The imaging was conducted under an electric field with voltages ranging from 0.5 to 30 kV and a tungsten filament current of 72 μA. Zircon laser ablation–inductively coupled plasma mass spectrometry (LA-ICP-MS) U-Pb isotope dating was performed at the State Key Laboratory of Geological Processes and Mineral Resources (GPMR) at the China University of Geosciences (Wuhan, China). For this purpose, a UP193FX 193 nm ArF excimer laser ablation system from Newwave Company (San Diego, CA, USA) was utilized. The laser that was used was from ATL Company (Wermelskirchen, Germany) and the ICP-MS instrument was Agilent 7500a (Santa Clara, CA, USA). The beam spot diameter used in this study was 32 μm. Both samples were subjected to external standard calibration using the zircon Standard 91500 [48]. The analytical method and process can be found in the literature [48]. The material's elemental composition and isotope ratio were computed via the GLITER_Ver4.0 program. The final zircon $^{207}Pb/^{206}Pb$ and $^{238}U/^{206}Pb$ diagrams of the Tera–Wasserburg concordia, along with the $^{206}Pb/^{238}U$ weighted average ages, were plotted via Isoplot 3.0 software [49].

3.2. Whole-Rock Geochemical Analysis

The testing of major, trace, and rare-earth samples was carried out by ALS Minerals-ALS Chemex (Guangzhou) Co., Ltd. (Guangzhou, China) The major element analysis was carried out via a PANalytical PW2424 X-ray fluorescence spectrometer (Almelo, The Netherlands) and analyzed using the ME-XRF26F method. The samples were prepared by mixing 0.5 g of the powdered sample with 5 g of $Li_2B_4O_7$ and $LiBO_2$ and then forming wafers. The range of study inaccuracies was 0.1~1% relative standard deviation (RSD).

Trace element analysis was carried out via an Agilent 5110 ICP emission spectrometer and an Agilent 7900 ICP mass spectrometer (Santa Clara, CA, USA), and the ME-MS61r method was used for analysis. Forty milligrams of sample powder was dissolved in a mixture of HF, HNO_3, and H_2SO_4 and then evaporated twice at temperatures between 150 and 200 °C. After evaporation, the residue was redissolved in 49 mL of 1% HNO_3 for analysis. As an internal standard, 1 mL of 500 ppb indium was inserted into the sample and used as an external standard.

The Agilent 7900 inductively coupled plasma mass spectrometer (Santa Clara, CA, USA) was used to analyze the rare-earth elements, employing the ME-MS81 analytical method. Next, 50 mg of lithium metaborate ($LiBO_2/Li_2B_4O_7$) flux was added to the sample, which was mixed thoroughly and fused in a furnace at 1025 °C. After cooling, the fused material was dissolved and diluted with nitric acid, hydrochloric acid, and hydrofluoric acid. The rare-earth element contents were subsequently analyzed with an inductively coupled plasma mass spectrometer. This method ensures analytical precision, with a relative deviation of less than 10%, and accuracy, with a relative error of less than 10%.

4. Results

4.1. Zircon U-Pb Dating

Zircons from the biotite granite porphyry sample ZK2105-1 were analyzed for U-Pb isotopes via LA-ICP-MS; the results for 14 analysis spots are presented in Table 1. Cathodoluminescence (CL) images reveal that the analyzed zircons are mostly euhedral to subhedral, with long-axis dimensions ranging from 55 μm to 185 μm and aspect ratios between 1:1 and 3:1 (Figure 4). The grains typically exhibit distinct internal structures, including oscillatory zoning and growth bands, characteristic of magmatic crystallization [50]. The uranium concentrations vary from 205 ppm to 2343 ppm, and the Th concentrations range from 55 ppm to 449 ppm. The resulting Th/U ratios span from 0.04 to 1.16, with an average of 0.34, consistent with a magmatic origin for most grains.

All 14 analytical spots yielded concordant or near-concordant U-Pb ages, plotting cohesively on the Tera–Wasserburg concordia diagram (Figure 5). Nine relatively concentrated isotopic ages provided a weighted mean $^{206}Pb/^{238}U$ age of 161.6 ± 2.1 Ma (MSWD = 1.5, n = 9). This age was interpreted as the crystallization age of the biotite granite porphyry. The other measurement points were relatively scattered, ranging from 606 to 879 Ma, which may indicate the age of captured Neoproterozoic zircons (e.g., ~847 Ma [51]).

Table 1. LA-ICP-MS zircon U-Pb isotope dating results of the biotite granite porphyry in the Zhuxi district.

No.	Content (ppm)			Th/U	Isotope Ratio						Isotope Age (Ma)	
	^{206}Pb	^{232}Th	^{238}U		$^{207}Pb/^{206}Pb$	1σ	$^{207}Pb/^{235}U$	1σ	$^{206}Pb/^{238}U$	1σ	$^{206}Pb/^{238}U$	1σ
1	22.7	77.4	806	0.10	0.0525	0.0016	0.1801	0.0056	0.0248	0.0003	158	2
2	77.1	152	647	0.23	0.0669	0.0016	0.9271	0.0285	0.1001	0.0022	615	13
3	27.0	52.7	949	0.06	0.0532	0.0019	0.1901	0.0067	0.0259	0.0004	165	2
4	7.56	152	223	0.68	0.0498	0.0037	0.1769	0.0129	0.0259	0.0005	165	3
5	11.1	228	325	0.70	0.0498	0.0029	0.1755	0.0101	0.0256	0.0003	163	2
6	29.5	192	996	0.19	0.0518	0.0017	0.1827	0.0062	0.0255	0.0004	162	2
7	61.3	99.9	2207	0.05	0.0493	0.0012	0.1715	0.0044	0.0250	0.0002	159	1
8	65.7	132	504	0.26	0.0659	0.0015	0.9962	0.0272	0.1093	0.0022	669	13
9	20.4	82.8	689	0.12	0.0526	0.0020	0.1861	0.0076	0.0255	0.0004	162	3
10	154	311	858	0.36	0.0669	0.0014	1.3558	0.0319	0.1460	0.0017	879	10
11	72.7	371	471	0.79	0.0657	0.0016	0.9971	0.0301	0.1093	0.0021	669	12
12	14.4	444	385	1.15	0.0499	0.0019	0.1753	0.0066	0.0257	0.0004	163	2
13	103	133	875	0.15	0.0654	0.0013	0.8958	0.0264	0.0985	0.0021	606	12
14	75.0	351	2456	0.14	0.0512	0.0013	0.1843	0.0057	0.0259	0.0005	165	3

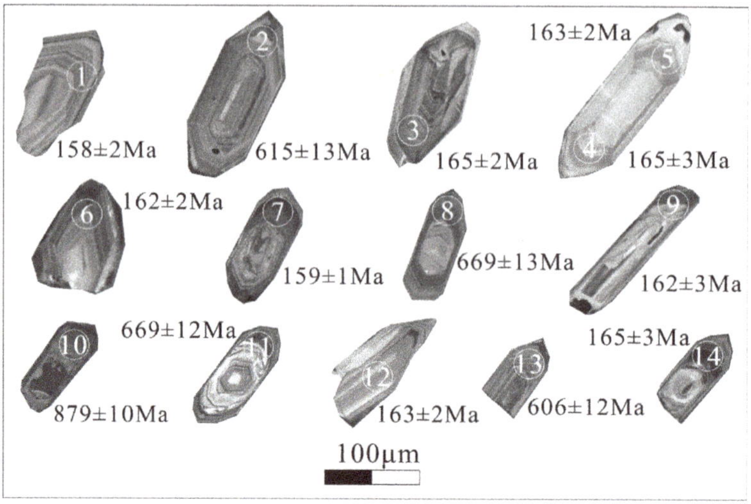

Figure 4. Zircon cathodoluminescence (CL) images of the biotite granite porphyry in the Zhuxi district.

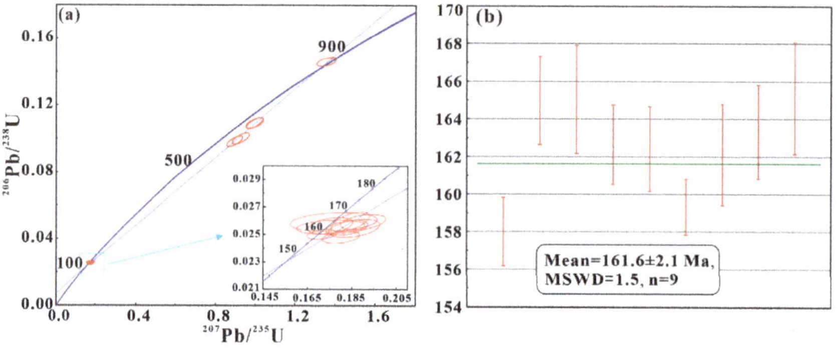

Figure 5. Zircon U-Pb isotope concordia diagram of the biotite granite porphyry in the Zhuxi district: (**a**) concordia diagram of the 14 LA-ICP-MS U-Pb ages; (**b**) weighted mean age plot of the 9 LA-ICP-MS U-Pb ages.

4.2. Characteristics of Major Elements

Whole-rock major element compositions for the newly analyzed biotite granite porphyry (ZK2105 samples) and previously published data for diorite porphyrite (ZX15-01-06 [42]) and biotite quartz monzonite porphyry (ZL-YQ11-15 [52]) are listed in Table 2. The three rock types show increasing SiO_2 contents, corresponding to increasing acidity, from the diorite porphyrite (avg. 59.13 wt%) to the biotite quartz monzonite porphyry (avg. 64.91 wt%) to the biotite granite porphyry (avg. 68.46 wt%). The TiO_2 content is highest in the diorite porphyrite (avg. 0.62 wt%), lowest in the biotite quartz monzonite porphyry (avg. 0.33 wt%), and intermediate in the biotite granite porphyry (avg. 0.41 wt%). The Al_2O_3 contents are relatively high and comparable across the three types (avg. 15.46 wt%, 15.28 wt%, and 14.94 wt%, respectively). With increasing SiO_2, the MgO (avg. 4.22, 1.90, and 0.96 wt%) and TFe_2O_3 (total iron as Fe_2O_3; avg. 4.49, 2.50, and 2.38 wt%) contents decrease, while the total alkali content ($Na_2O + K_2O$; avg. 4.76, 5.35, and 6.48 wt%) increases. Notably, the biotite quartz monzonite porphyry exhibits a significantly higher K_2O content compared to the other two rock types at similar SiO_2 levels.

The Rittman index values ($\sigma = (Na_2O + K_2O)^2/(SiO_2 - 43)$) for all three rock types are less than 3.3, indicating a calc-alkaline affinity. On the SiO_2 vs. K_2O classification diagram (Figure 6a), the samples plot within the calc-alkaline and high-K calc-alkaline fields. In the A/NK vs. A/CNK diagram (Figure 6b; A/NK = molar $Al_2O_3/(Na_2O + K_2O)$, A/CNK = molar $Al_2O_3/(CaO + Na_2O + K_2O)$), all samples plot in the peraluminous field (A/CNK > 1). Thus, all three intrusive types are classified as peraluminous, calc-alkaline to high-K calc-alkaline rocks. The Harker diagrams (Figure 7) show that inter-element correlations with SiO_2 do not define simple linear trends across the three rock types, and apparent compositional gaps or discontinuities exist. This suggests that they are unlikely to be related solely by fractional crystallization from a single parental magma.

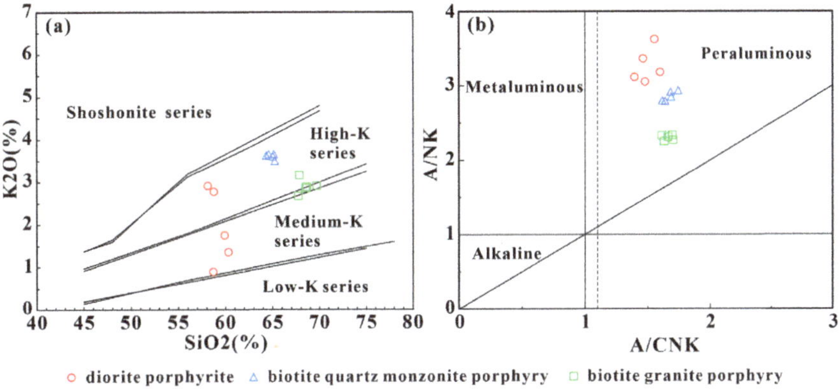

Figure 6. Petrological diagrams of the adakitic rocks in the Zhuxi district: (**a**) Relationships between K_2O and SiO_2 (the base map is from [53]). (**b**) A/NK vs. A/CNK diagram (the base map is from [54]).

4.3. Characteristics of Rare-Earth Elements

Rare-earth element (REE) and trace element data for the three rock types are presented in Table 2. Chondrite-normalized REE patterns (Figure 8a) are characterized by enrichment in light REEs (LREEs) relative to heavy REEs (HREEs), resulting in moderately to steeply right-dipping patterns. The LREE segments are steeper than the relatively flat HREE segments. The total REE contents (ΣREEs) are similar for the diorite porphyrite (avg. 202.8 ppm) and the biotite quartz monzonite porphyry (avg. 208.8 ppm) but lower in the more evolved biotite granite porphyry (avg. 173.0 ppm). The ΣREEs contents follow the same trend (avg. 191.2, 199.0, and 166.2 ppm, respectively). The ΣREEs contents decrease systematically with increasing SiO_2, from the diorite porphyrite (avg. 11.6 ppm) through the biotite quartz monzonite porphyry (avg. 9.8 ppm) to the biotite granite porphyry (avg. 6.8 ppm). Consequently, the degree of LREE/HREE fractionation, indicated by ratios such as $(La/Yb)_N$, increases with differentiation (the average $(La/Yb)_N$ values are approximately 16.5, 20.4, and 24.4 for the three rock types, respectively). All samples exhibit slight negative Eu anomalies, with δEu (Eu/Eu*) values ranging from 0.83 to 0.90 for the diorite porphyrite, from 0.74 to 0.81 for the biotite quartz monzonite porphyry, and from 0.77 to 0.81 for the biotite granite porphyry.

Table 2. Major element and trace element contents of biotite granite porphyry (ZK2105-1-5), diorite porphyrite (ZK15-01-06) [42], and biotite quartz monzonite porphyry (ZL-YQ11-15) [52].

Elements	ZK2105-2-1	ZK2105-2-2	ZK2105-1-5	ZK2105-2-4	ZK2105-2-5	ZX15-01	ZX15-02	ZX15-03	ZX15-04	ZX15-06	ZL-YQ11	ZL-YQ12	ZL-YQ13	ZL-YQ14	ZL-YQ15
SiO_2	68.58	69.66	67.72	67.83	68.5	58.66	58.08	60.3	59.89	58.73	65.2	65.04	64.7	65.12	64.48
Al_2O_3	14.77	14.77	14.86	15.16	15.11	15.46	15.25	15.83	15.25	15.52	14.85	15.67	15.11	15.37	15.39
TiO_2	0.4	0.41	0.41	0.41	0.41	0.63	0.62	0.65	0.59	0.63	0.35	0.35	0.33	0.32	0.31
TFe_2O_3	2.33	2.35	2.46	2.38	2.4	4.51	4.52	4.56	4.26	4.6	2.5	2.52	2.53	2.49	2.45
MgO	0.95	0.94	0.96	0.99	0.96	4.19	4.34	4.32	3.97	4.29	1.81	1.95	1.97	1.94	1.85
MnO	0.03	0.03	0.03	0.03	0.03	0.08	0.07	0.08	0.08	0.08	0.03	0.03	0.03	0.03	0.03
CaO	2.55	2.18	2.84	2.58	2.43	5.68	5.88	4.93	5.31	6.13	3.89	3.66	3.84	3.73	3.87
Na_2O	3.59	3.59	3.7	3.38	3.64	3.37	1.63	3.62	3.25	2.21	1.84	1.75	1.77	1.74	1.7
K_2O	2.88	2.91	2.67	3.16	2.86	0.9	2.91	1.36	1.75	2.78	3.47	3.59	3.65	3.65	3.61
P_2O_5	0.12	0.12	0.12	0.12	0.12	0.2	0.2	0.21	0.19	0.2	0.1	0.1	0.1	0.12	0.08
LOI	3.42	2.75	3.94	3.57	3.48	4.86	4.91	2.89	4.05	3.43	3.11	2.48	3.08	2.64	3.4
Li	88.5	69.3	84.5	79.7	69.4	202	165	116	137	92.4	10.92	9.79	9.45	11.53	9.68
Be	2.85	2.82	2.78	2.86	2.86	2.8	3.08	2.47	2.73	2.4	4.63	2.75	2.69	4.19	3.27
Sc	5.58	5.73	5.83	5.32	5.32	17	182	18	16.1	17.6	5.27	5.97	5.77	5.68	5.71
V	45.8	45.5	48.8	44	44.3	136	135	140	121	136	54.98	56.36	57.02	55.58	53.01
Cr	17.4	15.7	16.7	16.1	13.3	182	184	198	180	188	14.92	16.98	18.37	18.7	12.28
Co	5.95	5.81	6.64	5.75	5.59	21.8	21.3	21.3	19.3	20.5	6.77	7.09	6.68	6.55	5.97
Ni	8.23	5.61	5.4	5.88	4.87	73.8	74.7	783	71	75.8	7.09	7.08	6.55	6.24	4.71
Cu	6.97	8.09	13.9	4.79	10.1	22	20.3	25.2	25.4	21.7	4.78	5.79	6.74	2.47	1.98
Zn	137	43.7	62.6	59.2	59.9	74.5	79.1	69.5	90.5	80.3	66.37	70.4	64.5	57.01	50.43
Ga	23.1	22.8	24.4	21.9	22.3	22.6	23.1	23.8	23.3	22.9	22.87	23.57	23.38	23.51	22.4
As	22.1	4.74	40.3	29.9	184	229	354	91.9	159	138	2.6	12.6	3.01	6.9	7.54
Rb	139	134	141	144	130	74.3	238	104	121	185	129.9	138.6	142.3	143.8	133.6
Sr	513	401	468	464	483	1164	437	1238	856	630	474.16	366.05	437.88	408.94	383.67
Y	7.68	7.57	8.12	6.97	7.42	14.9	14.6	14.8	13.9	142	8.89	7.46	7.55	7.26	8
Zr	194	200	186	187	184	182	183	184	179	178	306	314	303	317	278
Nb	7.15	7.07	7.5	7.62	7.12	9.6	9.78	9.91	9.5	953	12.15	11.42	11.42	11.04	9.55
Cs	36.5	33.2	49	41.4	33.8	93.1	254	95.3	112	167	17.32	19.11	18.43	18.59	18.02

Table 2. *Cont.*

Elements	ZK2105-2-1	ZK2105-2-2	ZK2105-1-5	ZK2105-2-4	ZK2105-2-5	ZX15-01	ZX15-02	ZX15-03	ZX15-04	ZX15-06	ZL-YQ11	ZL-YQ12	ZL-YQ13	ZL-YQ14	ZL-YQ15
Ba	719	628	572	896	680	378	398	658	659	591	999	855	1150	974	1099
Hf	4.89	4.76	4.61	4.61	4.53	5.03	4.99	4.96	4.99	5	7.91	7.7	7.88	7.76	7.19
Ta	0.662	0.656	0.639	0.628	0.622	0.6	0.59	0.58	0.58	0.59	1.75	1.59	1.95	0.69	0.56
Pb	42	16.8	38.3	37.6	32.5	20	17.7	222	392	323	35.57	32.98	34.01	31.85	27.02
Th	18.7	18	18.4	17.7	17.4	17.7	17.2	18.1	17.7	17.4	17.86	18.12	19.02	19.24	17.89
U	4.17	4.09	3.93	3.75	3.83	2.94	2.86	3.01	2.95	2.91	4	4.17	4.21	4.17	4
Mo	-	-	0.225	-	-	0.23	0.5	0.19	0.22	0.12	0.75	8.46	0.78	0.42	0.37
Sb	2.52	2.09	2.91	2.26	1.91	3.68	15.4	4.5	15.9	11.3	0.53	1.14	0.92	0.82	0.7
W	0.213	0.37	0.516	0.847	0.12	1.45	2.24	1.43	0.85	0.6	1.55	1.74	1.74	1.89	1.25
Bi	0.174	0.085	0.051	0.0419	0.204	0.14	0.24	0.12	0.54	0.16	0.45	0.38	0.46	0.09	0.06
La	42.5	38.3	45.1	41.8	38.8	50	47.2	49.2	49.1	46.8	50.16	50.1	48.68	49.33	52.03
Ce	83.4	84.1	87.1	81.7	78.5	91.6	86	92	89.1	86.5	93.24	94.38	95.24	93.25	95.7
Pr	7.88	7.41	8.24	7.45	7.33	10.1	9.62	9.95	9.79	9.46	9.83	9.88	10.21	9.7	10.2
Nd	28.7	28.3	31.4	27.1	27.8	37.6	35.1	37.4	36.4	34.7	37.2	37.5	37.89	36.47	37.93
Sm	4.54	4.54	4.76	4.53	4.31	6.31	5.87	6.4	6.2	6.06	5.68	5.63	6.01	5.58	5.9
Eu	1.05	1.02	1.1	1.03	1.01	1.53	1.55	1.5	1.56	1.49	1.48	1.4	1.39	1.38	1.5
Gd	3.19	3.24	3.33	3	3.08	4.32	4.32	4.24	4.34	431	5.39	5.17	5.32	5.13	5.33
Tb	0.402	0.402	0.425	0.385	0.387	0.57	0.55	0.59	0.56	0.56	0.57	0.53	0.54	0.53	0.54
Dy	1.56	1.59	1.54	1.54	1.55	3.1	2.87	3.18	3.01	3.07	2.21	1.94	2	1.98	2.01
Ho	0.26	0.247	0.259	0.245	0.241	0.52	0.48	0.51	0.49	0.5	0.35	0.3	0.3	0.3	0.32
Er	0.793	0.717	0.711	0.695	0.697	1.57	1.56	1.64	1.56	1.65	0.88	0.74	0.75	0.74	0.78
Tm	0.0859	0.087	0.0857	0.0792	0.0803	0.2	0.19	0.19	0.19	0.2	0.111	0.093	0.09	0.094	0.099
Yb	0.562	0.568	0.582	0.526	0.536	1.21	1.2	1.26	1.19	1.2	0.73	0.64	0.62	0.65	0.68
Lu	0.0842	0.0809	0.0825	0.0794	0.0764	0.19	0.18	0.19	0.17	0.18	0.107	0.091	0.091	0.087	0.097
ΣREE	175.01	170.60	184.72	170.16	164.40	208.82	196.69	208.25	203.66	196.68	207.94	208.39	209.13	205.22	213.12
ΣLREE	168.07	163.67	177.70	163.61	157.75	197.14	185.34	196.45	192.15	185.01	197.59	198.89	199.42	195.71	203.26
ΣHREE	6.94	6.93	7.02	6.55	6.65	11.68	11.35	11.80	11.51	11.67	10.35	9.50	9.71	9.51	9.86
ΣLREE/ΣHREE	24.23	23.61	25.33	24.98	23.73	16.88	16.33	16.65	16.69	15.85	19.09	20.93	20.54	20.58	20.62

Table 2. *Cont.*

Elements	ZK2105-2-1	ZK2105-2-2	ZK2105-1-5	ZK2105-2-4	ZK2105-2-5	ZX15-01	ZX15-02	ZX15-03	ZX15-04	ZX15-06	ZL-YQ11	ZL-YQ12	ZL-YQ13	ZL-YQ14	ZL-YQ15
δEu	0.80	0.77	0.80	0.80	0.81	0.85	0.90	0.83	0.87	0.85	0.81	0.78	0.74	0.77	0.80
Sr/Y	66.80	52.97	57.64	66.57	65.09	78.12	29.93	83.65	61.58	4.44	53.34	49.07	58.00	56.33	47.96
Rb/Sr	0.27	0.33	0.30	0.31	0.27	0.06	0.54	0.08	0.14	0.29	0.27	0.38	0.32	0.35	0.35
Nb/Ta	10.80	10.78	11.74	12.13	11.45	16.00	16.58	17.09	16.38	16.15	6.94	7.18	5.86	16.00	17.05
Zr/Hf	39.67	42.02	40.35	40.56	40.62	36.18	36.67	37.10	35.87	35.60	38.69	40.78	38.45	40.85	38.66
Nb/U	1.71	1.73	1.91	2.03	1.86	3.27	3.42	3.29	3.22	3.27	3.04	2.74	2.71	2.65	2.39
Data sources	Data analyzed in this paper					Data from paper [42]					Data from paper [52]				

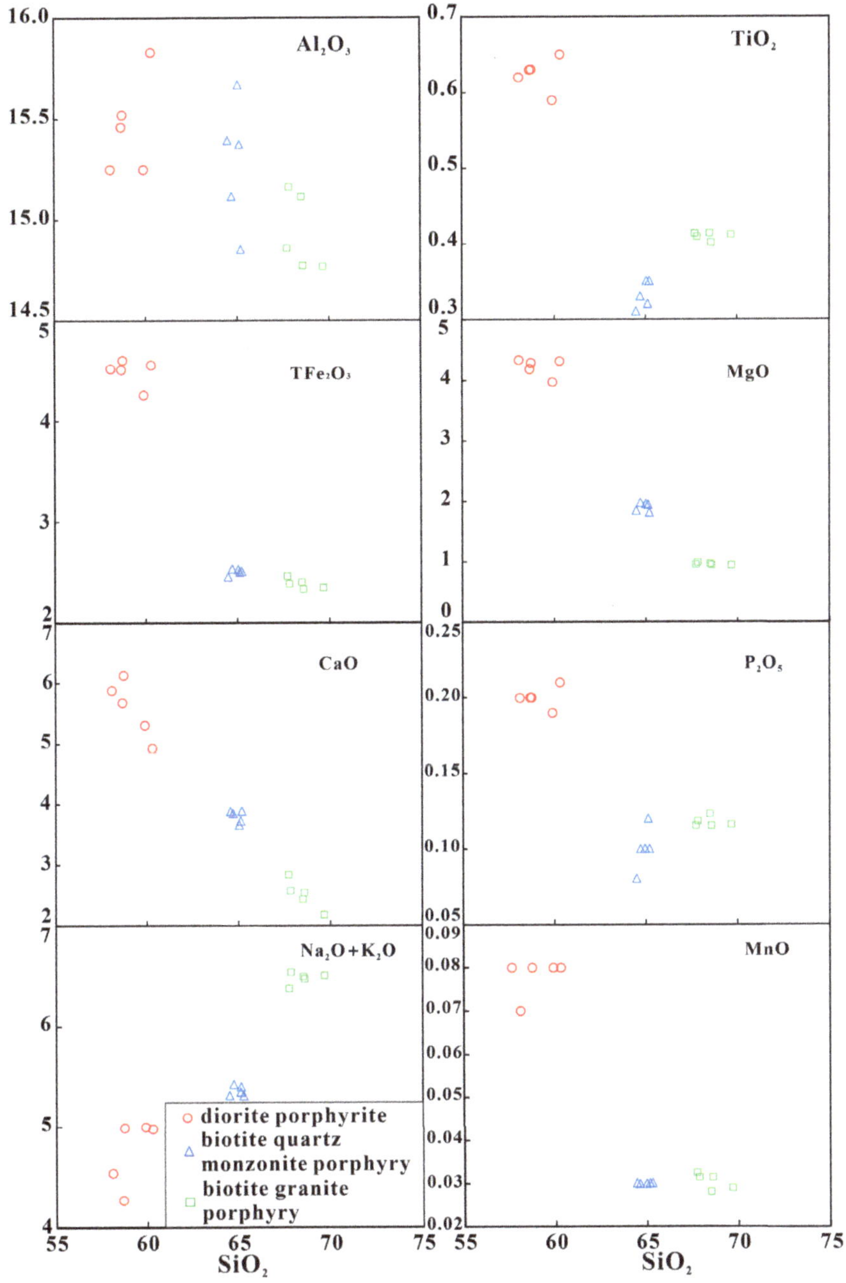

Figure 7. Diagram of SiO_2 and major oxides in the adakitic rocks in the Zhuxi district.

The primitive mantle-normalized trace element patterns (Figure 8b) for all three rock types display enrichment in large-ion lithophile elements (LILEs, e.g., Ba, U, K, Pb, Sr) and depletion in high-field-strength elements (HFSEs, e.g., Nb, Ta, Ti), as well as P. These features, particularly the pronounced negative Nb-Ta-Ti anomalies, are characteristic of arc-related magmas. Crucially, all samples exhibit low Y contents (range 6.97–14.9 ppm, avg. 9.95 ppm) and low Yb contents (range 0.53–1.26 ppm, avg. 0.81 ppm). Combined with high Sr contents, this results in high Sr/Y ratios (range 29.9–83.7, avg. 58.1). On the $(La/Yb)_N$-Yb_N and Sr/Y vs. Y discrimination diagrams (Figure 9a,b), the samples plot predominantly within the adakite field. Combined with their high Al_2O_3 contents (Table 2), these geochemical features (high Al_2O_3, high Sr/Y, low Y, low Yb) confirm that all three intrusive rock types possess adakitic affinities.

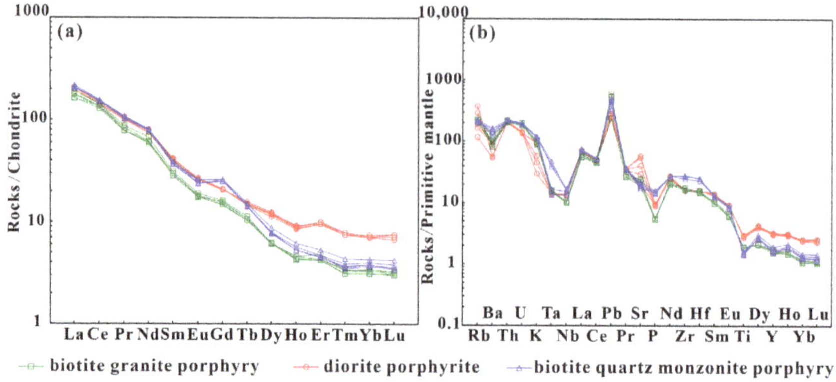

Figure 8. Distribution patterns of REEs and trace elements in adakitic rocks in the Zhuxi district: (**a**) chondrite-normalized REE patterns; (**b**) primitive mantle-normalized trace element patterns (the standard values can be founded in the related paper [55]).

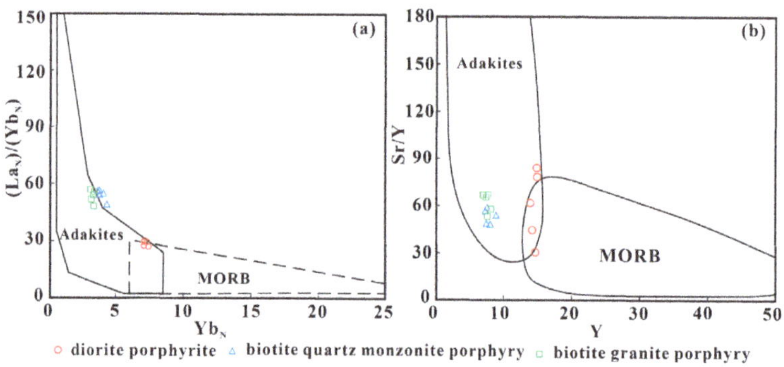

Figure 9. Discrimination diagram of the adakitic rocks in the Zhuxi district: (**a**) $(La_N)/(Yb_N)$ vs. Yb_N diagram; (**b**) (Sr/Y) vs. Y diagram (the base map is from [56]).

5. Discussions

5.1. Sources of Diorite Porphyrite, Biotite Quartz Monzonite Porphyry, and Biotite Granite Porphyry

Petrographic observations indicate that samples from all three intrusive types investigated (biotite granite porphyry, diorite porphyrite, and biotite quartz monzonite porphyry) have experienced variable degrees of post-emplacement alteration (e.g., chloritization of biotite, sericitization of feldspar; Figure 3). This is consistent with their relatively high Loss-on-Ignition (LOI) values, which generally exceed 3 wt% (Table 2), suggesting interactions with hydrothermal fluids or weathering processes. To evaluate the potential influence of alteration on whole-rock geochemistry and ensure the robustness of subsequent petrogenetic interpretations, the behavior of elements considered to be relatively immobile during low–moderate-temperature alteration, such as high-field-strength elements (HFSEs, e.g., Th, Zr, Hf, Ti, Nb, Ta) and rare-earth elements (REEs), was assessed.

Correlation diagrams for pairs of geochemically coherent, relatively immobile elements, such as Th vs. La and Zr vs. Hf (Figure 10), show strong positive linear correlations for samples from all three intrusive suites. This coherence suggests that these elements, and likely other HFSEs and REEs, largely preserved their primary magmatic ratios and were not significantly mobilized or fractionated relative to each other by the observed alteration. Furthermore, the overall consistency and systematic variations observed in the chondrite-normalized REE patterns (Figure 8a) and primitive mantle-normalized trace element patterns (Figure 8b), particularly for the HFSEs, lend further support to their relative immobility and utility in reflecting primary magma characteristics [57].

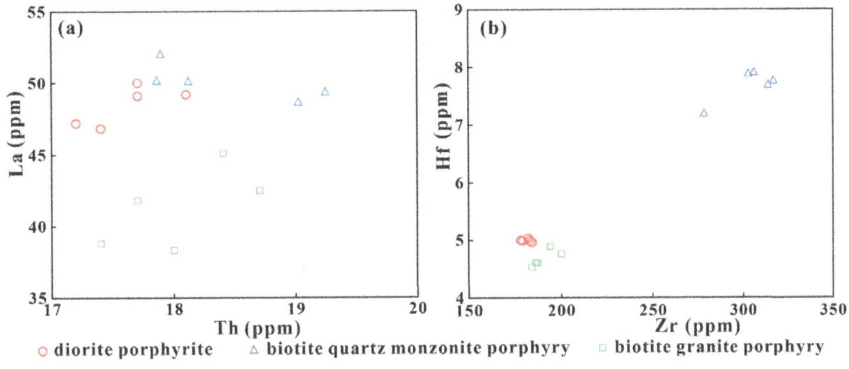

Figure 10. La vs. Th and Hf vs. Zr diagrams of the adakitic rocks in the Zhuxi district: (**a**). La vs. Th diagram; (**b**). Hf vs. Zr diagram.

Therefore, while acknowledging the presence of alterations, the HFSE and REE geochemical data are considered to be reliable indicators of primary magmatic processes and source characteristics for these intrusive rocks. These elements will be utilized preferentially in the following sections to discuss the petrogenesis of the diorite porphyrite, biotite quartz monzonite porphyry, and biotite granite porphyry, and to constrain their tectonic setting.

Three distinct types of adakitic intrusions have been identified within the Zhuxi district: the diorite porphyrite identified within altered granite (borehole ZK4212 [42]), the biotite quartz monzonite porphyry from the Zaolin ore block [52], and the biotite granite porphyry newly characterized in this study (borehole ZK2105). These rocks display a compositional range from intermediate (diorite porphyrite) to intermediate–acidic (biotite quartz monzonite porphyry) to acidic (biotite granite porphyry). Although their zircon U-Pb crystallization ages are broadly contemporaneous, clustering within the late Jurassic (diorite porphyrite at 162.7 ± 1.4 Ma [42], biotite quartz monzonite porphyry at 157 ± 1 Ma [52], and biotite granite porphyry at 161.6 ± 2.1 Ma; this study), several lines of evidence argue against a simple comagmatic relationship linking them through fractional crystallization.

Geochemically, the three rock types fail to define linear trends on Harker diagrams (Figure 7). Instead, distinct compositional gaps and non-linear variations are observed, particularly for elements like TiO_2, MgO, and K_2O. Furthermore, they exhibit distinct REE fractionation patterns (Figure 8a); the diorite porphyrite shows relatively flatter HREE patterns (lower $(La/Yb)_N$), whereas the biotite granite porphyry displays the most fractionated HREE patterns (highest $(La/Yb)_N$). While sharing overall trace element characteristics typical of adakitic rocks (Figure 8b), subtle but significant differences exist—notably, the biotite quartz monzonite porphyry belonging distinctly to the high-K calc-alkaline series, whereas the others are predominantly medium-K calc-alkaline (Figure 6a).

Spatially, the three rock types also exhibit distinct distributions. Both the diorite porphyrite and the biotite granite porphyry were intersected by drilling within the main Zhuxi W-Cu deposit area, whereas the biotite quartz monzonite porphyry occurs in the separate Zaolin ore block. Geological interpretations based on drill cores suggest that the biotite granite porphyry may have been tectonically emplaced via thrust faults or nappe structures, obscuring its original intrusive relationships. The diorite porphyrite was intersected nearby, hosted within altered granite [42], further complicating the interpretation of primary spatial relationships. Crucially, no direct field evidence (e.g., observed cross-cutting relationships, gradational contacts) linking the three distinct intrusive types as part of a single zoned pluton or evolutionary sequence has been documented.

Considering the combined geochemical discrepancies (non-linear major element trends, differing REE fractionation patterns, varying K contents) and the lack of clear spatial or temporal field relationships indicative of a shared, evolving magmatic system, we conclude that these three adakitic intrusions are unlikely to be comagmatic. They likely represent discrete magma batches, potentially derived from different source materials, under varying physical conditions (e.g., pressure, temperature, water content), or involving different petrogenetic processes.

5.2. Characteristics of the Magmatic Source

Several petrogenetic models have been proposed for the formation of adakitic rocks, including (1) partial melting of subducted oceanic crust [1], (2) mixing between felsic crustal melts and mantle-derived mafic magmas [58], (3) assimilation and fractional crystallization (AFC) processes involving mantle-derived basaltic magmas interacting with crustal materials [59–61], (4) partial melting of delaminated lower crust [62,63], and (5) partial melting of thickened lower crust [8,62–64].

The Nb/Ta ratios for the three Zhuxi intrusive types range from 6.94 to 17.09 (avg. 12.81), generally approaching typical continental crustal values (~11–13 [65,66]). Their Zr/Hf ratios (range 35.87–42.02, avg. 38.81) are slightly greater than that of crustal magma (~33 [67]). Crucially, the Nb/U ratios (range 1.71–3.42, avg. 2.62) are significantly lower than the primitive mantle values (~30–47), strongly indicating substantial crustal input or modification during magma genesis. Collectively, these ratios suggest significant crustal involvement in the petrogenesis of all three rock types.

Derivation solely through fractional crystallization of parental basaltic magma appears unlikely. Such a process typically requires voluminous associated mafic intrusions [68], and although mafic rocks occur regionally, no voluminous, coeval (~160 Ma) mafic intrusions have been documented in direct association with the Zhuxi adakitic rocks.

Magma mixing processes often leave characteristic petrographic evidence, such as mafic microgranular enclaves (MMEs) or mineral disequilibrium textures [69,70]. While MMEs were not observed in the studied samples of the three rock types, microscopic examination revealed the presence of resorbed or rounded quartz phenocrysts within the biotite quartz monzonite porphyry (Figure 11). This texture can be interpreted as mineralogical evidence of disequilibrium, potentially resulting from magma mixing. In contrast, such direct petrographic evidence for magma mixing is absent in the diorite porphyrite and the biotite granite porphyry. This suggests that magma mixing may have played a specific role in the petrogenesis of the biotite quartz monzonite porphyry. However, considering that its overall geochemical signature (e.g., low Nb/U, Nb/Ta ratios) still points towards a dominantly crustal character, we can infer that the mixing likely involved a predominantly crust-derived felsic melt incorporating a relatively minor component of mantle-derived mafic magma, rather than being a 50:50 mixture or mafic-dominated hybrid.

Figure 11. Rounded quartz in the biotite quartz monzonite porphyry (Bt—biotite; Pl—plagioclase; Qz—quartz) (revised from [52]).

The magma formed by the partial melting of delaminated crust often exhibits a relatively high contents of MgO, Cr, and Ni due to its interaction with upwelling mafic magma [61,71]. However, the observed low MgO, Cr, and Ni contents of the biotite granite porphyry (Table 2), appear inconsistent with melting of a mafic-dominated source associated with delamination. Conversely, these geochemical characteristics are more consistent with the magma originating from partial melting of a thickened crust [62–64].

The diorite porphyrite exhibits relatively higher MgO, Cr, and Ni contents [42]. This composition could be consistent with partial melting of delaminated lower crust. Petrogenetic models involving delamination and subsequent melting are often linked to intraplate extensional tectonic settings [15]. At the same time, a key feature of all three intrusive types is their prominent arc-like geochemical signature, including negative Nb-Ta-Ti anomalies (Figure 8b) [72] and their plotting within volcanic-arc fields on tectonic discrimination diagrams (e.g., Rb vs. Y + Nb, Figure 12a; Rb-Hf-Ta ternary diagram, Figure 12b). Since the diagenetic age of these three rock types is ~160 Ma, extensive research indicates that South China experienced significant lithospheric extension and intraplate deformation during the Jurassic–Cretaceous period, following earlier collisional orogenies. This establishes a post-orogenic extensional setting, which is conducive to lithospheric thinning, asthenospheric upwelling, and potentially lower crustal delamination—processes that can trigger the melting of both metasomatized lithospheric mantle and lower crust. Therefore, we believed that the diorite porphyrite originated from partial melting involving delaminated lower crustal materials within this extensional regime. The required heat could be supplied by the upwelling asthenosphere following delamination. The arc-like geochemical signature can be interpreted as an inherited feature, reflecting pervasive metasomatism of the South China lithosphere (both mantle and lower crust) during earlier Paleozoic subduction events, rather than indicating active arc magmatism at ~160 Ma. This interpretation aligns the geochemical data with the inferred post-orogenic extensional tectonic setting conducive to delamination-related magmatism.

In summary, the petrogenesis of the ~160 Ma adakitic suite in the Zhuxi district appears to be intrinsically linked to a dynamic late Jurassic tectonic regime characterized by post-orogenic lithospheric extension following earlier crustal thickening. This setting likely facilitated complex crust–mantle interactions, potentially involving delamination of the thickened lithosphere. According to this study, we propose a scenario initiated by asthenospheric upwelling, possibly triggered by extension and/or delamination, which induced partial melting of the overlying metasomatized subcontinental lithospheric mantle (SCLM), generating primary mantle-derived magmas. Underplating and/or advection of heat by these mantle magmas likely induced partial melting of the pre-existing thickened lower crust, generating the silicic melts parental to the low-MgO biotite granite porphyry. Concurrently, influenced by mantle-derived magmas, the delaminated lower crustal material underwent partial melting, ultimately resulting in the formation of diorite porphyrite. Finally, the biotite quartz monzonite porphyry likely represents a hybrid product, formed through mixing or assimilation processes involving ascending mantle-derived magmas (or melts derived from metasomatized SCLM) and a significant component of crust-derived silicic melt, consistent with its intermediate composition, high-K character, and subtle petrographic evidence of disequilibrium.

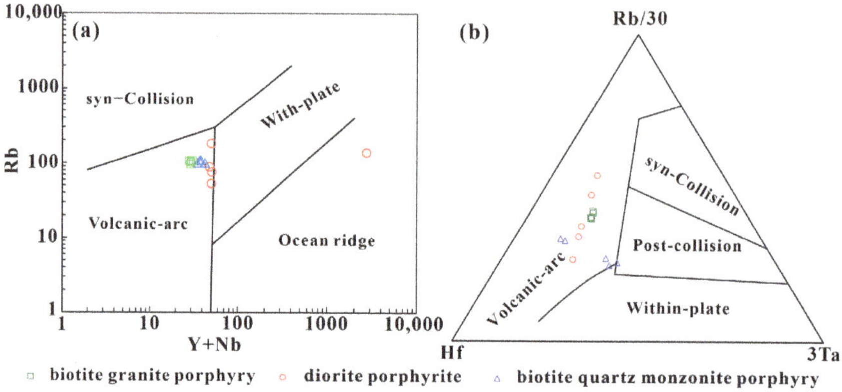

Figure 12. Tectonic discrimination diagram of the adakitic rocks in the Zhuxi district: (**a**) Rb vs. (Y + Nb) tectonic discrimination diagram (the base map is from [73]); (**b**) Rb-Hf-Ta tectonic discrimination diagram (the base map is from [74]).

5.3. Implications for Copper Mineralization in the Zhuxi District

Yanshanian magmatism and associated mineralization in the Zhuxi district occurred in several pulses, with significant activity recognized around ~160 Ma, possibly followed by events around ~150 Ma and ~130 Ma. The ~160 Ma event involved the emplacement of I-type granitoids and related porphyries exhibiting adakitic affinities, as documented in this study and previous work [42,52]. Mineralization associated with this stage includes significant W-Mo systems, exemplified by the Taqian deposit where molybdenite yields Re-Os ages of 162 ± 2 Ma [75]. This timing highlights the potential for broadly contemporaneous Cu-Mo mineralization during this period.

A strong spatial and temporal association between adakitic magmatism and porphyry Cu(-Mo-Au) deposits is widely recognized globally and within South China, such as those documented in the Middle–Lower Yangtze River Metallogenic Belt e.g., [9–17]. The giant Dexing porphyry Cu deposit, situated adjacent to the Zhuxi district, is a prominent regional example linked to adakitic intrusions [18–21]. While the specific adakitic intrusions studied here (particularly the biotite granite porphyry from ZK2105) may be tectonically displaced and are not observed in direct intrusive contact with the main Taqian ore zones, their ~160 Ma crystallization ages are essentially contemporaneous with the ~162 Ma mineralization event documented at Taqian [75]. This temporal coincidence strongly suggests a genetic link between the ~160 Ma adakitic magmatism and associated polymetallic (including Cu-Mo) mineralization potential in the district.

Classic models often associate the fertility of subduction-related adakites for porphyry Cu deposits with high water contents, high oxygen fugacity (fO_2), and efficient transport of sulfur and metals derived from the subducting slab or overlying mantle wedge [11,76]. However, it is increasingly recognized that adakitic magmas generated in non-subduction settings (e.g., post-collisional, intraplate extension) can also be fertile, potentially involving different mechanisms for metal enrichment or volatile behavior. Processes such as crust–mantle mixing (inferred for the biotite quartz monzonite porphyry) or partial melting involving metasomatized delaminated lower crust (inferred for the diorite porphyrite), as proposed for the Zhuxi adakites, can influence melt oxidation states, volatile contents, and metal budgets in ways that are potentially favorable for Cu-Mo-Au mineralization [77].

As discussed previously (Section 5.2), the inferred tectonic setting for the Zhuxi district at ~160 Ma is likely one of post-orogenic/intraplate extension. The specific petrogenetic processes inferred here—including partial melting involving delaminated lower crust and mixing between crust- and mantle-derived components—occurred within this dynamic setting. These processes, associated with the ~160 Ma adakitic magmatism, likely established favorable physicochemical conditions for the formation of associated porphyry-type Cu-Mo mineralization during this important metallogenic episode in the Zhuxi district.

6. Conclusions

(1) Zircon U-Pb geochronology indicates a crystallization age of 161.6 $\pm$ 2.1 Ma for the newly identified biotite granite porphyry within the Zhuxi W-Cu polymetallic deposit.

(2) The three investigated intrusive types (diorite porphyrite, biotite quartz monzonite porphyry, and biotite granite porphyry) display a compositional progression from intermediate to acidic. All are peraluminous and belong to the calc-alkaline to high-K calc-alkaline series, with the biotite quartz monzonite porphyry being distinctly potassic. They share adakitic geochemical affinities, characterized by LREE enrichment relative to depleted HREEs ($(LREE/HREE)_N$ increases with SiO_2), slight negative Eu anomalies, enrichment in LILEs (e.g., Ba, K, Sr), depletion in HFSEs (e.g., Nb, Ta, Ti) and P, high Sr/Y ratios, and low Y and Yb contents.

(3) Despite their broadly contemporaneous emplacement ages (~160 Ma), geochemical data (e.g., non-linear Harker trends, distinct REE fractionation patterns) and spatial distributions suggest that the three adakitic intrusive types are not comagmatic. Their petrogenesis is linked to complex crust–mantle interactions within a post-orogenic setting: The biotite granite porphyry likely originated from partial melting of thickened felsic lower crust, induced by heat from underplated mantle-derived magmas. The diorite porphyrite was interpreted to result from partial melting involving delaminated lower crustal material, potentially reflecting a more mafic or metasomatized source component compared to the granite porphyry source. The biotite quartz monzonite porphyry likely formed through mixing between crust-derived silicic melt and a minor component of ascending mantle-derived magma (or melt from metasomatized SCLM).

(4) The formation of these ~160 Ma adakitic rocks occurred within a dynamic tectonic regime involving post-orogenic lithospheric extension, potentially accompanied by delamination following earlier crustal thickening. The associated petrogenetic processes, including partial melting of diverse crustal sources (thickened and delaminated lower crust) and crust–mantle mixing, likely established favorable physicochemical conditions for contemporaneous porphyry-type Cu-Mo mineralization in the Zhuxi district.

Author Contributions: Writing—Original Draft, Y.O.; Methodology, Q.C.; Visualization, R.Z.; Writing—Review and Editing, T.L. All authors have read and agreed to the published version of the manuscript.

Funding: This study was jointly supported by the "Deep Earth Probe and Mineral Resources Exploration-National Science and Technology Major Project, grant number 2024ZD1002204", "Geological Exploration Project funded by Jiangxi Provincial Finance, grant number 20250020", "Key R&D Plan Project of Science and Technology Department of Jiangxi Province, grant number 20212BBG73045", "Jiangxi Natural Science Foundation General Program, grant number 20224BAB203037", "The Youth Science and Technology Leaders Training Program Project of Jiangxi

Geological Bureau, grant numbers 2022JXDZKJRC02 and 2023JXDZKJRC05", and the "Science and Technology Plan Project of Yingtan City, grant numbers 20233-185656, 202412-28880 and 2024SYD017".

Data Availability Statement: The data presented in this study are available upon request from the corresponding author.

Acknowledgments: We are grateful to the editors and the anonymous reviewers for handling our manuscript and providing valuable suggestions for improving it.

Conflicts of Interest: The authors declare no conflicts of interest.

References

1. Defant, M.J.; Drummond, M.S. Derivation of some modern arcmagmas by melting of young subducted lithosphere. *Nature* **1990**, *347*, 662–665. [CrossRef]
2. Defant, M.J.; Maury, R.C.; Ripley, E.M.; Feigenson, M.D.; Jacques, D. An example of island-arc petrogenesis: Geochemistry and petrology of the southern Luzon Arc, Philippines. *J. Petrol.* **1991**, *32*, 455–500. [CrossRef]
3. Zhang, Q.; Xu, J.F.; Wang, Y.; Xiao, L.; Liu, H.T.; Wang, Y.L. Diversity of adakite. *Geol. Bull. China* **2004**, *23*, 959–965.
4. Cai, L.M.; Huang, S.S.; Yang, Y.B.; Wang, L.J. On the adakite-like features of Zhoutieqiao rock mass in Yixing-Liyang area of eastern China. *J. Geol.* **2017**, *41*, 548–554.
5. Wang, Q.; Xu, J.F.; Wang, J.X.; Zhao, Z.H.; Wang, R.J.; Qiu, J.X.; Xiong, X.L.; Sang, L.K.; Peng, L.H. Determination of adakite type gray gneiss in the North Dabie Mountains and its relationship with ultra-high pressure metamorphism. *Chin. Sci. Bull.* **2000**, *45*, 1017–1024.
6. Wang, Q.; Xu, J.F.; Zhao, Z.H. The summary and comment on research on a new kind of igneous rock: Adakite. *Adv. Earth Sci.* **2001**, *16*, 201–208.
7. Zhang, Q.; Qian, Q.; Wang, E.Q.; Wang, Y.; Zhao, T.P.; Hao, J.; Guo, G.J. An east China plateau in Mid-Late Yanshanian period: Implication from adakites. *Chin. J. Geol.* **2001**, *36*, 248–255.
8. Zhang, Q.; Wang, Y.; Qian, Q.; Yang, J.H.; Wang, Y.L.; Zhao, T.P.; Guo, G.J. The characteristics and tectonic-metallogenic significances of the adakites in Yanshan period from eastern China. *Acta Petrol. Sin.* **2001**, *17*, 236–244.
9. Leng, C.B.; Zhang, X.C.; Chen, Y.J.; Wang, S.X.; Gou, T.Z.; Chen, W. Discussion on the relationship between Chinese porphyry copper deposits and adakitic rocks. *Earth Sci. Front.* **2007**, *14*, 199–210.
10. Li, Y.; Ling, M.X.; Ding, X.; Liu, J.; Han, F.; Sun, W.D. Adakites or adakitic rocks and associated metallogenesis in eastern China. *Geotecton. Metallog.* **2009**, *33*, 448–464.
11. Liu, H.T.; Zhang, Q.; Liu, J.M.; Ye, J.; Zeng, Q.D.; Yu, C.M. Adakite versus porphyry copper and epithermal gold deposits: A possible metallogenetic specialization of magmatism required in-deep assessment. *Acta Petrol. Sin.* **2004**, *20*, 205–218.
12. Wang, F.Y.; Liu, S.A.; Li, S.G.; He, Y.S. Genesis and metallogenic relationships of early Cretaceous high magnesium adakites in eastern China: A study on zircon O-Hf isotopes and trace elements. *Acta Mineral. Sin.* **2011**, *31*, 93.
13. Wang, Q.; Tang, G.J.; Jia, X.H.; Zi, F.; Jiang, Z.Q.; Xu, J.F.; Zhao, Z.H. The metalliferous mineralization associatied with adakitic rocks. *Geol. J. China Univ.* **2008**, *14*, 350–364.
14. Wang, Q.; Xu, J.F.; Zhao, Z.H. Intermediate-acid igneous rocks strongly depleted in heavy rare earth elements (or adakitic rocks) and copper-gold metallogenesis. *Earth Sci. Front.* **2003**, *10*, 561–572.
15. Wang, Q.; Xu, J.F.; Zhao, Z.H.; Zi, F.; Tang, G.J.; Jia, X.H.; Jiang, Z.Q. Adakites or adakite rocks and associated metal metallogenesis in China. *Bull. Mineral. Petrol. Geochem.* **2007**, *26*, 336–349.
16. Wang, Y.; Cheng, S.H. Discussion on the relationship between C-type adakite and Mesozoic metal mineralization in eastern China. *Acta Mineral. Sin.* **2009**, *S1*, 27–28.
17. Wang, Y.L.; Zhang, Q.; Wang, Q.; Liu, H.T.; Wang, Y. Study on adakitic rock and Cu-Au mineralization. *Acta Petrol. Sin.* **2003**, *19*, 543–550.
18. Hou, Z.Q.; Meng, X.J.; Qu, X.M.; Gao, Y.F. Copper ore potential of adakitic intrusives in Gangdese porphyry copper belt: Constrains from rock phase and deep melting process. *Miner. Depos.* **2005**, *24*, 108–121.
19. Wang, Q.; Xu, J.F.; Jian, P.; Bao, Z.W.; Zhao, Z.H.; Li, C.F.; Xiong, X.L.; Ma, J.L. Petrogenesis of adakitic porphyries in an extensional tectonic setting, Dexing, South China: Implications for the genesis of porphyry copper mineralization. *J. Petrol.* **2006**, *47*, 119–144. [CrossRef]
20. Wang, Q.; Zhao, Z.H.; Jian, P.; Xu, J.F.; Bao, Z.W.; Ma, J.L. SHRIMP zircon geochronology and Nd-Sr isotopic eochemistry of the Dexing granodiorite porphyries. *Acta Petrol. Sin.* **2004**, *20*, 315–324.
21. Zhang, J.F.; Li, Z.T.; Jin, C.Z. Adakites in northeastern China and their mineralized implications. *Acta Petrol. Sin.* **2004**, *20*, 361–368.
22. Wang, X.G.; Ouyang, Y.P.; Chen, G.H.; Liu, J.G.; Wei, J.; Rao, J.F. *W-Cu Deposit in Zhuxi, Jiangxi Province*; Geological Publishing House: Beijing, China, 2022.

23. Chen, G.H.; Shu, L.S.; Shu, L.M.; Zhang, C.; Ouyang, Y.P. Geological characteristics and mineralization setting of the Zhuxi tungsten (copper) polymetallic deposit in the Eastern Jiangnan Orogen. *Sci. China Earth Sci.* **2016**, *59*, 803–823. [CrossRef]
24. Chen, G.H.; Wan, H.Z.; Shu, L.S.; Zhang, C.; Kang, C. An analysis on ore-controlling conditions and geological features of the Cu-W polymetallic ore deposit in the Zhuxi area of Jingdezhen, Jiangxi Province. *Acta Petrol. Sin.* **2012**, *28*, 3901–3914.
25. Liu, M.; Song, S.W.; Cui, Y.R.; Chen, G.H.; Rao, J.F.; Ouyang, Y.P. In-situ U-Pb geochronology and trace element analysis for the scheelite and apatite from the deep seated stratiform-like W(Cu) ore of the Zhuxi tungsten deposit, northeastern Jiangxi Province. *Acta Petrol. Sin.* **2021**, *37*, 717–732.
26. He, X.L. Geological Characteristics and Origin Discussion of the Zhuxi Tungsten-Copper Deposit. Ph.D. Thesis, China University of Geosciences, Beijing, China, 2017.
27. He, X.L.; Zhang, D.; Wu, G.G.; Di, Y.J.; Zhang, Z.H.; Li, F.; Hu, B.J.; Huo, H.L.; Li, N.; Zhang, X.M.; et al. Control of interaction between stress and fluid in tectonic transition background on metallogenesis of giant Zhuxi W-Cu deposit, South China. *Miner. Depos.* **2021**, *40*, 1135–1159.
28. Hu, Z.H. The Formation Conditions and Matallogenic Regularity of Zhuxi Tungsten Polymentallic Deposit in Northeast of Jiangxi Province. Ph.D. Thesis, Chengdu University of Technology, Chengdu, China, 2015.
29. Ouyang, Y.P.; Zeng, R.L.; Shu, L.M.; Deng, Y.G.; Rao, J.F.; Wei, J. Genesis and geological significance of quartz vein-type wolframite in the deep of Zhuxi W-Cu deposit. *J. Guilin Univ. Technol.* **2023**, *43*, 184–192.
30. Ouyang, Y.P.; Wei, J.; Lu, Y.; Zhang, W.; Yao, Z.Y.; Rao, J.F.; Chen, J.; Pan, X. Muscovite ^{40}Ar–^{39}Ar age and its geological significance in Zhuxi W(Cu) deposit, northeastern Jiangxi. *J. Cent. South Univ.* **2019**, *26*, 3488–3501. [CrossRef]
31. Ouyang, Y.P.; Rao, J.F.; Liao, S.P.; He, X.R.; Hu, Q.H.; Wei, J.; Yang, M.G. Rock- and ore-controlling structure in the Zhuxi ore concentration area in the northeastern Jiangxi Province. *Geol. China* **2019**, *46*, 878–893.
32. Pan, X.F.; Hou, Z.Q.; Li, Y.; Chen, G.H.; Zhao, M.; Zhang, T.F.; Zhang, C.; Wei, J.; Kang, C. Dating the giant Zhuxi W–Cu deposit (Taqian–Fuchun Ore Belt) in South China using molybdenite Re–Os and muscovite Ar–Ar system. *Ore Geol. Rev.* **2017**, *86*, 719–733. [CrossRef]
33. Pan, X.F.; Hou, Z.Q.; Zhao, M.; Chen, G.H.; Rao, J.F.; Li, Y.; Wei, J.; Ouyang, Y.P. Geochronology and geochemistry of the granites from the Zhuxi W-Cu ore deposit in South China: Implication for petrogenesis, geodynamical setting and mineralization. *Lithos* **2018**, *304–307*, 155–179. [CrossRef]
34. Pan, X.F.; Hou, Z.Q.; Zhao, M.; Li, Y.; Ouyang, Y.P.; Wei, J.; Yang, Y.S. Fluid inclusion and stable isotope constraints on the genesis of the world-class Zhuxi W(Cu) skarn deposit in South China. *J. Asian Earth Sci.* **2020**, *190*, 104192. [CrossRef]
35. Pan, X.F.; Ren, Y.F.; Hou, Z.Q.; Ouyang, Y.P.; Gong, X.J.; Li, Q.Y.; Yang, Y.S. Petrogenesis and geodynamic implications of a newly discovered basanite dike in Zaolin, Jingdezhen City, South China. *Lithosphere* **2021**, *1*, 9732167. [CrossRef]
36. Song, S.W.; Mao, J.W.; Zhu, Y.F.; Yao, Z.Y.; Chen, G.H.; Rao, J.F.; Ouyang, Y. Partial-melting of fertile metasedimentary rocks controlling the ore formation in the Jiangnan porphyry-skarn tungsten belt, south China: A case study at the giant Zhuxi W-Cu skarn deposit. *Lithos* **2018**, *304–307*, 180–199. [CrossRef]
37. Romer, R.L.; Kroner, U. Phanerozoic tin and tungsten mineralization—Tectonic controls on the distribution of enriched protoliths and heat sources for crustal melting. *Gondwana Res.* **2016**, *31*, 60–95. [CrossRef]
38. Romer, R.L.; Kroner, U. Sediment and weathering control on the distribution of Paleozoic magmatic tin-tungsten mineralization. *Miner. Depos.* **2015**, *50*, 327–338. [CrossRef]
39. Richards, J.P. The oxidation state, and sulfur and Cu contents of arc magmas: Implications for metallogeny. *Lithos* **2015**, *233*, 27–45. [CrossRef]
40. Richards, J.P. Tectonic, magmatic, and metallogenic evolution of the Tethyan orogen: From subduction to collision. *Ore Geol. Rev.* **2015**, *70*, 323–345. [CrossRef]
41. Mungall, J.E. Roasting the mantle: Slab melting and the genesis of major Au and Au-rich Cu deposits. *Geology* **2002**, *30*, 915–918. [CrossRef]
42. Rao, J.F.; Ouyang, Y.P.; Chen, Q.; Wei, J.; Chen, G.H.; Zeng, X.H. Geochronology of the diorite porphyrites in Zhuxi W-Cu deposit and its metallogenic significances. *China Tungsten Ind.* **2020**, *35*, 43–52.
43. Yang, M.G.; Wang, G.H.; Xu, G.M.; Hu, Q.H. Basic characteristics of the Marina Pacific tectonic activities in Jiangxi Province and its adjacent areas. *East China Geol.* **2016**, *37*, 10–18.
44. Yang, M.G.; Zhu, P.J.; Wang, G.H. Division of tectonic-metallogenetic units in South China. *Shanghai Land Resour.* **2018**, *39*, 13–24.
45. Ouyang, Y.P.; Zeng, R.L.; Wei, J.; Chen, Q.; Rao, J.F.; Chen, G.H.; Zeng, X.H.; Xu, Y.X.; Deng, Y.G.; Luo, L.C.; et al. *Research Report of the Supernormal Enrichment Mechanism and Deep Prospecting Technology of the Zhuxi Tungsten-Copper Deposit in Northeast Jiangxi Province;* The Tenth Geological Brigade of Jiangxi Geological Bureau: Yingtan, China, 2023.
46. Ouyang, Y.P.; Rao, J.F. Using the catchment-based fractal model to delineate geochemical anomalies associated with Cu-W polymetallic deposits in the Zhuxi, Jiangxi Province. *Arab. J. Geosci.* **2018**, *11*, 797. [CrossRef]
47. Rao, J.F.; Yao, Z.Y.; Ouyang, Y.P. Tectonization-magmation-mineralization of the Taqian-Fuchun W-Cu polymetallic mineralization concentration area. *Adv. Geosci.* **2017**, *7*, 632–644. [CrossRef]

48. Hou, K.J.; Li, Y.H.; Tian, Y.R. In situ U-Pb zircon dating using Laser Ablation-multi Ion Counting-ICP-MS. *Miner. Depos.* **2009**, *28*, 481–492.
49. Ludwig, K.R. *Isoplot 3.0: A Geochronological Toolkit for Microsoft Excel*; Geochronology Center Special Publication: Berkeley, CA, USA, 2003.
50. Rubatto, D. Zircon trace element geochemistry: Partitioning with garnet and the link between U–Pb ages and metamorphism. *Chem. Geol.* **2002**, *184*, 123–138. [CrossRef]
51. Wan, H.Z.; Liu, X.L.; Liu, Z.Q.; Liu, S.B.; Chen, Y.C.; Wang, C.H.; Chen, G.H.; Liang, L.J.; Li, S.S.; Zhang, S.D.; et al. LA-ICP-MS zircon U-Pb dating of granodioritic porphyry located Zhuxi copper-tungsten mine in Northeast Jiangxi and its geological significance. *Rock Miner. Anal.* **2015**, *34*, 494–502.
52. Li, Y.H.; Tang, W.; Chen, Q.; Zou, J.; Rao, J.F.; Ouyang, Y.P.; Wu, B. Geochronological and geochemical characteristics of biotite quartz monzonite porphyry in Zhuxi ore concentration area, and its geological implications. *Miner. Depos.* **2021**, *40*, 1285–1298.
53. Peccerillo, A.; Taylor, S.R. Geochemistry of eocene calc–alkaline volcanic rocks from the Kastamonu area, Northern Turkey. *Contrib. Mineral. Petrol.* **1976**, *58*, 63–81. [CrossRef]
54. Maniar, P.D.; Piccoli, P.M. Tectonic discrimination of granitoids. *Geol. Soc. Am. Bull.* **1989**, *101*, 635–643. [CrossRef]
55. Sun, S.S.; Mcdonough, W.F. *Chemical and Isotopic Systematics of Oceanic Basalts: Implications for Mantle Composition and Processes*; Geological Society Special Publications: London, UK, 1989.
56. Defant, M.J.; Xu, J.F.; Kepezhinskas, P.; Wang, Q.; Zhang, Q.; Xiao, L. Adakites: Some variations on a theme. *Acta Petrol. Sin.* **2002**, *18*, 129–142.
57. Hou, T.; Zhang, Z.C.; Santosh, M.; Encarnacion, J.; Wang, M. The Cihai diabase in the Beishan region, NW China: Isotope geochronology, geochemistry and implications for Cornwall-style iron mineralization. *J. Asian Earth Sci.* **2013**, *70–71*, 231–249. [CrossRef]
58. Chen, B.; Jahn, B.M.; Suzuki, K. Petrological and Nd-Sr-Os isotopic constraints on the origin of high-Mg adakitic rocks from the North China Craton: Tectonic implications. *Geology* **2013**, *41*, 91–94. [CrossRef]
59. Martin, H.; Smithies, R.H.; Rapp, R.; Moyen, J.-F.; Champion, D. An overview of adakite, tonalite trondhjemite granodiorite (TTG), and sanukitoid: Relationships and some implications for crustal evolution. *Lithos* **2005**, *79*, 1–24. [CrossRef]
60. Moyen, J.F.; Martin, H. Forty years of TTG research. *Lithos* **2012**, *148*, 312–336. [CrossRef]
61. Rapp, R.P.; Shimizu, N.; Norman, M.D.; Applegate, G.S. Reaction between slab-derived melts and peridotite in the mantle wedge: Experimental constraints at 3.8 GPa. *Chem. Geol.* **1999**, *160*, 335–356. [CrossRef]
62. Atherton, M.P.; Petford, N. Generation of sodium-rich magmasfrom newly underplated basaltic crust. *Nature* **1993**, *362*, 144–146. [CrossRef]
63. Gao, S.; Rudnick, R.L.; Yuan, H.L.; Liu, X.M.; Liu, Y.S.; Xu, W.L.; Ling, W.L.; Ayers, J.; Wang, X.C.; Wang, Q.H. Recycling lower continental crust in the North China Craton. *Nature* **2004**, *432*, 892–897. [CrossRef]
64. Xu, J.F.; Shinjo, R.; Defant, M.J.; Wang, Q.; Rapp, R.P. Origin of Mesozoic adakitic intrusive rocks in the Ningzhen area of east China: Partial melting of delaminated lower continental crust? *Geology* **2002**, *30*, 1111–1114. [CrossRef]
65. Barth, M.G.; Mcdonough, W.F.; Rudnick, R.L. Tracking thebudget of Nb and Ta in the continental crust. *Chem. Geol.* **2000**, *165*, 197–213. [CrossRef]
66. Green, T.H. Significance of Nb/Ta as an indicator of geochemical processes in the crust-mantle system. *Chem. Geol.* **1995**, *120*, 347–359. [CrossRef]
67. Taylor, S.R.; Mclennan, S.M. *The Continental Crust: Its Composition and Evolution*; Blackwell Scientific Publication: Oxford, UK, 1985.
68. Frost, C.D.; Frost, B.R.; Bell, J.M.; Chamberlain, K.R. The relationship between A-type granites and residual magmas from anorthosite: Evidence from the northern Sherman batholith, Laramie Mountains, Wyoming, USA. *Precambrian Res.* **2002**, *119*, 45–71. [CrossRef]
69. Bao, B.; Jiang, J.S.; Yu, Y.S.; Qin, Z.W.; Zheng, Z.C.; Liu, X.; Shan, L.; Dong, Y.Y.; Li, F.L. Petrogenesis and metallogeny significance of quartz diorite porphyries from the Dexing porphyry Cu deposit, Jiangxi Province. *Acta Petrol. Mineral.* **2023**, *42*, 483–501.
70. Yang, Z.M.; Lu, Y.J.; Hou, Z.Q.; Chang, Z.S. High-Mg diorite from Qulong in Southern Tibet: Implications for the genesis of adakite-like intrusions and associated porphyry Cu deposits in collisional orogens. *J. Petrol.* **2015**, *56*, 227–254. [CrossRef]
71. Prouteau, G.; Scaillet, B.; Pichavant, M.; Maury, R.C. Evidence for mantle metasomatism by hydrous silicic melts derived from subducted oceanic crust. *Nature* **2001**, *410*, 197–200. [CrossRef]
72. Jahn, B.M.; Wu, F.Y.; Lo, C.H.; Tsai, C.H. Crust-mantle interaction induced by deep subduction of the continental crust: Geochemical and Sr-Nd isotopic evidence from post-collisional mafic-ultramafic intrusions of the northern dabie complex, Central China. *Chem. Geol.* **1999**, *157*, 119–146. [CrossRef]
73. Pearce, J.A.; Harris, N.B.W.; Tindle, A.G. Trace element discrimination diagrams for the tectonic interpretation of granitic rocks. *J. Petrol.* **1984**, *25*, 956–983. [CrossRef]
74. Harris, N.B.W.; Pearce, J.A.; Tindle, A.G. Geochemical characteristics of collision-zone magmatism. *Geol. Soc. Lond. Spec. Publ.* **1986**, *19*, 67–81. [CrossRef]

75. Huang, A.J.; Wen, Z.G.; Liu, S.B.; Liu, X.Q.; Liu, X.M.; Zhang, J.J.; Shi, H.G.; Liu, Z.Q. Re-Os isotopic dating of molybdenite from the Taqian W-Mo deposit in Leping County, Jiangxi Province and its geological implications. *Acta Petrol. Mineral.* **2013**, *32*, 496–504.
76. Reich, M.; Parada, M.A.; Palacios, C.; Dietrich, A.; Schultz, F.; Lehmann, B. Adakite-like signature of Late Miocene intrusions at the Los Pelambres giant porphyry copper deposit in the Andes of Central Chile: Metallogenic implications. *Miner. Depos.* **2003**, *38*, 876–885. [CrossRef]
77. Fan, J.J.; Wang, Q.; Li, J.; Wei, G.J.; Wyman, D.; Zhao, Z.H.; Liu, Y.; Ma, J.L.; Zhang, L.; Wang, Z.L. Molybdenum and boron isotopic compositions of porphyry Cu mineralization-related adakitic rocks in central-eastern China: New insights into their petrogenesis and crust-mantle interaction. *J. Geophys. Res. Solid Earth* **2020**, *125*, e2020JB020474. [CrossRef]

minerals

MDPI

Article

Petrogenesis, Geochemistry, and Geological Significance of the Kongco Granitic Porphyry Dykes in the Northern Part of the Central Lhasa Microblock, Tibet

Anping Xiang [1], Hong Liu [1,2,*], Wenxin Fan [2,*], Qing Zhou [1], Hong Wang [1] and Kaizhi Li [2]

[1] Chengdu Center, China Geological Survey, Chengdu 610081, China; 18008084525@163.com (A.X.); zhouq556@163.com (Q.Z.)

[2] College of Earth Sciences, Chengdu University of Technology, Chengdu 610059, China; 19827555352@163.com

* Correspondence: liuh@mail.cgs.gov.cn (H.L.); fwx@stu.cdut.edu.cn (W.F.)

Abstract: The Kongco area of Nima in the northern part of the Lhasa terrane has a suite of alkaline granitic porphyry dykes associated with Early Cretaceous granites and accompanied by Cu/Mo mineralization. LA-ICP-MS $^{206}Pb/^{238}U$ zircon geochronology performed on the dykes produced an age of 104.15 ± 0.94 Ma (MSWD = 0.98), indicating the Early Cretaceous emplacement of the dykes. The dykes exhibit high silica (SiO_2 = 76.22~77.90 wt.%), high potassium (K_2O = 4.97~6.21 wt.%), high alkalinity ($K_2O + Na_2O$ = 8.07~8.98 wt.%), low calcium (CaO = 0.24~0.83 wt.%), low magnesium (MgO = 0.06~0.20 wt.%), and moderate aluminum content (Al_2O_3 = 11.93~12.45 wt.%). The Rieterman index (σ) ranges from 1.93 to 2.34. A/NK (molar ratio $Al_2O_3/(Na_2O + K_2O)$) and A/CNK (molar ratio $Al_2O_3/(CaO + Na_2O + K_2O)$) values of the dykes range from 1.06 to 1.18 and 0.98 to 1.09, respectively. The dykes are relatively enriched in Rb, Th, U, K, Ta, Ce, Nd, Zr, Hf, Sm, Y, Yb, and Lu, and they show a noticeable relative depletion in Ba, Nb, Sr, P, Eu, and Ti, as well as an average differentiation index (DI) of 96.42. The dykes also exhibit high FeO_T/MgO ratios (3.60~10.41), Ga/Al ratios (2.22×10^{-4}~3.01×10^{-4}), Y/Nb ratios (1.75~2.40), and Rb/Nb ratios (8.36~20.76). Additionally, they have high whole-rock Zr saturation temperatures (884~914 °C), a pronounced Eu negative anomaly (δEu = 0.04~0.23), and a rightward-sloping "V-shaped" rare earth element pattern. These characteristics suggest that the granitic porphyry dykes can be classified as A2-type granites formed in a post-collisional tectonic environment and that they are weakly peraluminous, high-potassium, and Calc-alkaline basaltic rocks. Positive $\varepsilon_{Hf}(t)$ values = 0.43~3.63 and a relatively young Hf crustal model age (T_{DM2} = 826~1005 Ma, $^{87}Sr/^{86}Sr$ ratios = 0.7043~0.7064, and $\varepsilon_{Nd}(t)$ = −8.60~−2.95 all indicate lower crust and mantle mixing. The lower crust and mantle mixing model is also supported by $(^{206}Pb/^{204}Pb)t$ = 18.627~18.788, $(^{207}Pb/^{204}Pb)t$ = 15.707~15.719, $(^{208}Pb/^{204}Pb)t$ = 39.038~39.110). Together, the Hf, Sr and Pb isotopic ratios indicate that the Kongco granitic porphyry dykes where derived from juvenile crust formed by the addition of mantle material to the lower crust. From this, we infer that the Kongco granitic porphyry dykes are related to a partial melting of the lower crust induced by subduction slab break-off and asthenospheric upwelling during the collision between the Qiangtang and Lhasa terranes and that they experienced significant fractional crystallization dominated by potassium feldspar and amphibole. These dykes are also accompanied by significant copper mineralization (five samples, copper content 0.2%), suggesting a close relationship between the magmatism associated with these dykes and regional metallogenesis, indicating a high potential for mineral exploration.

https://doi.org/10.3390/min15030283

Keywords: Kongco; granitic porphyry dykes; zircon U-Pb; Sr-Nd-Pb; mineral exploration

1. Introduction

The Tibetan Plateau, the youngest plateau in the world, is an ideal location for geodynamic research and serves as a natural laboratory for studying mineral deposits. It consists of a series of nearly east–west-trending blocks or microblocks and suture zones characterized by a serpentinite mélange (Figure 1a) [1–9]. From north to south (Figure 1a), these include the Songpan–Ganzi block (SG), Qiangtang block (QT), Lhasa block (LS), and Himalayan block (HM). These blocks and suture zones, influenced by the subduction of the Tethys Ocean and subsequent continental collisions, host extensive magmatic rocks and mineral resources [10–33], attracting significant attention from geologists. The LS, also known as the Gangdese block or the Nyainqentanglha block, is bordered to the north by the Bangong–Nujiang Suture Zone (BNSZ) and to the south by the Indus–Yarlung Zangbo Suture Zone (IYSZ) (Figure 1b) [19,28,34–41]. It is further subdivided by the Shiquanhe–Namucuo serpentinite mélange zone (SNSZ) and the Luobadui–Milashan Fault Zone (LMF) into three subunits: the Northern Lhasa Microblock (NL), the Central Lhasa Microblock (CL), and the Southern Lhasa Microblock (SL) (Figure 1b). The BNSZ is a significant suture zone in the northern part of the LS and extends over 1500 km from east to west. It is characterized by compound rocks, mélanges, and segmented ophiolite fragments (Figure 1b) [42]. It represents the remnants of the Tethys Ocean [43], providing a crucial window into the geodynamic history of the Tethys Ocean during the Mesozoic. Some researchers propose that the Bangong–Nujiang oceanic basin opened in the Late Permian, with oceanic crust subduction occurring during the Early to Middle Jurassic [44,45]. However, due to the complex tectonic features and large-scale features of the region, the exact timing of the closure of the Bangong–Nujiang Ocean remains contentious [2,46–51].

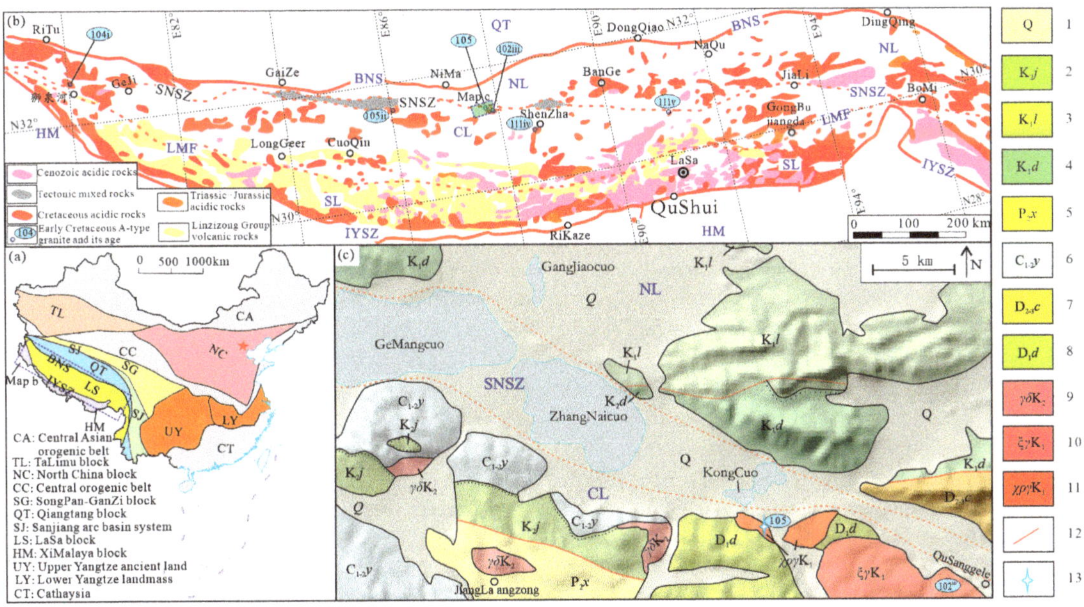

Figure 1. Geological maps of the research area. (**a**) Tectonic map of China (modified after Liu et al., 2015 [24]); (**b**) tectonic map of Gangdise (modified after Liu et al., 2019a, 2021b [47,52]); (**c**) geological

map of Kongco (modified after Liu et al., 2018 [26]). 1—quaternary; 2—sandstone and conglomerate of the Upper Cretaceous Jingzhushan Formation; 3—bioclastic limestone with siltstone of the Lower Cretaceous Langshan Formation; 4—sandstone, limestone and volcanic rock of the Lower Cretaceous Duoni Formation; 5—limestone of the Middle Permian Xiala Formation; 6—sandstone, shale, and limestone of the Carboniferous Yongzhu Formation; 7—sandstone intercalated with limestone of the Middle Upper Devonian Changshehu Formation; 8—limestone of the Lower Devonian dardong Formation; 9—Early Cretaceous intermediate acid rock; 10—Early Cretaceous syenogranite; 11—Early Cretaceous alkali feldspar granite; 12—fault; 13—sampling positions. NL: Northern Lhasa Subterrane; SNSZ: Shiquanhe–Namuco serpentinite mélange zone; CL: Central Lhasa Microblock; LMF: Luobadui–Milashan Fault Zone; SL: Southern Lhasa Microblock; IYSZ: Indus–Yarlung Zangbo Suture Zone; BNS: Bangong–Nujiang Suture Zone; SJ: Sanjiang arc basin system.

In recent years, a series of Cretaceous magmatic activities have been successively discovered in the northern part of the LS [53–63]. The study of these magmatic events provides material for exploring the evolutionary history of the BNSZ and its accompanying mineralization events. However, the geodynamic background of this series of magmatic activities remains controversial, and its contribution to regional mineralization processes remains unclear. The controversy is broadly based on two arguments. The first argument was provided by [29], who proposed that the series was related to a northward subduction of the Neo-Tethys Ocean, while Sun et al. (2024) [64] suggested that the series represented a tectonic–magmatic–mineralization event. Other arguments propose that the magmatic activities in the series were products of magmatic processes during the collisional stages between the QT and the LS [3,26,65–72]. In light of these debates, this study investigates the petrogenesis, geochemistry, and metallogenesis of the Kongco granitic porphyry dykes in the Northern Part of the Central Lhasa Microblock in Tibet. The systematic petrological, mineralogical, geochemical and mineralization study of these dykes may provide crucial material for resolving the tectonic–magmatic–mineralization events in the region.

2. Geology

The study area is located in the Kongco area of Nima County of Nagqu, Tibet, at the northern margin of the CL, bordering the SNSZ in the north (Figure 1c). The study area's geology is complex and is characterized by a predominant NEE-SEE-trending structure, along with NEE-SWW-trending structures formed by post-collisional extensional processes (Figure 1c). The study area comprises three structural units oriented along the northeast–southwest direction: the Northern Lhasa block (NL), the SNSZ, and the CL (Figure 1c). The exposed stratigraphy in the region primarily consists of a Mesozoic stratum, including the Upper Cretaceous Jingzhushan Formation (K_2j) conglomerates, Lower Cretaceous Langshan Formation (K_1l) bioclastic limestone interbedded with sandstones, Duoni Formation (K_1d) sandstones, limestone, and volcanic rocks, Permian Xiala Formation (P_2x) limestone, Carboniferous Yongzhu Formation ($C_{1\text{-}2}y$) sandstones, shales and limestone, Middle to Upper Devonian Changshehu Formation ($D_{2\text{-}3}c$) sandstones interbedded with limestone, and Devonian Daerdong Formation (D_1d) limestone (Figure 1c). The Daerdong Formation is characterized by thin-bedded dark gray limestone, bioclastic limestone and marl rich in fossils including brachiopods (Nowakia acuaria), crinoids, corals, bryozoans and trilobites, indicating deposition in a deep-water basin environment (Figure 1c). The Cretaceous magmatic activity in the study area was intense, with exposed intrusive rocks primarily comprising medium-acidic lithologies, including two phases: Early Cretaceous (~105 Ma) and Late Cretaceous (~90 Ma) (Figure 1c). These magmatites intruded into the strata of the Jingzhushan Formation and Xiala Formation in the form of complex plutons and dykes [26,73–75]. Additionally, reports indicate medium-acidic magmatic activity around ~90 Ma and ~102 Ma, suggesting crustal melting transitioning from deep to shallow depths, providing significant evidence for an extensional tectonic setting [76–79].

3. Granitic Porphyry Dykes and Copper Mineralization

In the study area, granitic porphyry dykes, ranging from 1 to 5 m in width, primarily intrude into the Jingzhushan Formation and Xiala Formation limestone with minor intrusion into the conglomerates. They are spatially adjacent to extensive exposures of Early Cretaceous granite, approximately 1–2 km away [52]. These dykes exhibit an irregular radial distribution, extending 500 m to 1000 m. Locally, the intrusion boundaries are distinct, displaying significant magmatic intrusion chill zones (Figure 2b), and are accompanied by weak copper mineralization (Figure 2b,e). The dyke width can reach up to 5 m or wider in some areas, with local occurrences as stocks. Along the contact zones between the dykes and limestone, skarn alteration has developed, primarily characterized by epidotization and silicification (Figure 2b,d). These zones also exhibit varying grades of pyrite and chalcopyrite mineralization, which developed along fracture surfaces. Under the influence of surface water and oxygenation, malachitization is prevalent.

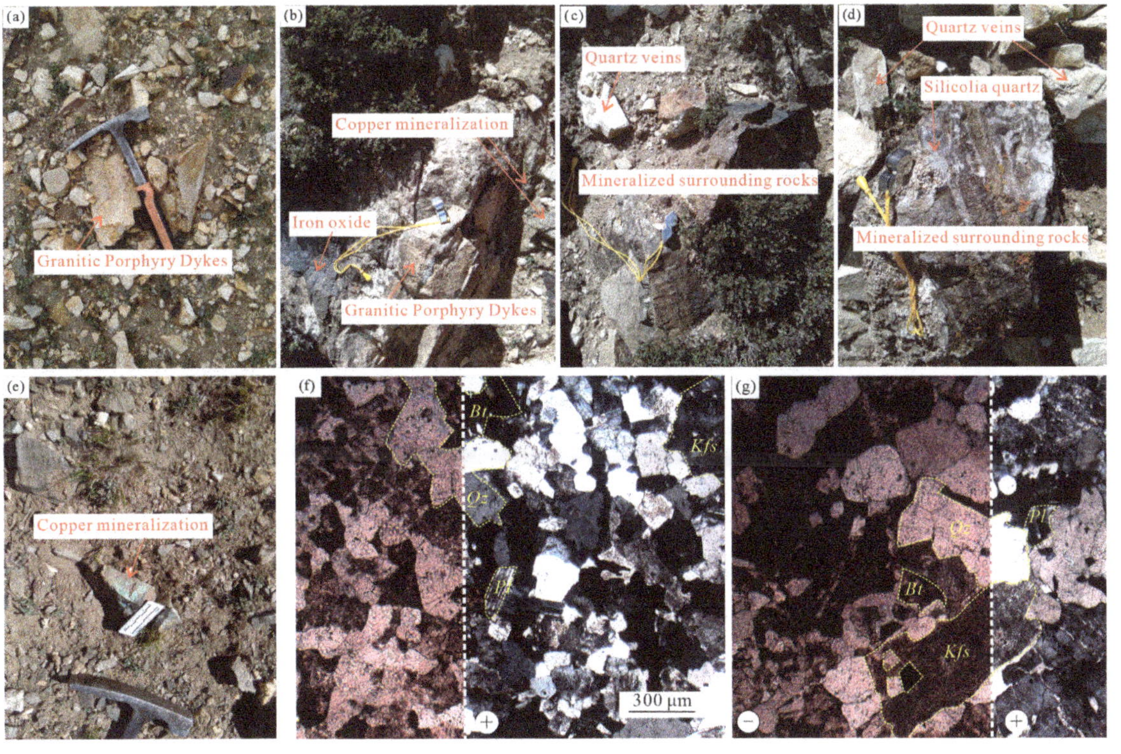

Figure 2. Petrographic characteristics of the rocks in Kongco. (**a**) Early Cretaceous Kongco granite; (**b**) fine-grained facies of Early Cretaceous Kongco granite; (**c**) medium-coarse-grained facies of Early Cretaceous Kongco granite; (**d**) invasion contact relationship and hydrothermal alteration; (**e**) peacock petrified altered rock; (**f**) micrograph of the fine-grained facies of Early Cretaceous Kongco granite; (**g**) microphotograph of medium-coarse-grained facies of Early Cretaceous Kongco granite. Qz: quartz; Kfs: K-feldspar; Pl: plagioclase; Bt: biotite.

Overall, granitic porphyry dykes exhibit uneven mineral grains (e.g., quartz and feldspar), transitioning from medium to fine grains (Figure 2a–c). The main minerals present are quartz, plagioclase, and potassium feldspar, with minor occurrences of pyrite as accessory minerals. These dykes are predominantly porphyritic, with quartz comprising approximately 60%, K-feldspar 30%, and plagioclase 10% of the phenocrysts. Quartz phenocrysts are colorless, euhedral to subhedral, and some grains are rounded. K-feldspar phenocrysts are typically perthitic, subhedral–tabular, and some exhibit graphic intergrowth with quartz. Some perthitic feldspars contain inclusions of plagioclase. Plagioclase phenocrysts are colorless, subhedral–tabular, exhibiting slight sericitization on the surface, with a calculated anorthite content of approximately 5%, primarily belonging to sodium-rich plagioclase (acidic plagioclase). The matrix is predominantly granophyric cryptocrystalline or microcrystalline, comprising approximately 70% of the rock, with phenocrysts making up the remaining 30%. Spatially, there is significant variation in phenocryst grain size, ranging locally from 5 mm to 10 mm, with rapid changes in grain size and in some areas as small as 1mm, displaying a blocky structure.

Mineralization is notably significant: within and along the irregular contact zones between the granitic porphyry dykes and the surrounding limestone formations, there are varying grades of pyrite and chalcopyrite mineralization. Additionally, a common feature observed in the outer zones of these contact areas is the presence of "iron–manganese crusts". The chalcopyrite (copper mineralization) and pyrite (iron mineralization) appear as uneven disseminations and veinlets within and along the contact zones between the intrusive veins and the limestone (Figure 2d). Abundant malachite is visible on the surface (Figure 2e). The mineralized alteration is predominantly characterized by epidotization, silicification and sericitization. Epidotization is widespread, occurring in clusters and irregularly distributed chicken-wire patterns along the contact zones between the veins and the surrounding rock. Strong silicification is prominent in highly mineralized areas, typically filling irregularly shaped masses in fractures or veins of varying widths. Both pyrite and chalcopyrite mineralization (malachitization) can be observed within and around the silicified alteration zones. Accompanying the silicification, varying degrees of sericitization are present, manifesting as fine scales of sericite often associated with quartz and pyrite, forming a type of alteration known as beresitization [71]. These observations suggest that the granitic porphyry dykes likely provided the necessary source materials and heat sources for mineralization. The limestone host rock provided conditions and space for the deposition and unloading of minerals. Nearby occurrences of skarn copper deposits such as Qusanggele and Kongco suggest a geological context conducive to mineralization.

4. Samples and Analysis Methods

In this study, the granite porphyry was taken as the research object, and a total of 9 samples for petrology, mineralogy, geochronology and geochemistry were collected in the area with mineralization. All samples were collected from granite porphyry related to mineralization or the contact zone of granite porphyry and mineralization (e.g., Figure 2d,e). The sampling location was near 87°31′15″ E and 31°28′01″ N and the sample numbers ranged from ZQ01~ZQ09. All samples were subjected to observation, identification and major and trace element analysis. One sample (ZQ05) was subjected to zircon U-Pb chronology and Lu-Hf isotope analysis, while three samples were subjected to Rb-Sr, Sm-Nd and Pb isotope testing.

Major and trace element analysis was performed by X-Ray Fluorescence spectroscopy (XRF) and Agilent 7700e ICP-MS, performed at the State Key Laboratory of Geological Processes and Mineral Resources (GPMR) at the China University of Geosciences (Wuhan, China), using AGV-2, BHVO-2, BCR-2 and RGM-2 as international standards. Zircon sorting, backscattered electron photography, zircon cathodoluminescence image analysis, zircon U-Pb isotope dating and trace element content were simultaneously analyzed by LA-ICP-MS at the Wuhan Shangpu Analysis Technology laboratory. The GeolasPro laser ablation system was composed of a COMPexPro 102 ArF 193 nm excimer laser and a MicroLas optical system and ICP-MS model Agilent 7500a. The internal standard used in the analysis was the international zircon standard 91500 (1062 $\pm$ 4 Ma) [80], and the detailed instrument parameters and analysis procedures can be found in [81]. The laser beam spot for this analysis was 32 μm. The offline processing of the analysis data was performed by the software ICPMSDataCal [82,83]. Zircon in situ micro-Lu-Hf isotope testing was performed using a combination of Neptune-LA-MC-ICP-MS and Geolas Pro laser ablation systems, and the detailed test procedure can be found in Meng et al. (2014) [84]. The diameter of the test beam spot was 32 μm, and all test positions were the same as or close to the U-Pb dating site. The international zircon standard GJ-1 (0.282030 $\pm$ 40(2σ); Jackson et al. [85] was used to monitor the Lu-Hf isotope testing. The isotope analysis of Rb-Sr, Sm-Nd, and Pb was completed at the Wuhan Shangpu Analysis Technology laboratory. Isotope analysis was performed using an ISOPROBE-T thermoionization mass spectrometer (Cheshire, UK), single-band, M+, tunable multi-Faraday receiver reception. Mass fractionation was corrected with $^{86}Sr/^{88}Sr = 0.1194$, standard measurement results: NbS987 (0.710250 $\pm$ 7), JMC ($^{143}Nd/^{144}Nd = 0.512109 \pm 3$), NBS981, ($^{208}Pb/^{206}Pb = 2.164940 \pm 15$, $^{207}Pb/^{206}Pb = 0.914338 \pm 7$, $^{204}Pb/^{206}Pb = 0.0591107 \pm 2$). the whole process background Pb < 100 pg. Laboratory process background: Rb 2×10^{-10} g, Sr 2×10^{-10} g, Sm, Nd less than 50 pg.

5. Results

5.1. Major and Trace Element Analysis

The major and trace element analytical results (Table 1) of the Kongco granitic porphyry dykes showed a high silica content (SiO_2 = 76.22~77.90 wt.%, average 77.09 wt.%), high potassium (K_2O = 4.97~6.21 wt.%, average 5.30 wt.%), high alkali ($K_2O + Na_2O$ = 8.07~8.98 wt.%, average 8.49 wt.%), low calcium (CaO = 0.24~0.83 wt.%, average 0.41 wt.%), low magnesium (MgO = 0.06~0.20 wt.%, average 0.10 wt.%), and a moderate aluminum content (Al_2O_3 = 11.93~12.45 wt.%, average 12.16 wt.%). The Rittmann index (σ) ranged from 1.93 to 2.34, averaging 2.11, indicating a transition from high potassium calc-alkaline to alkaline series granites (Figure 3a,b). The A/NK(molar ratio $Al_2O_3/(Na_2O + K_2O)$) and A/CNK(molar ratio $Al_2O_3/(CaO + Na_2O + K_2O)$) ratios ranged from 1.06 to 1.18 and 0.98 to 1.09, respectively, suggesting weak peraluminous characteristics in the granitic porphyry dykes (Figure 3c). This alkaline-to-aluminous character is also observed in other Early Cretaceous A-type granites within the CL (Figures 1b and 3c) [66,75,86–88]. The differentiation index (DI) had values ranging from 95.1 to 97.15 with an average of 96.42, indicating a relatively high degree of rock differentiation. The calculated zircon saturation temperatures for the Kongco granitic porphyry dykes ranged from 884 to 914 °C (average 893 °C), suggesting crystallization under relatively high-temperature conditions (Table 1) [89–92].

Table 1. Major and trace element compositions of the rocks in Kongco.

Sample	ZQ01	ZQ02	ZQ03	ZQ04	ZQ05	ZQ06	ZQ07	ZQ08	ZQ09
SiO_2 (wt.%)	76.61	77.45	77.90	77.64	77.44	77.51	76.22	76.31	76.73
TiO_2 (wt.%)	0.10	0.06	0.05	0.10	0.10	0.11	0.11	0.12	0.06
Al_2O_3 (wt.%)	12.23	11.93	11.96	12.21	12.42	12.45	12.19	12.00	12.04
Fe_2O_3 (wt.%)	0.70	0.35	0.38	0.48	0.26	0.21	0.42	0.62	0.48
FeO (wt.%)	0.19	0.32	0.22	0.08	0.13	0.11	0.16	0.54	0.35
MnO (wt.%)	0.02	0.02	0.01	0.01	0.01	0.01	0.03	0.02	0.02
MgO (wt.%)	0.17	0.06	0.07	0.08	0.09	0.08	0.09	0.20	0.09
CaO (wt.%)	0.48	0.43	0.41	0.28	0.26	0.29	0.83	0.46	0.24
Na_2O (wt.%)	2.94	2.77	3.46	3.24	3.09	3.49	2.76	3.41	3.56
K_2O (wt.%)	5.13	6.21	4.99	5.09	5.37	4.98	5.94	4.97	4.98
P_2O_5 (wt.%)	0.06	0.01	0.01	0.02	0.02	0.03	0.02	0.02	0.01
Fe_2O_3T (wt.%)	0.91	0.71	0.62	0.57	0.40	0.33	0.60	1.22	0.87
FeOT (wt.%)	0.82	0.63	0.56	0.51	0.36	0.30	0.54	1.10	0.78
LOI (wt.%)	1.26	0.35	0.50	0.74	0.77	0.69	1.20	0.60	0.52
Total (wt.%)	99.89	99.96	99.97	99.97	99.96	99.95	99.97	99.28	99.08
A/NK	1.18	1.06	1.08	1.13	1.14	1.12	1.11	1.09	1.07
A/CNK	1.09	0.99	1.01	1.08	1.09	1.07	0.98	1.01	1.03
δ43	1.93	2.34	2.04	2.00	2.07	2.08	2.27	2.10	2.15
DI	95.1	97.04	96.9	96.84	96.92	97.15	95.29	95.49	97.1
Q	38.99	36.27	37.47	38.62	38.29	37.45	35.76	36.3	36.4
Kfs	53.04	58.3	55.65	55.82	56.57	57.03	55.42	55.29	58.25
Pl	4.96	3.86	5.4	3.4	2.95	3.66	7.14	5.89	3.31
La (ppm)	/	/	20.2	18.6	9.9	13.3	31.1	/	/
Ce (ppm)	/	/	47.8	42.7	29.8	27.9	68.8	/	/
Pr (ppm)	/	/	5.51	4.74	2.22	3.05	7.74	/	/
Nd (ppm)	/	/	20.8	16.6	7.9	11.0	25.7	/	/
Sm (ppm)	/	/	5.78	3.81	1.77	2.43	5.99	/	/
Eu (ppm)	/	/	0.076	0.151	0.129	0.182	0.141	/	/
Gd (ppm)	/	/	6.21	4.25	2.00	2.42	5.70	/	/
Tb (ppm)	/	/	1.191	0.810	0.439	0.505	1.026	/	/
Dy (ppm)	/	/	8.40	5.79	3.54	3.77	7.01	/	/
Ho (ppm)	/	/	1.78	1.21	0.79	0.85	1.44	/	/
Er (ppm)	/	/	5.44	3.86	2.63	2.82	4.46	/	/
Tm (ppm)	/	/	0.918	0.669	0.484	0.485	0.715	/	/
Yb (ppm)	/	/	6.13	4.42	3.20	3.30	4.95	/	/
Lu (ppm)	/	/	0.903	0.665	0.483	0.497	0.697	/	/
Y (ppm)	/	/	56.6	39.3	26.3	28.8	45.5	/	/
Rb (ppm)	/	/	270	282	227	249	381	/	/
Ba (ppm)	/	/	31	84	122	120	56	/	/
Th (ppm)	/	/	44.2	39	31.4	32.9	46.3	/	/
U (ppm)	/	/	5.06	5.25	5.24	5.72	5.22	/	/
Nb (ppm)	/	/	32.3	20.1	11.6	12.0	24.4	/	/
Ta (ppm)	/	/	3.75	2.47	1.75	1.81	2.69	/	/
Sr (ppm)	/	/	12.1	24.2	36.8	33.4	14.4	/	/
Zr (ppm)	/	/	102	112	97.1	99.0	146	/	/
Hf (ppm)	/	/	5.19	4.42	3.59	3.56	5.31	/	/

Table 1. *Cont.*

Sample	ZQ01	ZQ02	ZQ03	ZQ04	ZQ05	ZQ06	ZQ07	ZQ08	ZQ09
Ga (ppm)	/	/	19.0	16.6	14.6	15.0	18.7	/	/
Sc (ppm)	/	/	2.12	2.53	2.37	2.49	3.04	/	/
Pb (ppm)	/	/	59.2	39.7	36.7	30.6	32.0	/	/
ΣREE	/	/	131	108	65	72	165	/	/
LREE/HREE	/	/	3.24	4.00	3.82	3.95	5.36	/	/
La_N/Yb_N	/	/	2.24	2.86	2.11	2.73	4.26	/	/
δEu	/	/	0.04	0.11	0.21	0.23	0.07	/	/
δCe	/	/	1.08	1.08	1.48	1.03	1.05	/	/
T_{Zr}(°C)	/	/	884	892	887	889	914	/	/

δ43 = $(Na_2O + K_2O)^2/(SiO_2 - 43)$, DI = Qz + Or + Ab + Ne + Lc + Kp (CIPW Mass Fraction Ratio); A/NK = $Al_2O_3/(CaO + K_2O + Na_2O)$, A/CNK = $Al_2O_3/(CaO + K_2O + Na_2O)$ (mol Concentration ratio); δEu = $2 \times Eu_N/(Sm_N + Gd_N)$; δCe = $2 \times Ce_N/(La_N + Pr_N)$; Tzr calculation methodology by Watson and Harrison (2005) [93]. T_{Zr} (°C) = 12,900/(2.95 + 0.85M + ln (496,000/Zrmelt) − 273 [90,92,94]. Note: ppm: 10^{-6}. "/" indicates that there is no test.

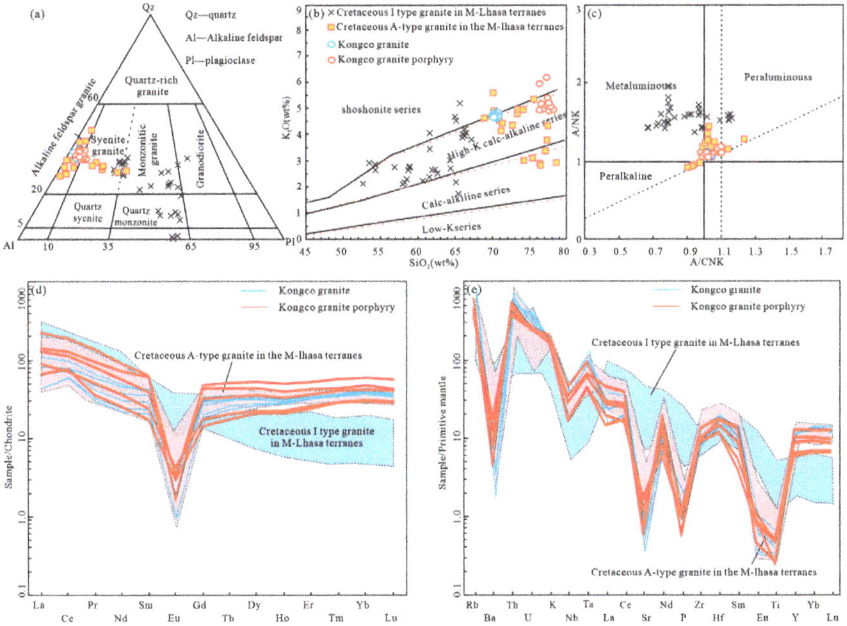

Figure 3. (**a**) QAP plots of rocks in Kongco; (**b**) K_2O vs. SiO_2 plots of rocks in Kongco (basemap after [95, 96]); (**c**) A/NK vs. A/CNK plots of rocks in KongCo (basemap after [97]); (**d**) chondrite-normalized REE patterns(normalizing values for REE are from Boynton, 1984 [98]); (**e**) mantle-normalized multi-element diagrams (normalizing values for REE and trace elements are from Sun 1989 [99]). Date according to these articles [3,26,60,73,75,86–88,100–104]. The solid and dashed lines represent different ranges determined, showing slight variations between them.

The Kongco granitic porphyry dykes exhibit similarities to other A-type granites in the CL. They display elevated total rare earth element content (ΣREE = 65×10^{-6}~165×10^{-6}, average 108.52×10^{-6}), moderate ratios of light rare earth elements to heavy rare earth elements (LREE/HREE = 3.24~5.36, average 4.07), and LaN/YbN ratios ranging from 2.11 to 4.26 (averaging 2.84). These values indicate slight enrichment in light rare earth elements and slight depletion in heavy rare earth elements, accompanied by a pronounced negative Eu anomaly (δEu = 0.04~0.23, average 0.13) and a weak positive Ce anomaly (δCe = 1.05~1.48, average 1.14). In the rare earth element chondrite-normalized distribution patterns (Figure 3d), all samples exhibited a consistent trend characterized by a gently sloping "V-shaped" curve, typical of A-type granites [105,106]. From the spider diagram of normalized to primitive mantle values (Figure 3e), the Kongco granitic-porphyry dykes are notably enriched in elements such as Rb, Th, U, K, Ta, Ce, Nd, Zr, Hf, Sm, Y, Yb, and Lu, while they are relatively depleted in Ba, Nb, Sr, P, Eu, and Ti. These geochemical characteristics reflect their origin and evolution within a magmatic setting [107,108].

5.2. Zircon U-Pb Age and Lu-Hf Isotopic Compositions

Zircon grains were extracted from the ZQ05 sample (granitic porphyry) and subsequently subjected to U-Pb chronology, trace element, and Lu-Hf isotope tests. The measurement points were specifically chosen from areas displaying clear rhythmic zoning structures. Detailed analysis locations and results are shown in Figures 4 and 5, Tables 2–4. These characteristics suggest that the zircons extracted from the Kongco ZQ05 sample are primarily magmatic zircons, characterized by their uniform texture, intact crystal morphology and relatively high Th/U ratios (ranging from 0.38 to 1.01 and averaging 0.65). The zircons exhibit dark CL imaging and dense oscillatory zoning under fluorescence microscopy (Figure 4), further confirming their origin within the crystallization history of the granite.

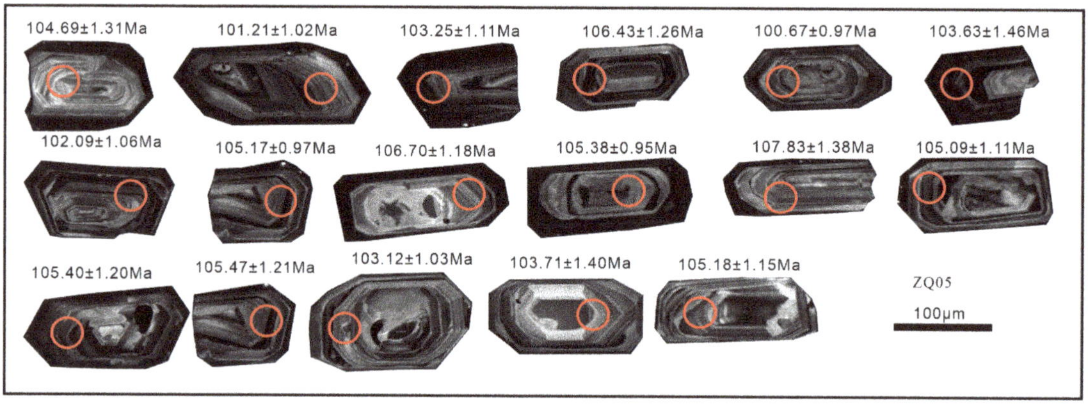

Figure 4. Cathodoluminescence (CL) images of zircon grains and LA-ICP-MS measurement points for zircon concordia diagrams. Note: The red circle represents the measurement point position of U-Pb.

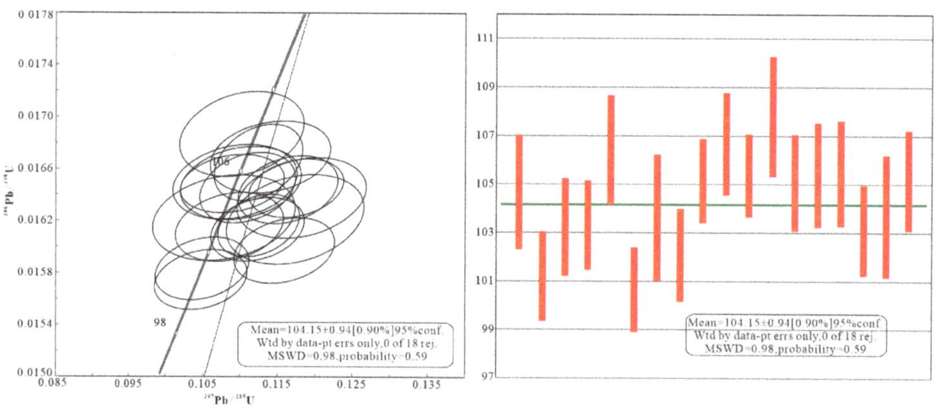

Figure 5. Zircon U-Pb age.

Table 2. LA-ICP-MS zircon U-Pb ages.

Sample	Pb	Th	U	Th/U	Isotope Ratio								Age (Ma)				Concordance
	$\times10^{-6}$	$\times10^{-6}$	$\times10^{-6}$		$^{207}Pb/^{206}Pb$	1σ	$^{207}Pb/^{235}U$	1σ	$^{206}Pb/^{238}U$	1σ	$^{208}Pb/^{232}Th$	1σ	$^{207}Pb/^{235}U$	1σ	$^{206}Pb/^{238}U$	1σ	%
ZQ05-1	22.25	252.60	387.71	0.65	0.0527	0.0027	0.1182	0.0057	0.0164	0.0002	0.0002	1.6119	113.42	5.16	104.69	2.36	91%
ZQ05-10	30.05	376.02	523.38	0.72	0.0482	0.0020	0.1049	0.0042	0.0158	0.0002	0.0001	1.4755	101.25	3.84	101.21	1.84	99%
ZQ05-11	31.58	386.60	515.62	0.75	0.0485	0.0022	0.1084	0.0048	0.0161	0.0002	0.0001	1.3910	104.53	4.41	103.25	2.01	98%
ZQ05-12	52.58	610.22	857.81	0.71	0.0507	0.0017	0.1124	0.0036	0.0162	0.0002	0.0001	1.4659	108.16	3.29	103.33	1.83	95%
ZQ05-13	50.99	539.06	1140.79	0.47	0.0487	0.0017	0.1124	0.0039	0.0166	0.0002	0.0001	2.2228	108.12	3.57	106.43	2.26	98%
ZQ05-14	48.95	540.05	951.95	0.57	0.0485	0.0021	0.1047	0.0041	0.0157	0.0002	0.0001	1.8378	101.06	3.81	100.67	1.76	99%
ZQ05-16	17.94	216.27	313.64	0.69	0.0534	0.0028	0.1173	0.0061	0.0162	0.0002	0.0002	1.5052	112.64	5.52	103.63	2.62	91%
ZQ05-17	48.93	633.05	629.36	1.01	0.0526	0.0020	0.1159	0.0044	0.0160	0.0002	0.0001	1.0351	111.35	4.05	102.09	1.91	91%
ZQ05-2	28.93	268.25	702.81	0.38	0.0521	0.0019	0.1175	0.0042	0.0164	0.0002	0.0002	2.7536	112.78	3.82	105.17	1.74	93%
ZQ05-21	22.87	235.89	431.89	0.55	0.0495	0.0023	0.1140	0.0054	0.0167	0.0002	0.0002	1.9441	109.61	4.88	106.70	2.13	97%
ZQ05-22	35.14	385.01	594.21	0.65	0.0480	0.0021	0.1088	0.0046	0.0165	0.0001	0.0001	1.6331	104.91	4.21	105.38	1.70	99%
ZQ05-23	17.84	200.74	377.43	0.53	0.0482	0.0026	0.1101	0.0055	0.0169	0.0002	0.0002	1.9355	106.05	5.05	107.83	2.49	98%
ZQ05-3	27.80	299.80	514.01	0.58	0.0517	0.0020	0.1164	0.0044	0.0164	0.0002	0.0001	1.7903	111.78	4.01	105.09	1.99	93%
ZQ05-4	23.75	280.40	403.35	0.70	0.0486	0.0025	0.1100	0.0054	0.0165	0.0002	0.0002	1.5069	105.95	4.91	105.40	2.15	99%
ZQ05-5	22.25	231.69	433.52	0.53	0.0483	0.0025	0.1092	0.0055	0.0165	0.0002	0.0002	1.9420	105.24	5.05	105.47	2.19	99%
ZQ05-6	32.63	390.71	528.91	0.74	0.0507	0.0020	0.1122	0.0041	0.0161	0.0002	0.0001	1.4169	108.01	3.78	103.12	1.85	95%
ZQ05-7	19.86	220.72	337.58	0.65	0.0492	0.0033	0.1077	0.0063	0.0162	0.0002	0.0002	1.6737	103.89	5.78	103.71	2.53	99%
ZQ05-8	54.77	623.64	749.34	0.83	0.0491	0.0020	0.1109	0.0042	0.0164	0.0002	0.0001	1.2927	106.79	3.88	105.18	2.07	98%

Table 3. LA-ICP-MS zircon REE compositions (ppm).

Sample	La	Ce	Pr	Nd	Sm	Eu	Gd	Tb	Dy	Ho	Er	Tm	Yb	Lu	Y	Ti	Hf
ZQ05-1	0.032	2.46	0.013	0.13	0.52	0.28	2.30	0.86	11.8	4.76	25.9	6.59	74.3	13.8	147	8.37	5776
ZQ05-10	0.62	14.9	0.17	1.76	2.87	0.27	14.2	5.90	82.7	33.6	170	39.7	422	76.0	1078	3.03	10,455
ZQ05-11	0.0000	15.8	0.10	1.37	2.54	0.39	17.4	6.82	93.9	38.2	191	44.2	465	83.2	1208	1.47	10,008
ZQ05-12	0.12	15.6	0.15	2.79	5.60	0.75	34.1	12.8	164	63.0	291	62.7	635	108	1911	3.65	10,100
ZQ05-13	13.8	57.8	5.48	26.7	10.0	0.69	28.4	9.66	121	47.8	229	51.4	532	93.8	1498	3.47	10,364
ZQ05-14	0.69	18.6	0.27	2.26	3.09	0.25	22.9	8.94	126	52.6	260	60.8	660	115	1635	2.93	11,175
ZQ05-16	0.064	9.09	0.12	1.37	3.54	0.21	22.1	7.90	105	40.1	185	40.7	436	78.2	1189	4.00	11,555
ZQ05-17	0.080	2.23	0.0000	0.17	0.57	0.27	2.08	0.75	11.1	4.35	24.1	6.10	66.3	12.7	136	9.89	5711
ZQ05-2	2.64	19.5	0.87	4.87	3.65	0.58	19.0	6.92	95.4	38.0	187	42.9	462	82.7	1214	6.59	9092
ZQ05-21	0.18	15.7	0.23	1.82	3.42	0.24	21.9	9.27	131	53.1	259	60.3	638	113	1658	4.35	10,850
ZQ05-22	5.52	25.6	1.53	7.94	3.16	0.43	15.2	5.79	75.6	30.8	152	35.0	373	67.1	981	3.89	9991
ZQ05-23	0.89	17.3	0.29	2.34	2.75	0.43	20.5	7.88	103	39.8	191	41.6	425	73.4	1222	5.45	9195
ZQ05-3	11.7	42.2	3.70	16.4	5.11	0.43	14.1	5.34	73.8	31.8	171	43.3	502	94.4	1072	0.000	10,177
ZQ05-4	0.26	15.1	0.11	1.22	2.17	0.48	16.5	6.39	91.4	37.3	189	44.6	484	89.6	1200	1.25	9694
ZQ05-5	0.034	11.0	0.16	2.53	5.10	0.87	28.4	10.1	130	51.0	241	53.7	565	99.3	1589	7.27	9453
ZQ05-6	0.066	12.6	0.053	1.11	3.05	0.28	17.8	7.13	99.8	41.2	205	47.1	506	92.3	1297	6.02	9516
ZQ05-7	2.20	22.4	0.80	5.35	5.08	0.57	31.1	11.5	149	57.7	261	55.4	545	92.0	1701	4.70	8986
ZQ05-8	0.25	8.59	0.13	1.89	3.78	0.33	24.0	8.93	114	43.9	196	42.9	419	72.7	1298	6.37	8529

Table 4. LA-ICP-MS zircon Lu-Hf isotope compositions.

Sample	Age (Ma)	$^{176}Hf/^{177}Hf$	$^{176}Lu/^{177}Hf$	$^{176}Yb/^{177}Hf$	$\varepsilon_{Hf}(0)$	$\varepsilon_{Hf}(t)$	T_{DM1}	T_{DM2}	$f_{Lu/Hf}$
ZQ05-01	104.7	0.2828	0.0015	0.0373	0.9	3.0	654.3	863.2	−0.96
ZQ05-02	105.2	0.2828	0.0012	0.0275	1.5	3.6	623.4	826.3	−0.96
ZQ05-03	105.1	0.2827	0.0013	0.0315	−1.7	0.4	754.5	1004.6	−0.96
ZQ05-04	105.4	0.2828	0.0012	0.0296	−0.4	1.8	698.8	930.9	−0.96
ZQ05-05	105.5	0.2828	0.0014	0.0340	0.4	2.5	670.1	886.7	−0.96
ZQ05-06	103.1	0.2827	0.0015	0.0373	−1.6	0.6	752.7	997.9	−0.95
ZQ05-07	103.7	0.2828	0.0018	0.0432	−0.4	1.7	710.8	934.3	−0.95
ZQ05-08	105.2	0.2828	0.0015	0.0362	−0.3	1.8	701.8	928.3	−0.96
ZQ05-10	101.2	0.2828	0.0017	0.0441	−0.2	1.9	703.0	924.1	−0.95
ZQ05-11	103.2	0.2827	0.0018	0.0433	−1.0	1.1	733.8	965.8	−0.95
ZQ05-12	103.3	0.2828	0.0017	0.0415	−0.7	1.4	719.6	948.4	−0.95

Notes: $\varepsilon_{Hf}(0) = (^{176}Hf/^{177}Hf/(^{176}Hf/^{177}Hf)_{CHUR,\ 0}-1) \times 10{,}000$; $\varepsilon_{Hf}(t) = (^{176}Hf/^{177}Hf - (^{176}Lu/^{177}Hf)_S \times (e^{\lambda t} - 1))/((^{176}Hf/^{177}Hf)_{CHUR,\ 0} - (^{176}Lu/^{177}Hf)_{CHUR} \times (e^{\lambda t} - 1)) - 1) \times 10{,}000$; $(^{176}Hf/^{177}Hf)_i = {}^{176}Hf/^{177}Hf - (^{176}Lu/^{177}Hf)_S \times (e^{\lambda t} - 1)) - 1$; $T_{DM2} = T_{DM1} - (T_{DM1} - t) \times (f_{CC} - f_{Lu/Hf})/(f_{CC} - f_{DM})$; $f_{Lu/Hf} = {}^{176}Lu/^{177}Hf/(^{176}Lu/^{177}Hf)_{CHUR} - 1$; $(^{176}Lu/^{177}Hf)_{CHUR} = 0.0332$, $(^{176}Hf/^{177}Hf)_{CHUR,\ 0} = 0.282772$ [109]; $(^{176}Lu/^{177}Hf)_{DM} = 0.0384$, $(^{176}Hf/^{177}Hf)_{DM} = 0.28325$ (Griffin et al., 2000 [110]); $f_{CC} = -0.55$, $f_{DM} = 0.16$, $\lambda = 1.867 \times 10^{-11}$ year^{-1} [111].

The ZQ05 sample was tested at 23 analysis points, yielding 18 valid data points for $^{206}Pb/^{238}U$ surface ages ranging from 101.21 Ma to 107.83 Ma. The weighted average age was calculated as 104.15 ± 0.94 Ma (MSWD = 0.89) (Figure 5 and Table 2). The zircon ages represent the crystallization age of the Kongco granitic porphyry dykes, which is consistent with the U-Pb zircon ages reported by Liu et al. (2022) [101] for Kongco granites. Therefore, the magmatic activity in the Kongco area likely occurred in the Early Cretaceous period, around 104 Ma.

In a similar manner, Lu-Hf isotope analysis was conducted on some selected dating points from the ZQ05 sample. The $^{176}Yb/^{177}Hf$ ratios for zircons ranged from 0.0275 to 0.0441, with an average of 0.0369 (Table 4). The $^{176}Lu/^{177}Hf$ ratios varied from 0.0012 to 0.0018, averaging 0.0015, while the $^{176}Hf/^{177}Hf$ ratios ranged from 0.282724 to 0.282814, with an average of 0.282763. The calculated $\varepsilon_{Hf}(0)$ values based on corresponding zircon ages primarily ranged from −1.69 to 1.49, averaging −0.31. The $\varepsilon_{Hf}(0)$ values were predominantly positive, ranging from 0.43 to 3.63, with an average of 1.80 (Table 4). The two-stage model ages (T_{DM2}) ranged from 826.3 Ma to 1004.6 Ma, averaging 928 Ma (Table 4, Figure 6).

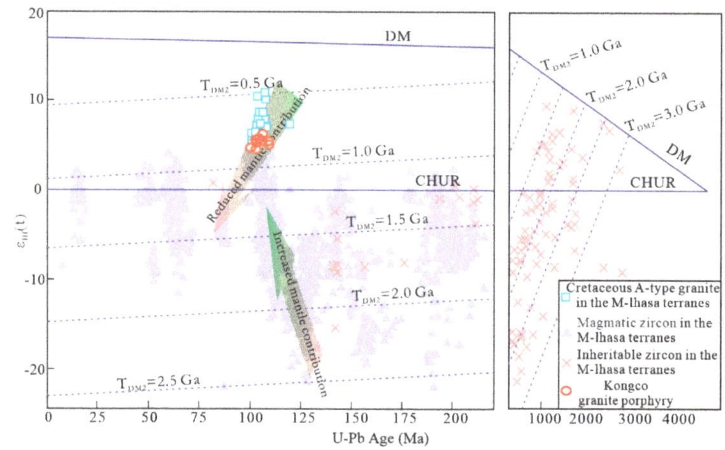

Figure 6. $\varepsilon_{Hf}(t)$ vs. U-Pb age plot of zircons from KongCo porphyry granite (base map after these articles [109–113].

5.3. Rb-Sr, Sm-Nd, Pb Isotopes

The Rb-Sr, Sm-Nd, and Pb isotope compositions of the granitic porphyry from Kongco are presented in Table 5. The $^{87}Rb/^{86}Sr$ and $^{87}Sr/^{86}Sr$ ratios range from 17.7352 to 76.1255 (average 52.0037) and from 0.7474 to 0.8182 (average 0.7945), respectively. The calculated $(^{87}Sr/^{86}Sr)t$ ranged from 0.7043 to 0.7064 (average 0.7054). The $^{147}Sm/^{144}Nd$ and $^{143}Nd/^{144}Nd$ ratios ranged from 0.1350 to 0.1528 (average 0.1413) and from 0.512155 to 0.512445 (average 0.512337), respectively. The calculated $(^{143}Nd/^{144}Nd)t$ and $\varepsilon_{Nd}(t)$ ranged from 0.5121 to 0.5124 (average 0.5123) and from −8.60 to −2.95 (average −5.13), respectively. The $^{206}Pb/^{204}Pb$, $^{207}Pb/^{204}Pb$ and $^{208}Pb/^{204}Pb$ ratios ranged from 18.779 to 18.947 (average 18.865), 15.714 to 15.723 (average 15.719), and 39.370 to 39.514 (average 39.427), respectively. The $(^{206}Pb/^{204}Pb)t$, $(^{207}Pb/^{204}Pb)t$, and $(^{208}Pb/^{204}Pb)t$ ranged from 18.627 to 18.788 (average 18.731), 15.707 to 15.719 (average 15.713), and 39.038 to 39.110 (average 39.082), respectively.

Table 5. Rb-Sr, Sm-Nd, and Pb isotope compositions.

Sample	t(Ma)	$^{87}Rb/^{86}Sr$	$^{87}Sr/^{86}Sr$	$(^{87}Sr/^{86}Sr)_t$	$^{147}Sm/^{144}Nd$	$^{143}Nd/^{144}Nd$	$f_{Sm/Nd}$	$\varepsilon_{Nd}(t)$	$(^{143}Nd/^{144}Nd)_t$	$^{206}Pb/^{204}Pb$	$(^{206}Pb/^{204}Pb)_t$	$^{207}Pb/^{204}Pb$	$(^{207}Pb/^{204}Pb)_t$	$^{208}Pb/^{204}Pb$	$(^{208}Pb/^{204}Pb)_t$	$\triangle\beta$	$\triangle\gamma$
ZQ03	104	62.1532	0.8182	0.7055	0.1528	0.512411	−0.22	−3.84	0.5123	18.868	18.777	15.723	15.719	39.370	39.110	25.4	43.8
ZQ04	104	17.7325	0.7474	0.7064	0.1350	0.512155	−0.31	−8.60	0.5121	18.779	18.627	15.714	15.707	39.396	39.098	24.6	43.4
ZQ05	104	76.1255	0.8179	0.7043	0.1362	0.512445	−0.31	−2.95	0.5124	18.947	18.788	15.720	15.712	39.514	39.038	24.9	41.8

Notes: $(^{87}Sr/^{86}Sr)_t = {}^{87}Sr/^{86}Sr - {}^{87}Rb/^{86}Sr\ (e^{\lambda t} - 1)$; $\lambda_{Rb} = 1.42 \times 10^{-11}t^{-1}$; $(^{143}Nd/^{144}Nd)_t = {}^{143}Nd/^{144}Nd - {}^{147}Sm/^{144}Nd\ (e^{\lambda t} - 1)$, $\lambda_{Sm} = 6.54 \times 10^{-12}t^{-1}$; $\varepsilon_{Nd}(0) = [^{143}Nd/^{144}Nd/(^{143}Nd/^{144}Nd)_{CHUR} - 1] \times 10000$; $\varepsilon_{Nd}(t) = \{[(^{143}Nd/^{144}Nd)_t/(^{143}Nd/^{144}Nd)_{(CHUR)t} - 1\} \times 10000$; $T_{DM2} = 1/\lambda \times \ln\{1+[(^{143}Nd/^{144}Nd)_S - ((^{147}Sm/^{144}Nd)_S - (^{147}Sm/^{144}Nd)_{cc})\ (e^{\lambda t}-1) - (^{143}Nd/^{144}Nd)_{DM}]/[(^{147}Sm/^{144}Nd)_{cc} - (^{147}Sm/^{144}Nd)_{DM}]\}$; $(^{143}Nd/^{144}Nd)_{(CHUR)t} = (^{143}Nd/^{144}Nd)_{CHUR} - (^{147}Sm/^{144}Nd)_{CHUR}\ (e^{\lambda t}-1)$; $f_{Sm/Nd} = (^{147}Sm/^{144}Nd)_S/(^{147}Sm/^{144}Nd)_{CHUR} - 1$; $(^{143}Nd/^{144}Nd)_{CHUR} = 0.512638$, $(^{147}Sm/^{144}Nd)_{CHUR} = 0, 1967$, $(^{147}Sm/^{144}Nd)_{cc} = 0.1180$, $(^{147}Sm/^{144}Nd)_{DM} = 0.2135$, $(^{143}Nd/^{144}Nd)_{DM} = 0.51315$. $\triangle\beta = (\beta - \beta M)/\beta M \times 1000$, $\triangle\gamma = (\gamma - \gamma M) / \gamma M \times 1000$; $\beta = (^{206}Pb/^{204}Pb)t$; $\gamma = (^{207}Pb/^{204}Pb)t$; $\beta_M = 15.33$; $\gamma_M = 37.47$. Calculation formula after DePaolo and Wasserburg (1976) [114].

6. Discussion

6.1. Types of Rocks

A-type granites are characterized by the presence of alkaline dark-colored minerals such as sodic plagioclase, biotite, and sodic amphibole. Chemically, they typically exhibit low Al_2O_3 (~12.40 wt.%), low CaO (~0.75 wt.%), low MgO (~0.20 wt.%), low A/NK value (~1.08), high Ca/Al ratio, and elevated concentrations of high-field-strength elements (HFSEs) [115]. Mineralogical analysis of the Kongco granitic porphyry dykes indicates that they are rich in potassium feldspar (average 56.15% calculated by CIPW norm), with abundant biotite and occasional hornblende. In terms of major element composition, they show a high potassium content (K_2O = 4.97~6.21 wt.%) and high alkalinity ($K_2O + Na_2O$ = 8.07~8.98 wt.%), along with low aluminum (Al_2O_3 = 11.93~12.45 wt.%), low calcium (CaO = 0.24~0.83 wt.%), and low magnesium (MgO = 0.06~0.20 wt.%) (Table 1). Based on the aluminum index, the Kongco granitic porphyry exhibit weakly peraluminous characteristics (A/NK value: 1.06~1.18, A/CNK value: 0.98~1.09) (Figure 3 and Table 1), which are typical features of A-type granites in terms of mineralogy and major element composition [106,115]. In terms of trace and rare earth elements (Figure 3, Table 1), the Kongco granitic porphyry shows a significant negative Eu anomaly (δEu = 0.04~0.23), with a slight positive Ce anomaly and light rare earth enrichment relative to heavy rare earths (LREE/HREE = 3.24~5.36). In the chondrite-normalized rare earth element distribution diagram (Figure 3d), it displays a characteristic "V-shaped" pattern with a rightward slope. Moreover, the Kongco granitic porphyry is enriched in Rb, Th, U, Pb, Zr and Hf and is relatively depleted in Nb, Ta, Ti, P, Ba, and Sr. These geochemical compositions align with those of A-type granites[48,106,116]. They are notably similar to Early Cretaceous A-type granites reported in the CL (Figure 3), distinguishing them from Late Cretaceous I-type granites in the same area (Figure 3). Criteria such as FeO^T/MgO (3.60~10.41) and Zr (97.1×10^{-6}~146×10^{-6}) further support their classification as A-type granites (Figure 7a,b). The formation temperature of A-type granites is generally higher than that of S-type and I-type granites [117,118]. Calculated using a whole-rock zirconium saturation thermometer [93,119], the formation temperature (Tzr) of the Kongco granitic porphyry ranged from 884 °C to 914 °C, falling within the typical range for A-type granite formation temperatures.

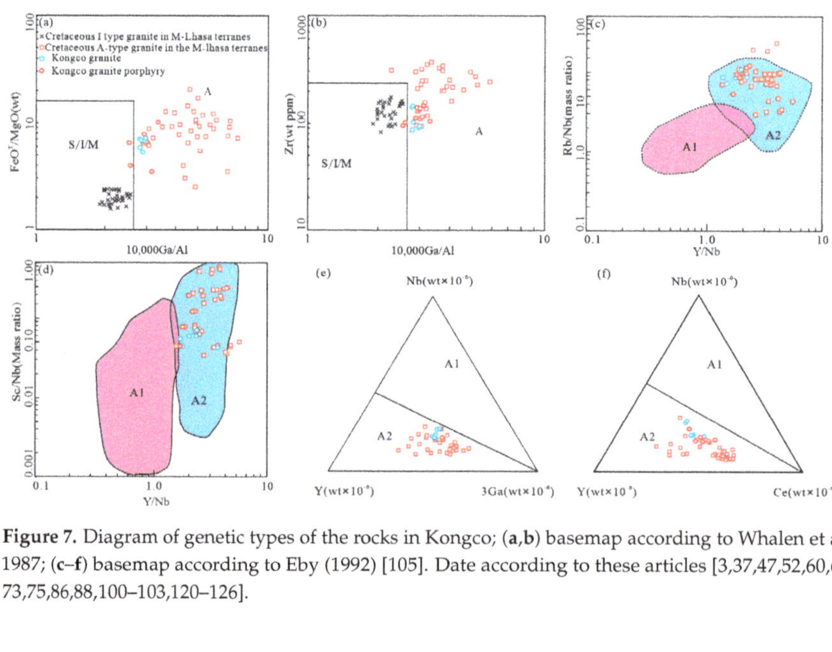

Figure 7. Diagram of genetic types of the rocks in Kongco; (**a**,**b**) basemap according to Whalen et al., 1987; (**c**–**f**) basemap according to Eby (1992) [105]. Date according to these articles [3,37,47,52,60,66,73,75,86,88,100–103,120–126].

A-type granites typically exhibit characteristics such as a non-orogenic, alkaline and anhydrous nature. However, with further research advancements, scholars have found that A-type granites can form not only in mantle plume or hotspot settings but also in post-orogenic tectonic environments [116]. Based on the petrological and geochemical characteristics of A-type granites, as well as differences in their material sources and tectonic backgrounds, petrologists classify them into two subtypes: A1 and A2 [116]. A1-type granites have trace element compositions similar to Ocean Island Basalts (OIBs) and are considered to form in non-orogenic environments (continental rifts or intraplate settings), while A2-type granites have trace element compositions similar to continental crustal material and Island Arc Basalts (IABs), indicating formation in post-collisional environments [64,127–130]. There is a transitional type between these two categories without clear boundaries [121,131]. A1-type granites typically exhibit very low Y/Nb and Rb/Nb ratios [105], whereas A2-type granites show the opposite with higher Y/Nb and Rb/Nb ratios [132–134]. Similarly to the Late Early Cretaceous (105 Ma) A-type granites reported in the CL, the Kongco granitic porphyry has elevated concentrations of Rb (227×10^{-6}~381×10^{-6}) and Y (26.3×10^{-6}~56.6×10^{-6}) and relatively lower concentrations of Nb (11.6×10^{-6}~32.3×10^{-6}), resulting in average Y/Nb and Rb/Nb values of 2.05 and 15.65, respectively. These values place it within the A2-type granite field on ternary diagrams (Table 1, Figure 7c–f) and in the post-collisional granite field on tectonic discrimination diagrams (Figure 8a,b).

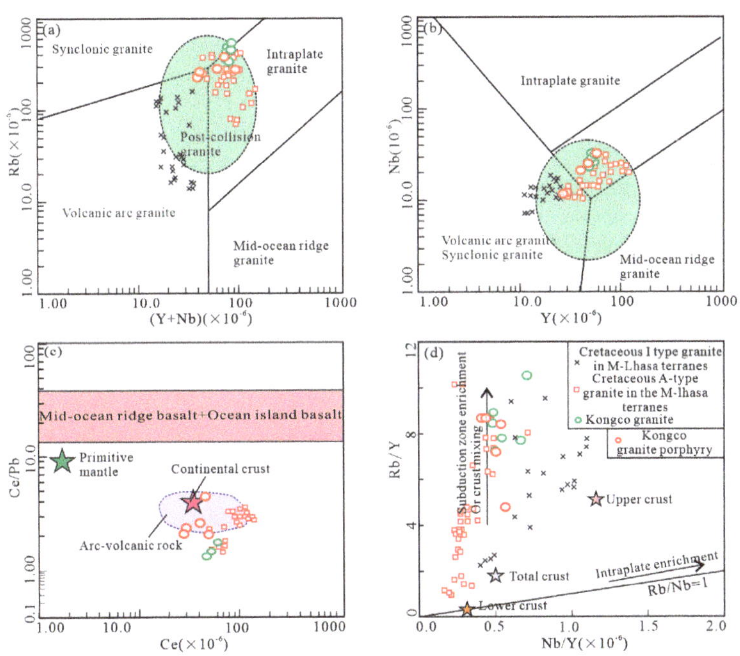

Figure 8. Structural environment diagram of the rocks in Kongco. (**a**,**b**) Basemap according to Pearce et al., 1984 [135]. (**c**,**d**) Basemap according to Boztuğ et al., 2007 [136].

From the above observations, the Kongco granitic porphyry is classified as a weakly peraluminous, high-K, calc-alkaline, potassium basalt series of A2 alkaline feldspar granite.

6.2. Genesis of the Kongco Granitic Porphyry

It is generally believed that A-type granites can form from various magmatic sources, primarily including (a) partial melting of residual F- or Cl-rich lower crustal granulites [106,117,137–139]; (b) low-pressure dehydration melting of shallow crustal igneous or granitic gabbroic (depth < 15 km) [140]; (c) partial melting of basaltic rocks [141–143]; (d) direct formation via mantle-derived mafic magma separation, crystallization, assimilation [144–146]; and (e) interaction between mantle-derived and crustal-derived melts [147]. Experimental petrology shows that melts derived from residual fluorine sources typically have higher MgO than TiO_2 content and exhibit strong peraluminous characteristics [148,149]. Similarly to other Early Cretaceous A-type granites reported in the CL, the Kongco granitic porphyry has elevated TiO_2/MgO ratios (0.59~1.33) and weakly peraluminous characteristics, excluding the possibility of partial melting of residual F- or Cl-rich lower crustal granulites (excluding possibilities "a" above). The Ce/Pb, Ce, Rb/Y and Nb/Y characteristics of Kongco granitic porphyry and other A-type granites reported in the CL show affinity with arc volcanic rocks (Figure 8c,d), and features in the Rb-Sr, Sm-Nd, and Pb isotopes indicate the presence of mantle-derived components (Table 5). Therefore, Kongco granitic porphyry cannot solely originate from pure crustal sources (including shallow crustal and lower crustal), excluding possibilities "b" and "c" above (Figure 8). The high silica rocks derived from mantle-derived mafic magma involve the separation and crystallization of mantle-derived mafic rocks on a large scale [150]. Although Kongco granitic porphyry exhibits a high silica content (SiO_2 = 76.22~77.90 wt.%), there are no reports of large-scale Late Cretaceous mafic magmatic activity in the Kongco region. This implies that the likelihood of Kongco granitic porphyry forming directly via magma separation, crystallization, and assimilation processes is minimal.

The geochemical data presented above strongly suggest that the Kongco granitic porphyry resulted from the interactions between mantle-derived melts and crustal melts. The main lines of evidence are as follows: (a) Kongco granitic porphyry is a weakly peraluminous alkaline feldspar granite. Similar coexistence from alkaline minerals to aluminous minerals also appears in other Early Cretaceous A-type granites in the CL (Figures 1b and 3c, [3,74,86,87,151,152]), as well as in A-type granites in southwestern China and northern Argentina [153]. Its formation may be related to crust–mantle interactions and sediment melting. (b) Kongco granitic porphyry and other Early Cretaceous A-type granites in the CL show a trend of crustal contamination in the Rb/Y-Nb/Y plot (Figure 7). The Lu, Yb, Hf, Rb and Sr isotopes of zircons in the Kongco granitic porphyry exhibit characteristics of crustal origin (Tables 4 and 5). The melting of crust enriched in high-field-strength element accessory minerals can produce magma with high concentrations of Zr, Ce, Y, and Ga characteristic of Kongco granitic porphyry [147]. Additionally, significant Nb depletion in the trace element spider diagram also indicates crustal components in this granite [154]). (c) In the t-ε_{Hf}(t) diagram (Figure 5), the Kongco granitic porphyry plots between the Chondritic Uniform Reservoir (CHUR) and the mantle evolution curve. It shares a similar source region with other Early Cretaceous granites in the CL, characterized by positive zircon ε_{Hf}(t) values (0.43~3.63), relatively young zircon Hf crustal model ages (T_{DM2} = 826~1005 Ma), relatively low whole-rock Rb/Y and Nb/Y ratios (Rb/Y = 4.77~8.65, Nb/Y = 0.42~0.57; Table 2), and higher whole-rock Ce/Pb values (0.81~2.15; Figure 7), indicating that the Kongco granitic porphyry originated from juvenile crustal material (Figure 7) with the involvement of mantle material. (d) The $(^{87}Sr/^{86}Sr)i$ and ε_{Nd}(t) isotopic characteristics (Table 5) suggest a mixture of mantle and lower crustal components for the Kongco granitic porphyry samples. In terms of Pb isotopes (Table 5), samples primarily fall within the field of mixing between Bangong–Nujiang ophiolites and crust–mantle hybrid lead zones.

It is generally believed that element depletion results from crystallization differentiation processes involving minerals that are enriched in those specific elements [124,155,156]. The Kongco granitic porphyry shows depletion in Ba, Nb, Sr, P, Eu, and Ti, indicating significant crystallization differentiation involving iron–magnesium minerals, titaniferous minerals, and phosphorus-rich minerals in its parental magma. Concurrently, the negative correlations observed between major oxides (FeO^T, Al_2O_3, CaO, MgO, P_2O_5 and TiO_2) and elements like Sr with SiO_2 (Figure 9) suggest crystallization differentiation of feldspar minerals, iron–magnesium minerals (such as amphibole and biotite), phosphorus-bearing minerals (such as apatite), and titanium-bearing minerals (such as ilmenite, titanite, and rutile). This indicates significant crystal fractionation during the formation of this granite. The lines of evidence discussed above suggest that the Kongco granitic porphyry is an Early Cretaceous A-type mass, most likely formed through interaction between mantle-derived mafic magmas and the remelting of juvenile lower crust, and that it underwent significant crystallization differentiation before intrusion.

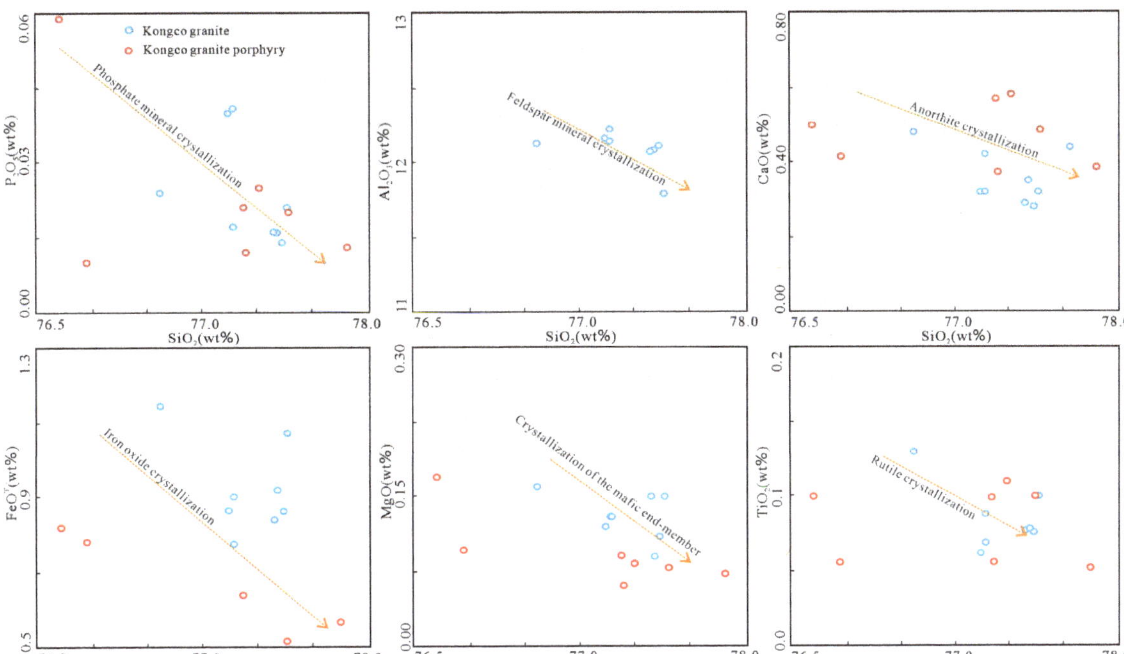

Figure 9. Harker diagrams indicating that significant crystal separation occurred during the formation of this granite. (Part of the data is sourced from Liu et al., 2022 [101]).

6.3. Structural Environment and Major Element Analysis

As analyzed earlier, the Kongco granitic porphyry, similar to the Late Early Cretaceous (~105 Ma) A-type granites reported in the CL, belongs to the A2-type granites associated with intracontinental collision. They are most likely formed through interactions between mantle-derived mafic magmas and the remelting of juvenile lower crust during post-collisional processes, where mantle material is thought to have been involved in petrogenesis through mechanisms such as slab tearing and lower crustal delamination [157]. There is controversy regarding the geodynamic context of the widespread Early Cretaceous magmatism in the northern and central parts of the Lhasa terrane [126,151,158–160]. (1) It has been attributed to the northward subduction of the Yarlung Tsangpo Neo-Tethyan oceanic crustal slab [17,161]; (2) it has been theorized to have formed due to southward subduction or slab detachment of the Bangong–Nujiang oceanic crustal slab [162–166]; (3) other have claimed that these magmatic activities are products of post-collisional processes between the Qiangtang terrane and the Lhasa terrane [167,168].

Zhang et al. (2004, 2012) [169,170] argue that the northward subduction of the Yarlung Tsangpo Neo-Tethys Ocean involved both low-angle subduction and increasing subduction angles, triggering magmatic eruptions in the central Tibetan during the Cretaceous period. In the low-angle subduction model, early subduction angles were small, resulting in relatively minor magmatic activity [169,170]. However, recent discoveries of extensive Jurassic–Cretaceous magmatism in the southern Lhasa terrane contradict the low-angle subduction model of oceanic crust slabs [72,122,123,126,171,172]. In contrast, Dai et al. (2015) [120] proposed that the northward subduction of the Yarlung Tsangpo Neo-Tethys was a normal deep subduction, and slab-tearing events triggered Early Cretaceous magmatic eruptions. Considering significant crustal shortening in the Lhasa terrane after the Cretaceous, during the late Early Cretaceous, the distance between the northern magmatic arc and the southern Yarlung Tsangpo Neo-Tethys subduction zone exceeded 600 km [44,173,174], which is difficult to explain with a normal subduction model. Therefore, we argue that the formation of the Early Cretaceous Kongco granitic porphyry was not related to the evolution of the Yarlung Tsangpo Neo-Tethys Ocean. Recent studies on the SNSZ indicate that it formed as a back-arc basin during the Bangong–Nujiang oceanic subduction [78,126,175]. Its small scale and rapid evolution suggest that the subduction processes in this basin were unrelated to the large-scale magmatic activity in the northern part of the Lhasa terrane [176].

The exact timing and subduction direction of the closure of the Bangong–Nujiang oceanic basin are still debated [13,19,23,51,52,54,67,177–182]. For example, Song et al. (2019) [183] proposed that during the Early Jurassic, the Bangong–Nujiang oceanic basin subducted northward beneath the Qiangtang block, while the collision between the Lhasa and Qiangtang blocks likely occurred around 140–130 Ma through arc–arc "soft" collision. Additionally, records from the Nima area in the northern part of the Lhasa block indicate a transition from marine to non-marine environments between 125 Ma and 118 Ma [174], suggesting a prior "soft" collision between the Lhasa and Qiangtang blocks. Chen et al. (2014) [86] also argued that the extensive magmatic activity around the Shenzha area in the central part of the Lhasa block at approximately 113 Ma resulted from southward subduction of the Bangong-Nujiang oceanic plate rupture. Moreover, evidence has been provided for the early Early Cretaceous (138~134 Ma) indicating that the Bangong–Nujiang ocean had not yet completely closed [184]. Furthermore, reports of extensive distribution of conglomerates from the K_2j Formation imply that since the late Early Cretaceous, the northern margin of the Lhasa block has entered a stage of intracontinental collision [43]. Despite significant discrepancies in these studies, it can be considered that during the late Early Cretaceous, the northern margin of the Lhasa block was in a collisional environment.

The extensive development of Mesozoic mildly acidic magmatism on both sides of the Bangong–Nujiang suture zone in Tibet exhibits a wide range of $\varepsilon_{Hf}(t)$ values during the Cretaceous (Table 4). From the early to the late Early Cretaceous (~105 Ma), $\varepsilon_{Hf}(t)$ abruptly increases from negative to positive values, indicating an increased mantle contribution to the magmas. Subsequently, from the late Early Cretaceous (~105 Ma) to the Late Cretaceous (~65 Ma), $\varepsilon_{Hf}(t)$ decreases to as much as negative values, suggesting a reduction in mantle contributions to the magmas. This implies an event during the late Early Cretaceous where mantle material was introduced into the magma source region, likely associated with post-collisional opening of slab windows. The presence of slab windows beneath the crust generates a suite of distinctive extension-related rock assemblages, including bimodal volcanic rocks, A-type granites, intraplate basalts, and adakites. In the northern part of the Lhasa microblock, collision-related magmatic events associated with extension have been reported during the late Early Cretaceous (~105 Ma), including basalts [79], adakite [185,186], bimodal volcanic rocks, and A-type granites [3,74,86,87,151]. These granitoids exhibit high Ce/Pb ratios, positive $\varepsilon_{Hf}(t)$ values, relatively young zircon Hf crustal model ages, and mantle and lower crustal mixed ($^{87}Sr/^{86}Sr$)i, $\varepsilon_{Nd}(t)$ and Pb isotope characteristics, providing evidence of sources from mantle-derived melts and juvenile crustal melts. This is consistent with contemporaneous extension-related magmatism in the region and further supports the existence of slab windows.

Therefore, the late Early Cretaceous magmatism in the northern part of the CL is unlikely to be a product of subduction of the Yarlung–Zangbo Neo-Tethyan oceanic crust slab or ridge subduction, nor is it directly related to the southward subduction and consumption of the Bangong–Nujiang oceanic slab. Instead, it is more likely the result of slab detachment during the collision process between the Bangong–Nujiang oceanic slab and the Qiangtang–Lhasa Block after closure of the subduction of the Bangong–Nujiang oceanic slab. During the late Early Cretaceous, under the collisional process of the Qiangtang-Lhasa Block, detachment of the subducted slab formed slab windows, causing mantle material to rise through the slab window and extension within the overlying lithosphere. Interaction between mantle-derived melts and juvenile crustal melts resulted in the formation of a series of A-type granites in the northern part of the Lhasa microblock within the Qiangtang-Lhasa Block (Figure 7).

6.4. Mineralization Analysis

As shown in Figure 2e, within and around the contact zones and mechanical weak planes of the Kongco granitic porphyry dykes, there are extensive occurrences of botryoidal textured malachite mineralization. Malachitization along microfracture surfaces is prominently developed and with varying grades of malachite patches within the Kongco granitic porphyry dykes. The close spatial relationship between this vein and mineralization suggests that the porphyry dykes had a genetic association with primary copper sulfide mineralization, which was later subjected to oxidation along the fractures in the Kongco granitic porphyry. What is more, within the mineralized alteration zone on both sides of the Kongco granite porphyry, our team collected five mineralized samples, and chemical analysis showed an average Cu content of 0.2% in the five samples (unpublished data). Combined with previous studies, these results indicate that this tectonic magmatic event during the Cretaceous was associated with varying grades of copper mineralization in the western section of the Gangdese belt. Therefore, the Cretaceous porphyry veins in the western section of the Gangdese belt exhibits significant mineralization potential, suggesting considerations for exploration efforts.

7. Conclusions

(1) The crystallization age of the Kongco granitic porphyry dykes is ~105 Ma, representing a late Early Cretaceous, weakly peraluminous, high-K, calc-alkaline, potassium basalt series of A2 alkaline feldspar granite.

(2) The Kongco granitic porphyry formed during the collision of the Qiangtang–Lhasa Block following the detachment of the subducted slab, inducing a partial melting of the crust due to asthenospheric upwelling. It subsequently underwent significant potassium feldspar- and biotite-dominated fractional crystallization.

(3) The close spatial relationship between this vein and mineralization suggests that the porphyry dykes had a genetic association with primary copper sulfide mineralization. This suggests that the porphyritic bodies exhibit significant mineralization potential and promising prospects for exploration.

Author Contributions: Conceptualization, H.L.; Software, K.L.; Writing—original draft, A.X.; Writing—review & editing, W.F.; Supervision, Q.Z.; Project administration, H.W. All authors have read and agreed to the published version of the manuscript.

Funding: Supporting projects include the National Key Research and Development Program of China (2024YFC2910102; 2023YFC2906805; 2021YFC2901803; 2023YFC2908400; 2021YFC2901903), the National Natural Science Foundation of China (92055314; 42272106; 42202105), the China Geological Survey Project (DD20230247; DD20221776), and the Yunnan Science and Technology Award–Outstanding Contribution Award Project (No. 2017001).

Data Availability Statement: The data that support the findings of this study are available.

Acknowledgments: The research work was supported by assistant Cao Huawen of Chengdu University of Technology, Wang Yiyun, assistant Huang Yong, and Zhang Tengjiao of Chengdu Geological Survey Center of China Geological Survey, and Zhang Jing, assistant Zeng Yunchuan, Wu Junyi, and Ji Xuan of China University of Geosciences (Beijing).

Conflicts of Interest: The authors declare no conflict of interest.

References

1. Li, W.C.; Wang, J.H.; He, Z.H.; Dou, S. Formation of Au-polymetallic ore deposits in alkaline porphyries at Beiya, Yunnan, Southwest China. *Ore Geol. Rev.* **2016**, *73*, 241–252. [CrossRef]
2. Li, W.C.; Yu, H.J.; Gao, X.; Liu, X.L.; Wang, J.H. Review of Mesozoic multiple magmatism and porphyry Cu–Mo (W) mineralization in the Yidun Arc, eastern Tibet Plateau. *Ore Geol. Rev.* **2017**, *90*, 795–812. [CrossRef]
3. Li, W.C.; Jiang, X.J. The Cenozoic tectono-magmatism-mineralization effect of the intracontinental tectonic transformation system in the western margin of Yangtze Block. *Earth Sci. Front.* **2020**, *27*, 151–164. (In Chinese with English Abstract)
4. Li, W.C.; Pan, G.T.; Zhang, X.F.; Wang, L.Q.; Zhou, J.X. Tectonic evolution and multi–episodic metallogenesis of the Sanjiang Paleo-Tethys multi-arc-basin-terrane system, SW Tibetan Plateau. *J. Asian Earth Sci.* **2021**, *221*, 104932. [CrossRef]
5. Mo, X.X. Growth and evolution of crust of Tibetan Plateau from perspective of magmatic rocks. *Earth Sci.* **2020**, *45*, 2245–2257. (In Chinese with English Abstract)
6. Pan, G.T.; Wang, L.Q.; Li, X.Z.; Wang, J.M.; Xu, Q. The tectonic framework and spatial allocation of the archipelagic arc-basin systems on the Qinghai-Xizang Plateau. *Sediment. Geol. Tethyan Geol.* **2001**, *21*, 1–26. (In Chinese with English Abstract)
7. Pan, G.T.; Zhu, D.C.; Wang, L.Q.; Liao, Z.L.; Geng, Q.R.; Jiang, X.S. Bangong Lake-Nu River suture zone-the northern boundary of Gondwanaland: Evidence from geology and geophysics. *Earth Sci. Front.* **2004**, *11*, 371–382. (In Chinese with English Abstract)
8. Pan, G.T.; Wang, L.Q.; Geng, Q.R.; Yin, F.G.; Wang, B.D.; Peng, Z.M.; Ren, F. Space-time structure of the Bangonghu-Shuanghu-Nujiang-Changning-Menglian Mega-suture zone: A discussion on geology and evolution of the Tethys Ocean. *Sediment. Geol. Tethyan Geol.* **2020**, *40*, 1–19. (In Chinese with English Abstract)
9. Zheng, Y.Y.; Ci, Q.; Gao, S.B.; Wu, S.; Jiang, X.J.; Chen, X. The Ag-Sn-Cu polymetallic minerogenetic series and prospecting direction in the western Gangdese belt, Tibet. *Earth Sci. Front.* **2021**, *28*, 379–402. (In Chinese with English Abstract)
10. Cai, Q.L.; Zheng, Z.W.; He, J. Age of zircon U-Pb and its geological significance of granite from western Gandese in Tibet. *J. East China Inst. Technol. Nat. Sci. Ed.* **2015**, *38*, 49–57. (In Chinese with English Abstract)
11. Cao, H.-W.; Li, G.-M.; Zhang, Z.; Zhang, L.-K.; Dong, S.-L.; Xia, X.-B.; Liang, W.; Fu, J.-G.; Huang, Y.; Xiang, A.-P.; et al. Miocene Sn polymetallic mineralization in the Tethyan Himalaya, southeastern Tibet: A case study of the Cuonadong deposit. *Ore Geol. Rev.* **2020**, *119*, 103403. [CrossRef]
12. Cao, H.W.; Li, G.M.; Zhang, R.Q.; Zhang, Y.H.; Zhang, L.K.; Dai, Z.W.; Zhang, Z.; Liang, W.; Dong, S.L.; Xia, X.B. Genesis of the Cuonadong tin polymetallic deposit in the Tethyan Himalaya: Evidence from geology, geochronology, fluid inclusions and multiple isotopes. *Gondwana Res.* **2021**, *92*, 72–101. [CrossRef]
13. Geng, Q.R.; Mao, X.C.; Zhang, Z.; Peng, Z.M.; Guan, J.L. New understanding in the middle and west part of Bangong Lake-Nujiang River metallogenic belt and its lmplication for prospecting. *Geol. Surv. China* **2015**, *2*, 1–11. (In Chinese with English Abstract)
14. Grimes, C.B.; John, B.E.; Kelemen, P.B.; Mazdab, F.K.; Wooden, J.L.; Cheadle, M.J.; Hanghøj, K.; Schwartz, J.J. Trace element chemistry of zircons from oceanic crust: A method for distinguishing detrital zircon provenance. *Geology* **2007**, *35*, 643–646. [CrossRef]
15. Hou, Z.Q.; Song, Y.C.; Li, Z.; Wang, Z.L.; Yang, Z.M.; Yang, Z.S.; Liu, Y.C.; Tian, S.H.; He, L.Q.; Chen, K.X.; et al. Thrust-controlled, sediments-hosted Pb-Zn-Ag-Cu deposits in eastern and northern margins of Tibetan orogenic belt: Geological features and tectonic model. *Miner. Depos.* **2008**, *27*, 123–144. (In Chinese with English Abstract)

16. Hou, Z.Q.; Yang, Z.M.; Lu, Y.J.; Kemp, A.; Zheng, Y.C.; Li, Q.Y.; Tang, J.X.; Yang, Z.S.; Duan, L.F. A genetic linkage between subduction- and collision-related porphyry Cu deposits in continental collision zones. *Geology* **2015**, *43*, 247–250. [CrossRef]
17. Huang, Y.; Li, G.M.; Ding, J.; Dai, J.; Yan, G.Q.; Dong, S.L.; Huang, H.X. Origin of the newly discovered Zhunuo porphyry Cu-Mo-Au deposit in the western part of the Gangdese porphyry copper belt in the southern Tibetan Plateau, SW China. *Acta Geol. Sin. Engl. Ed.* **2017**, *91*, 109–134. [CrossRef]
18. Huang, Y.; Cao, H.W.; Li, G.M.; Brueckner, S.M.; Zhang, Z.; Dong, L.; Dai, Z.W.; Lu, L.; Li, Y.B. Middle—Late Triassic bimodal intrusive rocks from the Tethyan Himalaya in South Tibet: Geochronology, petrogenesis and tectonic implications. *Lithos* **2018**, *318–319*, 78–90. [CrossRef]
19. Huang, Y.; Ren, M.H.; Jowitt, S.M.; Li, G.M.; Fu, J.G.; Zhang, Z.; Lang, X.H.; Liu, H.; Zhang, L. Middle Triassic arc magmatism in the southern Lhasa terrane: Geochronology, petrogenesis and tectonic setting. *Lithos* **2021**, *380–381*, 105857. [CrossRef]
20. Li, G.M.; Duan, Z.M.; Liu, B.; Zhang, H.; Dong, S.L.; Zhang, L. The discovery of Jurassic accretionary complexes in Duolong area, northern Bangong Co-Nujiang suture zone, Tibet, and its geologic significance. *Geol. Bull. China* **2011**, *30*, 1256–1260. (In Chinese with English Abstract)
21. Li, G.M.; Qin, K.Z.; Li, J.X.; Evans, N.J.; Zhao, J.X.; Cao, M.J.; Zhang, X.N. Cretaceous magmatism and metallogeny in the Bangong-Nujiang metallogenic belt, central Tibet: Evidence from petrogeochemistry, zircon U-Pb ages, and Hf-O isotopic compositions. *Gondwana Res.* **2017**, *41*, 110–127. [CrossRef]
22. Li, G.M.; Zhang, L.K.; Xia, X.B.; Liang, W.; Hou, C.Q. New exploration progresses, resource potentials and prospecting targets of strategic minerals in the southern Qinghai-Tibet Plateau. *Sediment. Geol. Tethyan Geol.* **2021**, *41*, 351–360. (In Chinese with English Abstract)
23. Li, W.C.; Zeng, P.S.; Hou, Z.Q.; White, N.C. The Pulang porphyry copper deposit and associated felsic intrusions in Yunnan province, southwest China. *Econ. Geol.* **2011**, *106*, 79–92.
24. Liu, H.; Huang, H.X.; Li, G.M.; Xiao, W.F.; Zhang, Z.L.; Liu, B.; Ma, D.F.; Dong, L.; Ma, D. Factor analysis in geochemical survey of the Shangxu gold deposit, northern Tibet. *Geol. China* **2015**, *42*, 1126–1136. (In Chinese with English Abstract)
25. Liu, H.; Li, G.M.; Huang, H.X.; Zhang, Z.L.; Xiao, W.F.; Jiao, Y.J.; Liang, S.X.; Guo, J.; Lan, S.S. Prospecting potential analysis of deep No. III ore section in the shangxu orogenic Gold deposit, bangong Co-Nujiang metallogenic belt, Tibet. *Acta Geol. Sin.* **2017**, *91*, 1245–1258. (In Chinese with English Abstract)
26. Liu, H.; Li, G.M.; Huang, H.X.; Cao, H.W.; Yuan, Q.; Li, Y.X.; Ouyang, Y.; Lan, S.S.; Lü, M.H.; Yan, G.Q. Petrogenesis of Late Cretaceous Jiangla'angzong I-Type granite in Central Lhasa Terrane, Tibet, China: Constraints from whole-rock geochemistry, zircon U-Pb geochronology, and Sr-Nd-Pb-Hf isotopes. *Acta Geol. Sin. Engl. Ed.* **2018**, *92*, 1396–1414. [CrossRef]
27. Song, Y.; Qu, X.M.; Xin, H.B.; Wang, Z.Y.; Du, D.D. The crustal carbon cycling in the Late Cretaceous extension of Bangong Lake arc zone in the Xizang (Tibetan) Plateau. *Geol. Rev.* **2013**, *59*, 225–233. (In Chinese with English Abstract)
28. Tang, J.X. Mineral resources base investigation and research status of the Tibet Plateau and its adjacent major metallogenic belts. *Acta Petrol. Sin.* **2019**, *35*, 617–624. (In Chinese with English Abstract)
29. Wang, X.; Lang, X.; Klemd, R.; Deng, Y.; Tang, J. Subduction initiation of the Neo-Tethys oceanic lithosphere by collision-induced subduction transference. *Gondwana Res.* **2022**, *104*, 54–69. [CrossRef]
30. Yang, Z.M.; Hou, Z.Q.; Xia, D.X.; Song, Y.C.; Li, Z. Relationship between Western Porphyry and mineralization in Qulong copper deposit of Tibet and its enlightenment to further exploration. *Miner. Depos.* **2008**, *27*, 28–36. (In Chinese with English Abstract)
31. Yang, Z.M.; Hou, Z.Q.; Jiang, Y.F.; Zhang, H.R.; Song, Y.C. Sr-Nd-Pb and zircon Hf isotopic constraints on petrogenesis of the Late Jurassic granitic porphyry at Qulong, Tibet. *Acta Petrol. Sin.* **2011**, *27*, 2003–2010. (In Chinese with English Abstract)
32. Zhang, K.X.; He, W.H.; Xu, Y.D.; Zhang, X.H.; Song, B.W.; Kou, X.H.; Wang, G.C. Reconstruction of main types for oceanic plate strata in the subduction-accretionary complex and feature of sequence for each type: An example from the Qinghai-Tibet Tethyan Permian strata. *Sediment. Geol. Tethyan Geol.* **2021**, *41*, 137–151. (In Chinese with English Abstract)
33. Zhao, L.X.; Gong, C.; He, J.; Peng, J.H.; Huang, S.C.; Yang, Z.L. Discovery of porphyry copper deposit and its significance of Dajiacuo in Cuoqin County, Tibet. *J. East China Inst. Technol. Nat. Sci. Ed.* **2013**, *36*, 13–20. (In Chinese with English Abstract)
34. Ding, L.; Lai, Q.Z. New geological evidence of crustal thickening in the Gangdese block prior to the Indo-Asian collision. *Chin. Sci. Bull.* **2003**, *48*, 1604–1610. [CrossRef]
35. Guynn, J.; Tropper, P.; Kapp, P.; Gehrels, G.E. Metamorphism of the amdo metamorphic complex, Tibet: Implications for the Jurassic tectonic evolution of the bangong suture zone. *J. Metamorph. Geol.* **2013**, *31*, 705–727. [CrossRef]
36. Huang, Y.; Ren, M.H.; Liang, W.; Li, G.M.; Heilbronn, K.; Dai, Z.W.; Wang, Y.Y.; Zhang, L. Origin of the Oligocene Tuolangla porphyry—Skarn Cu-W-Mo deposit in Lhasa terrane, southern Tibet. *China Geol.* **2020**, *3*, 369–384.
37. Liu, H.; Zhang, L.K.; Huang, H.X.; Li, G.M.; Ouyang, Y.; Yu, H.; Liang, W.; Zhang, H.M.; Chen, X.P. Evolution of ore-forming fluids in the Luobuzhen epithermal gold-silver deposit in western Gangdisi: Fluid inclusion and H-O isotope evidence. *Earth Sci. Front.* **2020**, *27*, 49–65. (In Chinese with English Abstract)
38. Pan, G.T.; Xiao, Q.H.; Lu, S.N.; Deng, J.F.; Feng, Y.M.; Zhang, K.X.; Zhang, Z.Y.; Wang, W.G.; Xing, G.F.; Hao, G.J.; et al. Subdivision of tectonic units in China. *Geol. China* **2009**, *36*, 1–28. (In Chinese with English Abstract)

39. Wu, H.; Li, C.; Hu, P.Y.; Zhang, H.Y.; Li, J. The discovery of Early Cretaceous bimodal volcanic rocks in the Dachagou area of Tibet and its significance. *Geol. Bull. China* **2014**, *33*, 1804–1814. (In Chinese with English Abstract)
40. Xu, Z.Q.; Yang, J.S.; Li, W.C.; Li, H.Q.; Cai, Z.H.; Yan, Z.; Ma, C.Q. Paleo-tethys system and accretionary orogen in the Tibet Plateau. *Acta Petrol. Sin.* **2013**, *29*, 1847–1860. (In Chinese with English Abstract)
41. Yang, J.S.; Xu, Z.Q.; Li, T.F.; Li, H.Q.; Li, Z.L.; Ren, Y.F.; Xu, X.Z.; Chen, S.Y. Oceanic subduction-type eclogite in the Lhasa block, Tibet, China: Remains of the Paleo-Tethys ocean basin? *Geol. Bull. China* **2007**, *26*, 1277–1287. (In Chinese with English Abstract)
42. Pan, G.T.; Wang, L.Q.; Li, R.S.; Yuan, S.H.; Ji, W.H.; Yin, F.G.; Zhang, W.P.; Wang, B.D. Tectonic evolution of the Qinghai-Tibet Plateau. *J. Asian Earth Sci.* **2012**, *53*, 3–14. [CrossRef]
43. Pan, G.T.; Mo, X.X.; Hou, Z.Q.; Zhu, D.C.; Wang, L.Q.; Li, G.M.; Zhao, Z.D.; Geng, Q.R.; Liao, Z.L. Spatial-temporal framework of the Gangdese Orogenic Belt and its evolution. *Acta Petrol. Sin.* **2006**, *22*, 521–533. (In Chinese with English Abstract)
44. Murphy, M.A.; Yin, A.; Harrison, T.M.; Dürr, S.B.; Chen, Z.; Ryerson, F.J.; Kidd, W.S.F.; Wang, X.; Zhou, X. Did the Indo-Asian collision alone create the Tibetan Plateau? *Geology* **1997**, *25*, 719–722. [CrossRef]
45. Tapponnier, P.; Xu, Z.Q.; Roger, F.; Meyer, B.; Arnaud, N.; Wittlinger, G.; Yang, J.S. Oblique stepwise rise and growth of the Tibet Plateau. *Science* **2001**, *294*, 1671–1677. [CrossRef]
46. Liu, H.; Zhang, H.; Li, G.M.; Huang, H.X.; Xiao, W.F.; You, Q.; Ma, D.F.; Zhang, H.; Zhang, H. Petrogenesis of the early cretaceous Qingcaoshan strongly peraluminous S-type granitic pluton, southern Qiangtang, northern Tibet: Constraints from whole-rock geochemistry and zircon U-Pb geochronology. *Acta Sci. Nat. Univ. Pekin.* **2016**, *52*, 848–860. (In Chinese with English Abstract)
47. Liu, H.; Zhang, L.K.; Huang, H.X.; Li, G.M.; Ouyang, Y.; Lü, M.H.; Liu, H.; Lan, S.S.; Yan, G.Q. Petrogenesis of Late Triassic Luerma Monzodiorite in western Gangdise, Tibet, China. *Earth Sci.* **2019**, *44*, 2339–2352. (In Chinese with English Abstract)
48. Shi, S.F.; Xiao, Y.F.; Yuan, H.W.; Wei, Y.F.; Qiu, C.Y.; Jiang, X.Q. Geochronology and geochemical characteristics of A-type granites in Boguorize area, the northern margin of Gangdese Belt, Tibet, China and their geological significance. *J. Earth Sci. Environ.* **2019**, *41*, 644–657. (In Chinese with English Abstract)
49. Xu, Z.Q.; Dilek, Y.; Cao, H.; Yang, J.S.; Robinson, P.; Ma, C.Q.; Li, H.Q.; Jolivet, M.; Roger, F.; Chen, X.J. Paleo-Tethyan evolution of Tibet as recorded in the East Cimmerides and West Cathaysides. *J. Asian Earth Sci.* **2015**, *105*, 320–337. [CrossRef]
50. Zhu, D.C.; Zhao, Z.D.; Niu, Y.L.; Dilek, Y.; Hou, Z.Q.; Mo, X.X. The origin and pre-Cenozoic evolution of the Tibetan Plateau. *Gondwana Res.* **2013**, *23*, 1429–1454. [CrossRef]
51. Zhu, D.C.; Li, S.M.; Cawood, P.A.; Wang, Q.; Zhao, Z.D.; Liu, S.A.; Wang, L.Q. Assembly of the Lhasa and Qiangtang terranes in central Tibet by divergent double subduction. *Lithos* **2016**, *245*, 7–17. [CrossRef]
52. Liu, H.; Huang, H.X.; Zhang, L.K.; Li, G.M.; Ouyang, Y.; Huang, Y.; Lu, M.H. Luerma, a newly discovered Late Triassic porphyry copper-gold ore spot in the western Gangdise metallogenic belt, Tibet. *Sediment. Geol. Tethyan Geol.* **2021**, *40*, 3502–3515. (In Chinese with English Abstract)
53. Cao, H.W.; Zhang, Y.H.; Santosh, M.; Li, G.M.; Hollis, S.P.; Zhang, L.K.; Pei, Q.M.; Tang, L.; Duan, Z.M. Petrogenesis and metallogenic implications of Cretaceous magmatism in Central Lhasa, Tibetan plateau: A case study from the Lunggar Fe skarn deposit and perspective review. *Geol. J.* **2019**, *54*, 2323–2346. [CrossRef]
54. Gao, S.B.; Zheng, Y.Y.; Tian, K.; Chen, X.; Jiang, X.J.; Gu, Y.R. Geochronology of magmatic intrusions and mineralization of lunggar iron deposit in Tibet and its implications for regional multi-stage iron mineralization: Geochemistry, zircon U-Pb and phlogopite Ar-Ar isotopic dating constraints. *Earth Sci.* **2021**, *46*, 1941–1959. (In Chinese with English Abstract)
55. Guo, Z.F.; Wilson, M.; Zhang, M.L.; Cheng, Z.H.; Zhang, L.H. Post-collisional ultrapotassic mafic magmatism in South Tibet: Products of partial melting of pyroxenite in the mantle wedge induced by roll-back and delamination of the subducted Indian continental lithosphere slab. *J. Petrol.* **2015**, *56*, 1365–1406. [CrossRef]
56. Huang, F.; Xu, J.; Zeng, Y.; Chen, J.; Wang, B.; Yu, H.; Chen, L.; Huang, W.; Tan, R. Slab breakoff of the Neo-Tethys Ocean in the Lhasa Terrane inferred from contemporaneous melting of the mantle and crust. *Geochem. Geophys. Geosystems* **2017**, *18*, 4074–4095. [CrossRef]
57. Huang, H.X.; Li, G.M.; Dong, S.L.; Liu, B.; Zhang, H.; Zhang, L.; Li, Y.X.; Shi, H.Z. SHRIMP zircon U-Pb age and geochemical characteristics of Qinglung granodiorite in Baingoin area, Tibet. *Geol. Bull. China* **2012**, *31*, 852–859. (In Chinese with English Abstract)
58. Peng, J.H.; Zhao, L.X.; He, J.; Huang, S.C.; Gong, C. Discovery of Indosinian magmatic rocks and its significance in western Gangdise, Tibet. *J. East China Inst. Technol. Nat. Sci. Ed.* **2013**, *36*, 21–26. (In Chinese with English Abstract)
59. Peng, Z.M.; Geng, Q.R.; Liu, S.S.; Zhang, Z.; Guan, J.L.; Cong, F. Zircon U-Pb ages and geochemical features of the diorite por-phyrite in Gajia Formation in the Nagqu area, Tibet. *Geol. Bull. China* **2015**, *34*, 354–363. (In Chinese with English Abstract)
60. Qu, X.M.; Hou, Z.Q.; Guo, L.J.; Xu, W.Y. Source compositions and crustal contaminations of adakitic ore-bearing porphyries in the Gangdise copper belt: Nd, Sr, Pb and O isotope constraints. *Acta Geol. Sin.* **2004**, *78*, 813–821. (In Chinese with English Abstract)
61. Wang, Q.; Zhu, D.C.; Zhao, Z.D.; Liu, S.A.; Chung, S.L.; Li, S.M.; Liu, D.; Dai, J.G.; Wang, L.Q.; Mo, X.X. Origin of the ca. 90 Ma magnesia-rich volcanic rocks in SE Nyima, central Tibet: Products of lithospheric delamination beneath the Lhasa-Qiangtang collision zone. *Lithos* **2014**, *198–199*, 24–37. [CrossRef]

62. Zhang, Z.; Song, J.L.; Tang, J.X.; Wang, L.Q.; Yao, X.F.; Li, Z.J. Petrogenesis, diagenesis and mineralization ages of Galale Cu-Au deposit, Tibet: Zircon U-Pb age, Hf isotopic composition and molybdenite Re-Os dating. *Earth Sci.* **2017**, *42*, 862–880. (In Chinese with English Abstract)
63. Zheng, Y.Y.; Sun, X.; Gao, S.B.; Wu, S.; Xu, J.; Jiang, J.S.; Chen, X.; Zhao, Z.Y.; Liu, Y. Metallogenesis and the minerogenetic series in the Gangdese polymetallic copper belt. *J. Asian Earth Sci.* **2015**, *103*, 23–39. [CrossRef]
64. Sun, G.; Sinclair, H.D.; Persano, C.; Stuart, F.M.; Hu, X. Late Cretaceous-Eocene exhumation of the northern Lhasa terrane and topographic implications for the Central Tibet. *Lithos* **2024**, *470*, 107528. [CrossRef]
65. Coulon, C.; Maluski, H.; Bollinger, C.; Wang, S. Mesozoic and Cenozoic volcanic rocks from central and southern Tibet: ^{39}Ar-^{40}Ar dating, petrological characteristics and geodynamical significance. *Earth Planet. Sci. Lett.* **1986**, *79*, 281–302. [CrossRef]
66. Li, H.; Wang, M.; Zeng, X.W.; Luo, A.B.; Yu, Y.P.; Zeng, X.J. Slab break-off origin of 105 Ma A-type porphyritic granites in the Asa area of Tibet. *Geol. Mag.* **2020**, *157*, 1281–1298. [CrossRef]
67. Li, J.X.; Qin, K.Z.; Li, G.M.; Richards, J.P.; Zhao, J.X.; Cao, M.J. Geochronology, geochemistry, and zircon Hf isotopic compositions of Mesozoic intermediate-felsic intrusions in central Tibet: Petrogenetic and tectonic implications. *Lithos* **2014**, *198–199*, 77–91. [CrossRef]
68. Matte, P. Tectonics of Western Tibet, between the Tarim and the Indus. *Earth Planet. Sci. Lett.* **1996**, *142*, 311–330. [CrossRef]
69. Qin, Z.; Yang, Z.J.; Tang, L.; Wang, M.Z.; She, P.T.; Han, K.; Zhang, D. Geochemical characteristics, zircon U-Pb age and metallogenic significance of Chalong granites in the Gangdise Belt of Tibet. *Geol. Bull. China* **2019**, *38*, 231–241. (In Chinese with English Abstract)
70. Wang, B.D.; Xu, J.F.; Liu, B.M.; Chen, J.L.; Wang, L.Q.; Guo, L.; Wang, D.B.; Zhang, W.P. Geochronology and ore-forming geological background of ~90 Ma porphyry copper deposit in the Lhasa terrane, Tibet Plateau. *Acta Geol. Sin.* **2013**, *87*, 71–80. (In Chinese with English Abstract)
71. Zhao, H.; Zhang, Y.; Xu, Y.; Shao, Y.; Chen, X.; Hao, J.; Zhao, L.; Shen, H.; Wang, X. Machine learning model for deep exploration: Utilizing short wavelength infrared (SWIR) of hydrothermal alteration minerals in the Qianchen gold deposit, Jiaodong Peninsula, Eastern China. *Ore Geol. Rev.* **2024**, *168*, 106060. [CrossRef]
72. Zhu, D.C.; Mo, X.X.; Niu, Y.L.; Zhao, Z.D.; Wang, L.Q.; Liu, Y.S.; Wu, F.Y. Geochemical investigation of Early Cretaceous igneous rocks along an east-west traverse throughout the central Lhasa Terrane, Tibet. *Chem. Geol.* **2009**, *268*, 298–312. [CrossRef]
73. Liu, H.; Li, G.M.; Huang, H.X.; Xiao, W.F.; Yan, G.Q.; Ma, D.F.; Zhang, H.; Zhang, H. Sources of Ore-forming materials in the Shangxu orogenic gold deposit, Northern Xizang (Tibet): Constraints form C, S, and Pb Isotopes. *Geol. Rev.* **2018**, *64*, 1285–1301. (In Chinese with English Abstract)
74. Wang, L.Q.; Wang, B.D.; Li, G.M.; Wang, D.B.; Peng, Z.M. Major progresses of geological survey and research in East Tethys: An overview. *Sediment. Geol. Tethyan Geol.* **2021**, *41*, 283–296. (In Chinese with English Abstract)
75. Wang, X.X.; Yan, G.Q.; Liu, H.; Huang, H.X.; Lai, Y.; Tian, E.Y.; Ouyang, Y. Genesis of Late Cretaceous Qusang'gele granitie in Central Lhasa Block, Tibet: Constraints by geochemistry, zircon U-Pb geochronology, and Sr-Nd-Pb-Hf isotopes. *Earth Sci.* **2021**, *46*, 2832–2849. (In Chinese with English Abstract)
76. Luo, A.B.; Wang, M.; Li, C.; Xie, C.M.; Fan, J.J.; Zhang, T.Y.; Liu, J.H.; Wang, W. Petrogenesis of early Late Cretaceous Asa-intrusive rocks in central Tibet, western China: Post-collisional partial melting of thickened lower crust. *Int. J. Earth Sci.* **2019**, *108*, 1979–1999. [CrossRef]
77. Ma, Y.; Yang, T.; Yang, Z.; Zhang, S.; Wu, H.; Li, H.; Chen, W.; Zhang, J.; Ding, J. Paleomagnetism and U-Pb zircon geochronology of Lower Cretaceous lava flows from the western Lhasa terrane: New constraints on the India-Asia collision process and intracontinental deformation within Asia. *J. Geophys. Res. Solid Earth* **2014**, *119*, 7404–7424. [CrossRef]
78. Zeng, X.W.; Wang, M.; Fan, J.J.; Li, C.; Xie, C.M.; Liu, Y.M.; Zhang, T.Y. Geochemistry and geochronology of gabbros from the Asa Ophiolite, Tibet: Implications for the Early Cretaceous evolution of the Meso-Tethys Ocean. *Lithos* **2018**, *320–321*, 192–206. [CrossRef]
79. Zeng, Y.C.; Xu, J.F.; Huang, F.; Li, M.J.; Chen, Q. Generation of the 105–100 Ma Dagze volcanic rocks in the north Lhasa Terrane by lower crustal melting at different temperature and depth: Implications for tectonic transition. *GSA Bull.* **2020**, *132*, 1257–1272. [CrossRef]
80. Wiedenbeck, M.; Alle, P.; Corfu, F.Y.; Griffin, W.L.; Meier, M.; Oberli, F.; Von Quadt, A.; Roddick, J.C.; Spiegel, W. Three natural zircon standards for U-Th-Pb, Lu-Hf, trace element and REE analyses. *Geostand. Newsl.* **1995**, *19*, 1–23. [CrossRef]
81. Zong, K.Q.; Klemd, R.; Yuan, Y.; He, Z.Y.; Guo, J.L.; Shi, X.L.; Liu, Y.S.; Hu, Z.C.; Zhang, Z.M. The assembly of Rodinia: The correlation of early Neoproterozoic (ca. 900 Ma) high-grade metamorphism and continental arc formation in the southern Beishan Orogen, southern Central Asian Orogenic Belt (CAOB). *Precambrian Res.* **2017**, *290*, 32–48. [CrossRef]
82. Liu, Y.S.; Hu, Z.C.; Gao, S.; Günther, D.; Xu, J.; Gao, C.G.; Chen, H.L. In situ analysis of major and trace elements of anhydrous minerals by LA-ICP-MS without applying an internal standard. *Chem. Geol.* **2008**, *257*, 34–43. [CrossRef]

83. Liu, Y.S.; Gao, S.; Hu, Z.C.; Gao, C.G.; Zong, K.Q.; Wang, D.B. Continental and oceanic crust recycling-induced melt-peridotite interactions in the Trans-North China Orogen: U-Pb dating, Hf isotopes and trace elements in zircons from mantle xenoliths. *J. Petrol.* **2010**, *51*, 537–571. [CrossRef]
84. Meng, F.Y.; Zhao, Z.D.; Zhu, D.C.; Mo, X.X.; Guan, Q.; Huang, Y.; Dong, G.C.; Zhou, S.; DePaolo, D.J.; Harrison, T.M.; et al. Late Cretaceous magmatism in Mamba area, central Lhasa subterrane: Products of back-arc extension of Neo-Tethyan Ocean? *Gondwana Res.* **2014**, *26*, 505–520. [CrossRef]
85. Jackson, S.E.; Pearson, N.J.; Griffin, W.L.; Belousova, E.A. The application of laser ablation-inductively coupled plasma-mass spectrometry to in situ U–Pb zircon geochronology. *Chem. Geol.* **2004**, *211*, 47–69. [CrossRef]
86. Chen, Y.; Zhu, D.C.; Zhao, Z.D.; Meng, F.Y.; Wang, Q.; Santosh, M.; Wang, L.Q.; Dong, G.C.; Mo, X.X. Slab breakoff triggered ca. 113 Ma magmatism around Xainza area of the Lhasa Terrane, Tibet. *Gondwana Res.* **2014**, *26*, 449–463. [CrossRef]
87. Qu, X.M.; Wang, R.J.; Xin, H.B.; Jiang, J.H.; Chen, H. Age and petrogenesis of A-type granites in the middle segment of the Bangonghu-Nujiang suture, Tibetan plateau. *Lithos* **2012**, *146–147*, 264–275. [CrossRef]
88. Xie, L.; Dun, D.; Zhu, L.D.; Ni, M.C.R.; Yang, W.G.; Yao, G.; Li, C.; He, B.; He, Y. Zircon U-Pb geochronology, geochemistry and geological significance of the Zhaduding A-type granites in northern Gangdise, Tibet. *Geol. China* **2015**, *42*, 1214–1227. (In Chinese with English Abstract)
89. Hanchar, J.M.; Watson, E.B. Zircon saturation thermometry. *Rev. Mineral. Geochem.* **2003**, *53*, 89–112. [CrossRef]
90. Miller, C.F.; McDowell, S.M.; Mapes, R.W. Hot and cold granites? Implications of zircon saturation temperatures and preservation of inheritance. *Geology* **2003**, *31*, 529–532. [CrossRef]
91. Teixeira, L.F.; Laurent, O.; Troch, J.; Siddoway, C.S.; Tavazzani, L.; Deering, C.; Bachmann, O. Tracking quartz and zircon provenance in sedimentary rocks using Ti distributions: Unlocking the volcanic-plutonic connection in old igneous systems. *Earth Planet. Sci. Lett.* **2024**, *643*, 118906. [CrossRef]
92. Watson, E.B.; Harrison, T.M. Zircon saturation revisited: Temperature and composition effects in a variety of crustal magma types. *Earth Planet. Sci. Lett.* **1983**, *64*, 295–304. [CrossRef]
93. Watson, E.B.; Harrison, T.M. Zircon thermometer reveals minimum melting conditions on earliest Earth. *Science* **2005**, *308*, 841–844. [CrossRef]
94. Miller, C.; Schuster, R.; Klötzli, U.; Frank, W.; Purtscheller, F. Post-collisional potassic and ultrapotassic magmatism in SW Tibet: Geochemical and Sr-Nd-Pb-O isotopic constraints for mantle source characteristics and petrogenesis. *J. Petrol.* **1999**, *40*, 1399–1424. [CrossRef]
95. Middlemost, E.A.K. *Magmas and Magmatic Rocks*; Longman: London, UK, 1985; pp. 1–266.
96. Peccerillo, R.; Taylor, S.R. Geochemistry of eocene calc-alkaline volcanic rocks from the Kastamonu area, Northern Turkey. *Contrib. Mineral. Petrol.* **1976**, *58*, 63–81. [CrossRef]
97. Maniar, P.D.; Piccoli, P.M. Tectonic discrimination of granitoids. *Geol. Soc. Am. Bull.* **1989**, *101*, 635–643. [CrossRef]
98. Boynton, W.V. Cosmochemistry of the rare earth elements: Meteorite studies. In *Developments in Geochemistry*; Henderson, P., Ed.; Elsevier: Amsterdam, The Netherlands, 1984; pp. 63–114.
99. Sun, S.S.; MchDonough, W.F. Chemical and isotopic systematics of oceanic basalts: Implications for mantle composition and processes. In *Magmatism in the Ocean Basins*; Saunders, A.D., Norry, M.J., Eds.; Geological Society: London, UK, 1989; Volume 42, pp. 303–345.
100. Dai, Z.W.; Huang, H.X.; Li, G.M.; Huizenga, J.M.; Santosh, M.; Cao, H.W.; Huang, C.M.; Ding, J. Formation of Late Cretaceous High-Mg granitoid porphyry in central Lhasa, Tibet: Implications for crustal thickening prior to India-Asia collision. *Geol. J.* **2020**, *55*, 6696–6717. [CrossRef]
101. Liu, H.; Li, G.M.; Li, W.C.; Zhang, J.H.; Li, Y.G.; Zhang, Z.L.; Guang, H.X.; Ouyang, Y.; Zhang, T.J. Epithermal mineralization at Budongla gold deposit in Zhongba county of Tibet: Evidence from fluid inclusions and H-O isotopes. *Miner. Depos.* **2021**, *40*, 311–328. (In Chinese with English Abstract)
102. Ouyang, Y.; Yang, W.N.; Huang, H.X.; Liu, H.; Zhang, J.L.; Zhang, J.H. Metallogenic dynamics background of Ga'erqiong Cu-Au deposit in Tibet, China. *Earth Sci. Res. J.* **2017**, *21*, 59–65. [CrossRef]
103. Yu, H.X.; Chen, J.L.; Xu, J.F.; Wang, B.D.; Wu, J.B.; Liang, H.Y. Geochemistry and origin of Late Cretaceous (~90 Ma) ore-bearing porphyry of Balazha in mid-northern Lhasa terrane, Tibet. *Acta Petrol. Sin.* **2011**, *27*, 2011–2022. (In Chinese with English Abstract)
104. Zhu, T.X.; Feng, X.T.; Wang, X.F.; Zhang, Y.J.; An, X.Y. Summary of the Late Triassic tectonic paleogeography in the Qinghai-Tibetan Plateau, China. *Sediment. Geol. Tethyan Geol.* **2020**, *40*, 59–71. (In Chinese with English Abstract)
105. Eby, G.N. Chemical subdivision of the A-type granitoids: Petrogenetic and tectonic implications. *Geology* **1992**, *20*, 641–644. [CrossRef]
106. Whalen, J.B.; Currie, K.L.; Chappell, B.W. A-type granites: Geochemical characteristics, discrimination and petrogenesis. *Contrib. Mineral. Petrol.* **1987**, *95*, 407–419. [CrossRef]
107. Niu, Y.; O'Hara, M.J. MORB mantle hosts the missing Eu (Sr, Nb, Ta and Ti) in the continental crust: New perspectives on crustal growth, crust–mantle differentiation and chemical structure of oceanic upper mantle. *Lithos* **2009**, *112*, 1–17. [CrossRef]

108. Roy, A.; Sarkar, A.; Jeyakumar, S.; Aggrawal, S.K.; Ebihara, M. Mid-Proterozoic plume-related thermal event in Eastern Indian craton: Evidence from trace elements, REE geochemistry and Sr-Nd isotope systematics of basic-ultrabasic intrusives from Dalma Volcanic Belt. *Gondwana Res.* **2002**, *5*, 133–146. [CrossRef]
109. Blichert-Toft, J.; Albarède, F. The Lu-Hf isotope geochemistry of chondrites and the evolution of the mantle-crust system. *Earth Planet. Sci. Lett.* **1997**, *148*, 243–258. [CrossRef]
110. Griffin, W.L.; Pearson, N.J.; Belousova, E.; Jackson, S.E.; van Achterbergh, E.; O'Reilly, S.Y.; Shee, S.R. The Hf isotope composition of cratonic mantle: LAM-MC-ICPMS analysis of zircon megacrysts in kimberlites. *Geochim. Et Cosmochim. Acta* **2000**, *64*, 133–147. [CrossRef]
111. Griffin, W.L.; Wang, X.; Jackson, S.E.; Pearson, N.J.; O'Reilly, S.Y.; Xu, X.S.; Zhou, X.M. Zircon chemistry and magma mixing, SE China: In-situ analysis of Hf isotopes, Tonglu and Pingtan igneous complexes. *Lithos* **2002**, *61*, 237–269. [CrossRef]
112. Amelin, Y.; Lee, D.C.; Halliday, A.N.; Pidgeon, R.T. Nature of the Earth's earliest crust from hafnium isotopes in single detrital zircons. *Nature* **1999**, *399*, 252–255. [CrossRef]
113. Wu, F.Y.; Li, X.H.; Zheng, Y.F.; Gao, S. Lu-Hf isotopic systematics and their applications in petrology. *Acta Petrol. Sin.* **2007**, *23*, 185–220. (In Chinese with English Abstract)
114. DePaolo, D.J.; Wasserburg, G.J. Nd isotopic variations and petrogenetic models. *Geophys. Res. Lett.* **1976**, *3*, 249–252. [CrossRef]
115. Deng, J.F.; Luo, Z.H.; Su, S.G. *Petrogenesis, Tectonic Environment and Mineralization*; Geological Publishing House: Beijing, China, 2009; pp. 1–381. (In Chinese)
116. Eby, G.N. The A-type granitoids: A review of their occurrence and chemical characteristics and speculations on their petrogenesis. *Lithos* **1990**, *26*, 115–134. [CrossRef]
117. King, P.L.; White, A.J.R.; Chappell, B.W.; Allen, C.M. Characterization and origin of aluminous A-type granites from the Lachlan fold belt, southeastern Australia. *J. Petrol.* **1997**, *38*, 371–391. [CrossRef]
118. Zhang, Q.; Wang, Y.; Pan, G.Q.; Li, C.D.; Jin, W.J. Sources of granites: Some crucial questions on granite study (4). *Acta Petrol. Sin.* **2008**, *24*, 1193–1204. (In Chinese with English Abstract)
119. Schiller, D.; Finger, F. Application of Ti-in-zircon thermometry to granite studies: Problems and possible solutions. *Contrib. Mineral. Petrol.* **2019**, *174*, 51. [CrossRef]
120. Dai, J.G.; Wang, C.S.; Zhu, D.C.; Li, Y.L.; Zhong, H.T.; Ge, Y.K. Multi-stage volcanic activities and geodynamic evolution of the Lhasa terrane during the Cretaceous: Insights from the Xigaze forearc basin. *Lithos* **2015**, *218–219*, 127–140. [CrossRef]
121. Jiang, X.Y.; Wu, K.; Luo, J.C.; Zhang, L.P.; Sun, W.D.; Xia, X.P. An A1-type granite that borders A2-type: Insights from the geochemical characteristics of the Zongyang A-type granite in the Lower Yangtze River Belt, China. *Int. Geol. Rev.* **2020**, *62*, 2203–2220. [CrossRef]
122. Liu, H.; Zhang, L.K.; Huang, H.X.; Li, G.M.; Lü, M.H.; Yan, G.Q.; Huang, Y.; Xie, H.; Lan, S.S. Sources of ore-forming materials of Luerma porphyry copper (gold) deposit, western Gangdise. *Miner. Depos.* **2019**, *38*, 631–643. (In Chinese with English Abstract)
123. Liu, H.; Zhang, L.K.; Huang, H.X.; Li, G.M.; Lü, M.H.; Yan, G.Q.; Huang, Y.; Lan, S.S.; Xie, H. Origin and Evolution of Ore-Forming Fluids in Luerma Porphyry Copper (Gold) Deposit from Western Gangdise. *Earth Sci.* **2019**, *44*, 1935–1956. (In Chinese with English Abstract)
124. Wu, F.Y.; Jahn, B.M.; Wilder, S.A.; Lo, C.H.; Yui, T.F.; Lin, Q.; Ge, W.C.; Sun, D.Y. Highly fractionated I-type granites in NE China (I): Geochronology and petrogenesis. *Lithos* **2003**, *66*, 241–273. [CrossRef]
125. Xie, C.M.; Li, C.; Li, G.M.; Zhang, L.K.; Wang, B.; Dong, Y.C.; Hao, Y.J. The research progress and problem of the Sumdo Paleo-Tethys Ocean, Tibet. *Sediment. Geol. Tethyan Geol.* **2020**, *40*, 1–13. (In Chinese with English Abstract)
126. Zhu, D.C.; Zhao, Z.D.; Niu, Y.; Mo, X.X.; Chung, S.L.; Hou, Z.Q.; Wang, L.Q.; Wu, F.Y. The Lhasa Terrane: Record of a microcontinent and its histories of drift and growth. *Earth Planet. Sci. Lett.* **2011**, *301*, 241–255. [CrossRef]
127. Martin, H. Adakitic magmas: Modern analogues of Archaean granitoids. *Lithos* **1999**, *46*, 411–429. [CrossRef]
128. Sun, G.C.; Gao, P.; Zhao, Z.F. Post-collisional reworking of subducted continental crust: Insights from late Paleozoic granites in the North Qaidam orogen, northeastern Tibet. *Lithos* **2022**, *432*, 106921. [CrossRef]
129. Sylvester, P.J. Post-collisional strongly peraluminous granites. *Lithos* **1998**, *45*, 29–44. [CrossRef]
130. Yang, H.; Zhang, H.; Luo, B.; Gao, Z.; Guo, L.; Xu, W. Generation of peraluminous granitic magma in a post-collisional setting: A case study from the eastern Qilian orogen, NE Tibetan Plateau. *Gondwana Res.* **2016**, *36*, 28–45. [CrossRef]
131. Zhang, J.J.; Sun, S.J.; Zhang, L.P.; Deng, J.H.; Li, S.; Sun, W.D. Origin of transitional IA-type syenite and its relationship to A-type intrusions in the Luzong Basin, the Lower Yangtze River Belt: Insights from geochemistry. *Chem. Geol.* **2023**, *626*, 121458. [CrossRef]
132. Chen, J.; Fu, L.; Wei, J.; Selby, D.; Zhang, D.; Zhou, H.; Zhao, X.; Liu, Y. Proto-Tethys magmatic evolution along northern Gondwana: Insights from Late Silurian—Middle Devonian A-type magmatism, East Kunlun Orogen, Northern Tibetan Plateau, China. *Lithos* **2020**, *356*, 105304. [CrossRef]
133. Wang, E.; Zhai, X.; Chen, W.; Wu, L.; Song, G.; Wang, Y.; Guo, Z.; Zhao, J.; Wang, J. Late devonian a-type granites from the beishan, southern central Asia Orogenic Belt: Implications for closure of the Paleo-Asia Ocean. *Minerals* **2023**, *13*, 565. [CrossRef]

134. Whalen, J.B.; Hildebrand, R.S. Trace element discrimination of arc, slab failure, and A-type granitic rocks. *Lithos* **2019**, *348*, 105179. [CrossRef]
135. Pearce, J.A.; Harris, N.B.W.; Tindle, A.G. Trace element discrimination diagrams for the tectonic interpretation of granitic rocks. *J. Petrol.* **1984**, *25*, 956–983. [CrossRef]
136. Boztuğ, D.; Harlavan, Y.; Arehart, G.; Satir, M.; Avci, N. K-Ar age, whole-rock and isotope geochemistry of A-type granitoids in the Divriği-Sivas region, eastern-central Anatolia, Turkey. *Lithos* **2007**, *97*, 193–218. [CrossRef]
137. Collins, W.J.; Beams, S.D.; White, A.J.R.; Chappell, B.W. Nature and origin of A-type granites with particular reference to southeastern Australia. *Contrib. Mineral. Petrol.* **1982**, *80*, 189–200. [CrossRef]
138. Huppert, H.E.; Sparks, R.S.J. The generation of granitic magmas by intrusion of basalt into continental crust. *J. Petrol.* **1988**, *29*, 599–624. [CrossRef]
139. Yang, J.H.; Wu, F.Y.; Chung, S.L.; Wilde, S.A.; Chu, M.F. A hybrid origin for the Qianshan A-type granite, northeast China: Geochemical and Sr-Nd-Hf isotopic evidence. *Lithos* **2006**, *89*, 89–106. [CrossRef]
140. Skjerlie, K.P.; Johnston, A.D. Vapor-absent melting at 10 kbar of a biotite- and amphibole-bearing tonalitic gneiss: Implications for the generation of A-type granites. *Geology* **1992**, *20*, 263–266. [CrossRef]
141. Wilson, M. *Igneous Petrogenesis*; Springer: Dordrecht, The Netherlands, 1989; pp. 1–446.
142. Wu, F.Y.; Sun, D.Y.; Li, H.M.; Jahn, B.M.; Wilde, S. A-type granites in northeastern China: Age and geochemical constraints on their petrogenesis. *Chem. Geol.* **2002**, *187*, 143–173. [CrossRef]
143. Wu, H.; Qiangba, Z.; Li, C.; Wang, Q.; Gesang, W.; Ciren, O.; Basang, D. Geochronology and geochemistry of Early Cretaceous granitic rocks in the Dongqiao Area, Central Tibet: Implications for magmatic origin and geological evolution. *J. Geol.* **2018**, *126*, 249–260. [CrossRef]
144. Foland, K.A.; Allen, J.C. Magma sources for Mesozoic anorogenic granites of the White Mountain magma series, New England, USA. *Contrib. Mineral. Petrol.* **1991**, *109*, 195–211. [CrossRef]
145. Sparks, R.S.J.; Marshall, L.A. Thermal and mechanical constraints on mixing between mafic and silicic magmas. *J. Volcanol. Geotherm. Res.* **1986**, *29*, 99–124. [CrossRef]
146. Turner, S.; Sandiford, M.; Foden, J. Some geodynamic and compositional constraints on "postorogenic" magmatism. *Geology* **1992**, *20*, 931–934. [CrossRef]
147. Kerr, A.; Fryer, B.J. Nd isotope evidence for crust-mantle interaction in the generation of A-type granitoid suites in Labrador, Canada. *Chem. Geol.* **1993**, *104*, 39–60. [CrossRef]
148. Creaser, R.A.; Price, R.C.; Wormald, R.J. A-type granites revisited: Assessment of a residual-source model. *Geology* **1991**, *19*, 163–166. [CrossRef]
149. Dooley, D.F.; Patiño Douce, A.E. Fluid-absent melting of F-rich phlogopite+rutile+quartz. *Am. Mineral.* **1996**, *81*, 202–212. [CrossRef]
150. Wilson, M. Magmatism and the geodynamics of basin formation. *Sediment. Geol.* **1993**, *86*, 5–29. [CrossRef]
151. Ji, W.Q.; Wu, F.Y.; Chung, S.L.; Liu, C.Z. The Gangdese magmatic constraints on a latest Cretaceous lithospheric delamination of the Lhasa terrane, southern Tibet. *Lithos* **2014**, *210*, 168–180. [CrossRef]
152. Zhu, B.Q. *Theories and Application of Isotopic System in Geoscience: Crustal and Mantle Evolution in China Continent*; Science Press: Beijing, China, 1998; pp. 1–330. (In Chinese)
153. Shellnutt, J.G.; Zhou, M.F. Permian peralkaline, peraluminous and metaluminous A-type granites in the Panxi district, SW China: Their relationship to the Emeishan mantle plume. *Chem. Geol.* **2007**, *243*, 286–316. [CrossRef]
154. Barth, M.G.; Mcdonough, W.F.; Rudnick, R.L. Tracking the budget of nb and ta in the continental crust. *Chem. Geol.* **2000**, *165*, 197–213. [CrossRef]
155. Chappell, B.W.; White, A.J.R. I- and S-type granites in the Lachlan Fold Belt. *Earth Environ. Sci. Trans. R. Soc. Edinb.* **1992**, *83*, 1–26.
156. Chappell, B.W. Aluminium saturation in I- and S-type granites and the characterization of fractionated haplogranites. *Lithos* **1999**, *46*, 535–551. [CrossRef]
157. Hou, Z.Q.; Gao, Y.F.; Qu, X.M.; Rui, Z.Y.; Mo, X.X. Origin of adakitic intrusives generated during mid-Miocene east-west extension in southern Tibet. *Earth Planet. Sci. Lett.* **2004**, *220*, 139–155. [CrossRef]
158. Li, S.M.; Wang, Q.; Zhu, D.C.; Stern, R.J.; Cawood, P.A.; Sui, Q.L.; Zhao, Z. One or two Early Cretaceous arc systems in the Lhasa Terrane, southern Tibet. *J. Geophys. Res. Solid Earth* **2018**, *123*, 3391–3413. [CrossRef]
159. Tang, W.L.; Huang, F.; Xu, J.F.; Zeng, Y.C.; Liu, X.J. Cretaceous magmatism in the northern Lhasa Terrane: Implications for the tectonic evolution and crustal growth tempos of central Tibet. *Geol. Soc. Am. Bull.* **2024**, *136*, 3440–3456. [CrossRef]
160. Tian, Y.; Huang, F.; Xu, J.; Zeng, Y.; Hu, P.; Yu, H.; Tian, Y.; Yang, Z.; Yang, X. Petrogenesis and geodynamic mechanisms of the Late Cretaceous magmatic 'flare-up' in the southern Lhasa Terrane, Tibet. *Lithos* **2022**, *424*, 106766. [CrossRef]
161. Zhang, C.; Liu, C.Z.; Xu, Y.; Ji, W.B.; Wang, J.M.; Wu, F.Y.; Liu, T.; Zhang, Z.Y.; Zhang, W.Q. Subduction re-initiation at dying ridge of Neo-Tethys: Insights from mafic and metamafic rocks in Lhaze ophiolitic mélange, Yarlung-Tsangbo Suture Zone. *Earth Planet. Sci. Lett.* **2019**, *523*, 115707. [CrossRef]

162. Chen, Q.; Xie, L.; Xiao, Z.J. Characteristics and tectonic evolution of Bangong Lake serpentinite mélange belt in the western Qinghai-Tibet Plateau. *J. East China Inst. Technol. Nat. Sci. Ed.* **2007**, *30*, 107–112. (In Chinese with English Abstract)
163. Hu, P.Y.; Zhai, Q.G.; Jahn, B.M.; Wang, J.; Li, C.; Chung, S.L.; Lee, H.Y.; Tang, S.H. Late Early Cretaceous magmatic rocks (118–113 Ma) in the middle segment of the Bangong-Nujiang suture zone, Tibetan Plateau: Evidence of lithospheric delamination. *Gondwana Res.* **2017**, *44*, 116–138. [CrossRef]
164. Hu, X.; Ma, A.; Xue, W.; Garzanti, E.; Cao, Y.; Li, S.-M.; Sun, G.; Lai, W. Exploring a lost ocean in the Tibetan Plateau: Birth, growth, and demise of the Bangong-Nujiang Ocean. *Earth-Sci. Rev.* **2022**, *229*, 104031. [CrossRef]
165. Tang, Y.; Zhai, Q.G.; Hu, P.Y.; Chung, S.L.; Xiao, X.C.; Wang, H.T.; Zhu, Z.C.; Wang, W.; Wu, H.; Lee, H.Y. Southward subduction of the Bangong-Nujiang Tethys Ocean: Insights from ca. 161–129 Ma arc volcanic rocks in the north of Lhasa terrane, Tibet. *Int. J. Earth Sci.* **2020**, *109*, 631–647. [CrossRef]
166. Wang, W.; Wang, M.; Zhai, Q.G.; Xie, C.M.; Hu, P.Y.; Li, C.; Liu, J.H.; Luo, A.B. Transition from oceanic subduction to continental collision recorded in the Bangong-Nujiang suture zone: Insights from Early Cretaceous magmatic rocks in the north-central Tibet. *Gondwana Res.* **2020**, *78*, 77–91. [CrossRef]
167. Chen, S.S.; Shi, R.D.; Gong, X.H.; Liu, D.L.; Huang, Q.S.; Yi, G.D.; Wu, K.; Zou, H.B. A syn-collisional model for Early Cretaceous magmatism in the northern and central Lhasa subterranes. *Gondwana Res.* **2017**, *41*, 93–109. [CrossRef]
168. Sun, M.; Tang, J.-X.; Chen, W.; Ma, X.-D.; Qu, X.-M.; Song, Y.; Li, X.-Y.; Ding, J.-S. Process of lithospheric delamination beneath the Lhasa—Qiangtang collision orogen: Constraints from the geochronology and geochemistry of Late Cretaceous volcanic rocks in the Lhasa terrane, central Tibet. *Lithos* **2020**, *356*, 105219. [CrossRef]
169. Zhang, K.J.; Xia, B.D.; Wang, G.M.; Li, Y.T.; Ye, H.F. Early Cretaceous stratigraphy, depositional environments, sandstone provenance, and tectonic setting of central Tibet, western China. *Geol. Soc. Am. Bull.* **2004**, *116*, 1202–1222. [CrossRef]
170. Zhang, K.J.; Zhang, Y.X.; Tang, X.C.; Xia, B. Late Mesozoic tectonic evolution and growth of the Tibetan Plateau prior to the Indo-Asian collision. *Earth-Sci. Rev.* **2012**, *114*, 236–249. [CrossRef]
171. Huang, H.X.; Liu, H.; Li, G.M.; Zhang, L.K.; Cao, H.W.; Zhou, Q.; Wang, X.X.; Yan, G.Q. Zircon U-Pb, molybdenite Re-Os and quartz vein Rb-Sr geochronology of the Luobuzhen Au-Ag and Hongshan Cu deposits, Tibet, China: Implications for the oligocene-miocene porphyry—Epithermal metallogenic system. *Minerals* **2019**, *9*, 476. [CrossRef]
172. Wen, D.R.; Chung, S.L.; Song, B.; Iizuka, Y.; Yang, H.J.; Ji, J.Q.; Liu, D.Y.; Gallet, S. Late Cretaceous Gangdese intrusions of adakitic geochemical characteristics, SE Tibet: Petrogenesis and tectonic implications. *Lithos* **2008**, *105*, 1–11. [CrossRef]
173. Hoskin, P.W.O. Trace-element composition of hydrothermal zircon and the alteration of Hadean zircon from the Jack Hills, Australia. *Geochim. Et Cosmochim. Acta* **2005**, *69*, 637–648. [CrossRef]
174. Kapp, P.; Yin, A.; Harrison, T.M.; Ding, L. Cretaceous-Tertiary shortening, basin development, and volcanism in central Tibet. *Geol. Soc. Am. Bull.* **2005**, *117*, 865–878. [CrossRef]
175. Xu, M.J.; Li, C.; Zhang, X.Z.; Wu, Y.W. Nature and evolution of the Neo-Tethys in central Tibet: Synthesis of ophiolitic petrology, geochemistry, and geochronology. *Int. Geol. Rev.* **2014**, *56*, 1072–1096. [CrossRef]
176. Zhu, D.C.; Pan, G.T.; Wang, L.Q.; Mo, X.X.; Zhao, Z.D.; Zhou, C.Y.; Liao, Z.L.; Dong, G.L.; Yuan, S.H. Tempo-spatial variations of Mesozoic magmatic rocks in the Gangdise belt, Tibet, China, with a discussion of geodynamic setting-related issues. *Geol. Bull. China* **2008**, *27*, 1535–1550. (In Chinese with English Abstract)
177. Gao, S.B.; Zheng, Y.Y.; Wang, J.S.; Zhang, Z.; Yang, C. The geochronology and geochemistry of intrusive rocks in Bange area: Constraints on the evolution time of the Bangong Lake-Nujiang ocean basin. *Acta Petrol. Sin.* **2011**, *27*, 1973–1982. (In Chinese with English Abstract)
178. Geng, Q.R.; Li, W.C.; Wang, L.Q.; Zeng, X.T.; Peng, Z.M.; Zhang, X.F.; Zhang, Z.; Cong, F.; Guan, J.L. Paleozoic tectonic framework and evolution of the central and western Tethys. *Sediment. Geol. Tethyan Geol.* **2021**, *41*, 297–315. (In Chinese with English Abstract)
179. Huang, H.X.; Dai, Z.W.; Liu, H.; Li, G.M.; Huizenga, J.M.; Zhang, L.K.; Huang, Y.; Cao, H.W.; Fu, J.G. Zircon U-Pb ages, geochemistry, and Sr-Nd-Pb-Hf isotopes of the Mugagangri monzogranite in the southern Qiangtang of Tibet, western China: Implications for the evolution of the Bangong Co-Nujiang Meso-Tethyan Ocean. *Geol. J.* **2021**, *56*, 3170–3186. [CrossRef]
180. Huang, H.X.; Luosang, J.C.; Dai, Z.W.; Liu, H.; Fu, J.G.; Li, G.M.; Cao, H.W.; Zhou, Q. Hydrothermal zircon geochronology in the Shangxu Gold Deposit and its Implication for the Early Cretaceous orogenic gold mineralization in the Middle Bangonghu-Nujiang suture zone. *Acta Geol. Sin. Engl. Ed.* **2021**, *95*, 1249–1259. [CrossRef]
181. Volkmer, J.E.; Kapp, P.; Guynn, J.H.; Lai, Q.Z. Cretaceous-Tertiary structural evolution of the north central Lhasa Terrane, Tibet. *Tectonics* **2007**, *26*, TC6007. [CrossRef]
182. Wang, B.D.; Wang, L.Q.; Chung, S.L.; Chen, J.L.; Yin, F.G.; Liu, H.; Li, X.B.; Chen, L.K. Evolution of the Bangong-Nujiang Tethyan ocean: Insights from the geochronology and geochemistry of mafic rocks within ophiolites. *Lithos* **2016**, *245*, 18–33. [CrossRef]
183. Song, Y.; Zeng, Q.G.; Liu, H.Y.; Liu, Z.B.; Li, H.F.; Dexi, Y.Z. An innovative perspective for the evolution of Bangong-Nujiang Ocean: Also discussing the Paleo-and Neo-Tethys conversion. *Acta Petrol. Sin.* **2019**, *35*, 625–641. (In Chinese with English Abstract)

184. Zeng, Y.C.; Xu, J.F.; Chen, J.L.; Wang, B.D.; Huang, F.; Xia, X.P.; Li, M.J. Early Cretaceous (∼138–134 Ma) forearc ophiolite and tectonomagmatic patterns in Central Tibet: Subduction termination and re-initiation of Meso-Tethys Ocean caused by collision of an Oceanic Plateau at the continental margin? *Tectonics* **2021**, *40*, e2020TC006423. [CrossRef]
185. Wu, H.; Li, C.; Hu, P.Y.; Li, X.K. Early Cretaceous (100–105 Ma) Adakitic magmatism in the Dachagou area, northern Lhasa terrane, Tibet: Implications for the Bangong-Nujiang Ocean subduction and slab break-off. *Int. Geol. Rev.* **2015**, *57*, 1172–1188. [CrossRef]
186. Wu, H.; Li, C.; Xu, M.J.; Li, X.K. Early Cretaceous adakitic magmatism in the Dachagou area, northern Lhasa terrane, Tibet: Implications for slab roll-back and subsequent slab break-off of the lithosphere of the Bangong-Nujiang Ocean. *J. Asian Earth Sci.* **2015**, *97*, 51–66. [CrossRef]

minerals

MDPI

Article

Mineralogy and Geochemistry of Titaniferous Iron Ores in El-Baroud Layered Gabbros: Fe-Ti Ore Genesis and Tectono-Metallogenetic Setting

Mohamed Zaki Khedr [1,*], Ahmed Moftah [1], N. H. El-Shibiny [1], Akihiro Tamura [2], Wei Tan [3], Yuji Ichiyama [4], Eiichi Takazawa [5], Ali Y. Kahal [6] and Kamal Abdelrahman [6]

1 Department of Geology, Faculty of Science, Kafrelsheikh University, Kafrelsheikh 33516, Egypt; ahmed.saad_a011@sci.kfs.edu.eg (A.M.)
2 Department of Earth Sciences, Kanazawa University, Ishikawa 920-1192, Japan
3 Key Laboratory of Mineralogy and Metallogeny, Guangzhou Institute of Geochemistry, Chinese Academy of Sciences, Guangzhou 510640, China; tanwei@gig.ac.cn
4 Graduate School of Science, Chiba University, 1-33 Yayoi-cho, Inage-ku, Chiba 263-8522, Japan; ichiyamay@chiba-u.jp
5 Geology Department, Faculty of Science, Niigata University, Niigata 950-2181, Japan; takazawa@geo.sc.niigata-u.ac.jp
6 Department of Geology and Geophysics, College of Science, King Saud University, P.O. Box 2455, Riyadh 11451, Saudi Arabia; khassanein@ksu.edu.sa (K.A.)
* Correspondence: mohamed.khader1@sci.kfs.edu.eg; Fax: +2-047-264-6545

Abstract: The Neoproterozoic pyroxene gabbros and gabbronorites in the El-Baroud mafic intrusion in the Northern Eastern Desert (NED) of Egypt host Fe-Ti oxide ore deposits. This study discusses the major and trace elements of both titaniferous iron ores and their host rocks, along with the mineral chemistry (major and in situ trace elements) of interstitial clinopyroxene (Cpx), to gain a deeper understanding of the Fe-Ti oxide genesis. These ores occur as disseminated (55–60 vol.% of Fe-Ti oxides) and massive types (85–95 vol.%) in the form of the dyke, layer, and lens. They are composed of titanomagnetite (80–87 vol.%) with subordinate ilmenite (10–15 vol.%) and magnetite (3–5 vol.%), in accordance with their high Fe_2O_3 (75.66 wt.% on average) and TiO_2 contents (16.30–17.60 wt.%). The Cpx in the investigated ores is diopside composition (Mg#; 0.72–0.83) and exhibits a nearly convex upward REE pattern, similar to Cpxs in the ferropicrite that originated from the primitive mantle. Melts in equilibrium with this Cpx resemble Greenstone ferropicrite melts; the parent melt of El-Baroud gabbros is possibly a ferropicritic melt that was derived from the lithospheric mantle during plume interaction. The El-Baroud gabbroic rocks were generated during the arc rifting and crystallized under a high oxygen fugacity at a temperature of 800–1000 °C and a pressure of 3 kbar with a depth of 12 km. The Fe-Ti oxide ores have been formed from ferropicritic parent melts by two processes, including in situ crystallization that leads to the formation of disseminated Fe-Ti oxides in the iron-rich gabbros at the bottom and liquid immiscibility that is responsible for the formation of thick Fe-Ti ore lenses and layers at the top of the gabbroic intrusion. Initially, titanomagnetite crystallized from the primary Ti-rich oxide melt. As cooling progressed, some of the excess titanium in this melt was exsolved in the form of the exsolution ilmenite lamellae within the titanomagnetite. The Fe-Ti oxide layers in the NED follow the trend of NW-SE (Najd trend), where their distribution is possibly controlled by the composition of parent melts (rich in Ti and Fe), high oxygen fugacity, and the structure related to the Najd fault system. The distribution of Fe-Ti oxide ores increases from the NED to the Southern Eastern Desert (SED), suggesting the dominant mantle plumes and/or shear zones in the SED relative to the NED.

Keywords: gabbroic intrusion; Fe-Ti oxide ores; mineral chemistry; liquid immiscibility; arc rifting; Arabian-Nubian Shield

1. Introduction

Magmatic iron ores from layered intrusions play an important role in hosting significant deposits of Fe, Ti, and V metals, which are essential for industrial processes and various applications [1,2]. Fe-Ti-V oxide ore-bearing layered intrusions are globally widespread, including Bushveld in South Africa [3], Skaergaard in East Greenland [4], Panzhihua in SW China [5], Duluth and Stillwater in America [6], Bjerkreim-Sokndal in Norway [7], and Sept-Iles in Canada [8]. Moreover, ferrogabbros that host significant Fe-Ti-(V) ore deposits are frequently associated with massif-type anorthosites [9–11]. These ore deposits comprise Ti-magnetite, ilmenite, and magnetite with a few apatite and Al-spinel, hosted in the mafic-ultramafic intrusions. They are also found as lenses and veins in anorthosite–gabbronorite–monzonite rocks [12].

Fe-Ti oxide ores in layered mafic intrusions are considered orthomagmatic deposits, and their parental melt is possibly ferrobasaltic and derived from more primitive (ferro-) picritic magmas [13]. Their genesis and formation mechanism in layered intrusions have been extensively debated and remain poorly understood [14]. Therefore, several theories have been suggested to explain the origin of the Fe-Ti oxide ores [15], including: (1) liquid immiscibility [16]; (2) magma addition and/or mixing [17]; (3) fractional crystallization by mineral settling and sorting [18]; (4) variations in oxygen content [19]; and (5) pressure changes [20]. Fractional crystallization is an essential genetic model for Fe-Ti oxide deposits, where this process may concentrate ore minerals from the basaltic magma, either (I) late-stage differentiation observed in the Bushveld Complex [21], the Kiglapait intrusion [22], and the Sonju Lake intrusion [23–25] or (II) relatively early-stage differentiation recorded in the Wuben deposit [26], through the process of the separation of immiscible Fe-Ti-(P)-rich liquids. Following this immiscible separation, the process proceeds with the settling and sorting of Fe-Ti oxides in the Khamal complex [27], the Panzhihua intrusion [5], and the Damiao complex [28].

In the Eastern Desert of Egypt, the Neoproterozoic mafic or mafic-ultramafic layered intrusions in the Arabian-Nubian Shield (ANS), formed in syn- to post-collisional settings such as Korab Kansi [29], Abu Ghalaga [12,30,31], Wadi Rahaba [32], and Gabal Akab El Negum [33], host a significant amount of Fe-Ti oxide ores. The Abu Ghalaga mine in gabbroic rocks possesses the biggest reserves (about 41 million tons) of these ores in the ANS, not only in Egypt [30]. The second-largest reserve is found in Korab Kansi [29,31], whereas the third-largest reserve may exist in the El-Baroud (this study) area. The minor Fe-Ti oxide ores are also recorded in Wadi Rahaba [32], Wadi Abu Fas, Gabal Kolminab [34,35], Gabal El Rokham, and Um Effein [36]. All these studies discussed Fe-Ti oxides in the layered intrusions based on the whole rock chemistry of gabbros and the mineral chemistry of ores such as ilmenite and magnetite. They described the type of Fe-Ti ores (massive or semi-massive) and their tectonic setting, but the source of high Fe-Ti and the mechanism of Fe-Ti oxide mineral accumulation are still debated. These studies suggested two possible parent melts of titaniferous iron ores: (1) ferropicritic (e.g., Korab Kansi [29] and Rahaba [32]) or (2) ferrobasaltic melts (e.g., Abu Ghalaga [31]), formed in either an island arc setting (e.g., Korab Kansi [29]) or a post-collisional setting (e.g., Rahaba [32]).

El-Baroud gabbroic rocks and their Fe-Ti oxide ores are located in the Northern Eastern Desert (NED) of Egypt and are considered one of the mineralized layered intrusions in the ANS. There is no study recording the existence of Fe-Ti oxide ores either in the El-Baroud mafic intrusion or in the NED. Although there are several studies discussing Fe-Ti oxide ores in the mafic intrusions in the Central and Southern Eastern Desert of Egypt, the sources of Fe and Ti and the genesis of these ores are still mysterious. This is the first record of the Fe-Ti-rich gabbros and their titaniferous iron ores in the El-Baroud gabbroic intrusion. This study discusses the field occurrence, micro-texture, mineral chemistry (major, trace, and REE), and whole-rock compositions of titaniferous iron ores and their host gabbroic rocks in the El-Baroud mafic intrusion. The main aim is to understand the genesis of titaniferous iron ores and the source of anomalously high Ti-Fe-rich melts, as well as shed light on their tectono-metallogenetic setting. This study may put constraints on the formation conditions (temperature, pressure, and fO_2) and distribution of these Fe-Ti oxide deposits in the Eastern Desert of Egypt.

2. Geological Setting of the El-Baroud Intrusion

The Arabian-Nubian Shield (ANS) includes three Neoproterozoic gabbroic rocks different in field occurrence, assemblages, origin, and geochemistry, including: (1) ophiolitic metagabbros mainly associated with serpentinites; (2) island-arc-related metagabbros (intrusive metagabbros) mainly associated with diorites/granodiorites; and (3) layered (previously named younger) gabbros [29,30,37–43]. The ophiolitic metagabbros are considered pieces of oceanic lithosphere and show tectonic contacts with the country rocks; they are free of Fe-Ti oxide ores. The gabbroic intrusions, which comprise intrusive metagabbros and layered gabbros, show intrusive contacts against surrounding rocks [29,31,32,37–39,43,44], while the former is highly metamorphosed with fewer Fe-Ti oxides [37–39], but the latter is sometimes rich in Fe-Ti oxide ores [30–33].

The El-Baroud mafic intrusion is found along the southern border of the Northern Eastern Desert (NED) of Egypt (Figure 1b). The studied area is located roughly 25 km southwest of Safaga city, at latitude 26° 45′ 00″ N and longitude 33° 45′ 00″ E, covering an area of about 26 Km^2 (Figure 1b). Its landscape is marked by abundant granitic rocks, which were intruded by the Fe-Ti-rich gabbroic intrusion that hosts Fe-Ti oxide veins (Figures 1b and 2a). The El-Baroud gabbroic rocks comprise pyroxene gabbro, gabbronorite, hornblende gabbro, and biotite hornblende gabbro, whereas pyroxene gabbro with subordinate gabbronorite is the main host of Fe-Ti oxide ores (Figure 1b). Contacts between the gabbroic rock units are gradational and show thermal evidence against country rocks.

The El-Baroud gabbroic rocks intruded in the tonalite–granodorite (previously named older granite or syn-collisional granite) and occurred as sporadic intrusions in this granite (Figure 1b). The granitoids found in the El-Baroud area have been classified into two separate rock units, known as the older granites and the younger granites, that are far away from the studied gabbroic intrusion [45]. The El-Baroud gabbroic rocks intruded in the syn-collisional tonalite–granodorite (Figure 1b), suggesting their younger age than older granites (Figures 1b and 2a) [45].

Like the Central Eastern Desert (CED) of Egypt, the NED of Egypt is dissected by NW-SE (Najd trend) and NE-SW (Barramiya-Mubarak trend) strike-slip faults; this Najd trend intersected with the NE-SW dextral strike fault, such as NE (to ENE) oriented shear zones. For instance, the Qena–Safaga NE-SW shear zone (Figure 1a) could be interpreted as either Najd-related conjugates or earlier fault systems [46,47]. In the El-Baroud area, various small-offset strike-slip faults are observed, trending in both NW-SE and NE-SW structural trends (Figure 1b); these faults show both left-lateral and right-lateral movements, respectively [48]. The foliations of El-Baroud pyroxene gabbro and Fe-Ti oxide ore veins follow the same NW-SE structural trend, parallel to the NW-SE Najid fault system. Specifically, the foliation of gabbro runs parallel to the Fe-Ti ore veins with a striking direction of N 40° W with dipping of 50° SW.

Fe-Ti oxide deposits increase in abundance from the NED (one mafic intrusion) through the Central (two intrusions) to the Southern Eastern Desert (SED) (five intrusions) of Egypt (Figure 1a). The El-Baroud Fe-Ti oxide deposits are only titaniferous iron ores in the NED of Egypt, compared with Wadi El-Miyah and Abu Ghalaga [12,30,31] in the CED, and the Korab Kansi [29], Wadi Rahaba [32], Abu Dahr, and Gabal Akab El Negum [33] in the SED (Figure 1a).

The El-Baroud Fe-Ti oxide ores are composed of titanomagnetite (80–87 vol.%), ilmenite (10–15 vol.%), and magnetite (3–5 vol.%); they occur as horizontal veins or layers, extending for over 60 m in length and approximately 1.5 m in thickness (Figure 2b). Most of these layers strike at the N 65° W structural trend (same as the Najd fault system) and show sharp contacts with the host pyroxene gabbro (Figure 2b). The Fe-Ti oxide ores occur in various morphological bodies, including: (1) the dyke or vertical vein of Ti-magnetite and ilmenite (Figure 2c); (2) horizontal layers of massive Fe-Ti oxide ores that are interlayered with the pyroxene gabbro (Figure 2d); (3) the cap of titanomagnetite, situated above the pyroxene gabbro (Figure 2e), which signifies the liquid immiscibility origin; and (4) the lens of Fe-Ti ore hosted by the pyroxene gabbro at the top of the gabbroic intrusion (Figure 2f). The horizontal ore layer extends for a distance of 30 m and is 3 m thick, but the vertical ore dyke is 30 m long and 4 m thick. All of the vertical Fe-Ti ore dykes are intersected by quartz veins, running parallel to the horizontal Fe-Ti oxide vein, with a thickness ranging from 10 to 20 cm. The titaniferous iron ores are mainly massive types with interstitial clinopyroxene (Figure 2g,h). These iron ores are stained with hematite–titanomagnetite ores (Figure 2h). The massive layered ore bodies exhibit distinct and sharp contacts with the host gabbro but show gradual contacts with disseminated ore bands in the lower parts of the mafic intrusion.

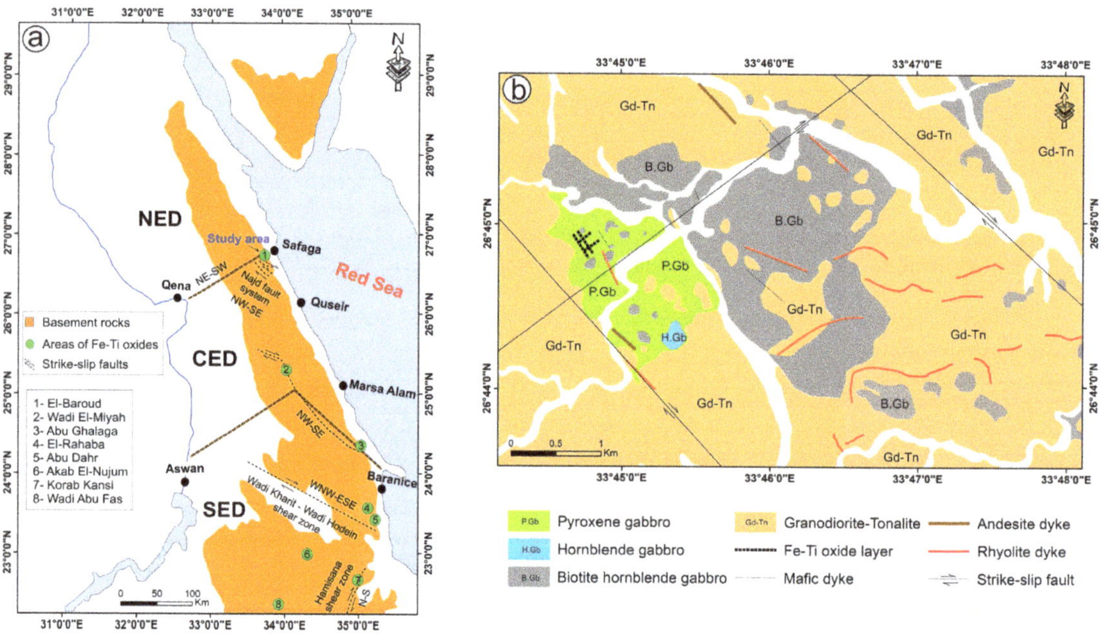

Figure 1. Regional and simplified geological map of the El-Baroud area, the Northern Eastern Desert (NED) of Egypt. (**a**) Regional geological map showing the location of the El-Baroud intrusion and other Fe-Ti oxide deposits in the Eastern Desert (modified after El-Ramly [37], where structures were compiled from Hamimi and Abd El-Wahed [46]). (**b**) A simplified geological map of the Wadi El-Baroud mafic intrusion, showing the distribution of the gabbroic rocks and associated Fe-Ti oxide ore deposits. Strike-slip faults and other structure trends are from Omran [48].

The layers/veins of titaniferous iron ores in the El-Baroud area resemble in shape and composition the Fe-Ti oxide ores in the Korab Kansi area [29,31]; both areas include the layered Fe-Ti-rich gabbroic intrusions. It is noted that the Fe-Ti oxide layers, foliation of the host pyroxene gabbro, dykes of bimodal compositions, and strike-slip faults in the El-Baroud area follow the structural trend of the Najd Fault System, which is a sinistral strike-slip with striking NW-SE or WNW-ESE (Figure 1b).

Figure 2. Photographs of the field relationship of the El-Baroud mafic intrusion hosting the massive Fe–Ti oxide ores. (**a**) A panoramic photograph of the El-Baroud intrusion. (**b**) Layer-like Fe-Ti oxide ores with sharp intrusive contacts with the host pyroxene gabbro. (**c**) A dyke-like ore body, cutting pyroxene gabbro. (**d**) Pyroxene gabbro interlayers between massive Fe-Ti oxide layers, showing sharp contacts against layers. (**e**) Occurrence of the Fe-Ti oxide ore lens as a cap in the top of pyroxene gabbro. (**f**) The lenticular Fe-Ti oxide body with sharp contacts against its host layered pyroxene gabbro. (**g**,**h**) Photographs of Fe-Ti oxide hand specimens, showing massive titanomagnetite with interstitial clinopyroxene (Cpx), where some altered titanomagnetite samples are stained with hematite, as in panel (**h**).

3. Sampling and Analytical Methods

Twenty samples of Fe-Ti oxide ores (Figure 2g,h) and fifteen samples of their host gabbros were collected from the El-Baroud mafic intrusion. All samples were prepared as thin sections (35) and polished at Niigata University, Japan. Ore microscopy (Figures 3 and 4) and scanning electron microscopy (SEM) coupled with energy dispersive X-ray spectroscopy (EDS) at Niigata University, Japan, were used to identify sulfide and some accessory minerals in the titaniferous iron ores. The SEM-EDS unit operated at 20 kV accelerating voltage, with a working distance (WD) of 10 mm. Micro-textural analysis of the Fe-Ti oxide samples was performed under the SEM.

Eight samples of the host gabbroic rocks and ten titaniferous iron ore samples were selected for bulk-rock chemical analysis. Their major and trace elements were analyzed using X-ray fluorescence (XRF) on lithium borate fused glass at the Mineral Division of the ALS Laboratory Group in Guangzhou, China. The analysis was conducted using the following codes: ME-XRF26 for major element fusion, ME-MS61r for trace element fusion, and ME-MS81g for rare earth element (REE) fusion. The detection limits for all elements were set at 0.01 wt.%. For precise primary and secondary analysis of Fe-Ti oxides using X-ray fluorescence spectrometry, two samples are first weighed. Then, a lithium borate–lithium nitrate melting flux containing lithium nitrate is added to one sample. The mixture is thoroughly mixed and melted at a high temperature. The molten material is poured into a platinum vessel. Once the molten material solidifies into a flat, glassy piece, it is analyzed using an X-ray fluorescence spectrometer. Simultaneously, another sample was placed in the muffle furnace and incinerated at 1000 °C. Once cooled, the sample is weighed again. The difference in weight between the sample before and after heating denotes the quantity lost during incineration. The total obtained from this method is the sum of the loss on ignition results and the elemental oxide results measured by XRF.

A flux of lithium borate ($LiBO_2/Li_2B_4O_7$) is added to the sample in order to perform a melting technique analysis of the REE content using inductively coupled plasma mass spectrometry (ICP-MS) at the Mineral Division of ALS Laboratory Group in Guangzhou, China. To analyze the REE content using the ICP-MS through the melting method, a flux of lithium borate is added to the sample. This flux is thoroughly mixed with the sample and then melted in a furnace at a temperature of 1025 °C. The molten liquid is left to cool until it solidifies. Subsequently, the solidified material was digested using nitric acid, hydrochloric acid, and hydrofluoric acid. The resulting solution was then diluted to the desired volume and analyzed using a plasma mass spectrometer.

The major-element contents of Cpx, magnetite, and ilmenite were determined using an electron probe microanalyzer (EPMA) (Table 2; Supplementary Table S1) with wavelength dispersive X-ray spectrometry (JEOL JXA-8230-XM160041-0041) at the Graduate School of Science, Chiba University, Japan. Standard samples, both natural and synthetic minerals, were used for calibration purposes. The following standards were used to determine and calibrate the values of Si, Cr, Fe, Ca, (K, Ti), Al, Mn, Na, Ni, and Mg: quartz, eskolaite, fayalite, wollastonite, $KTiPO_5$, corundum, manganosite, jadeite, nickel oxide, and periclase, in that order. For Si, Al, Na, and Mg, the diffracting crystals were TAP; for Cr, Ca, K, and Ti, PETJ; and for Fe, Mn, and Ni, LIF. The parameters for the analysis were as follows: an acceleration voltage of 15 kV, a beam current of 20 nA, and a probe diameter of 3 μm. The data were corrected by using oxide ZAF matrix correction.

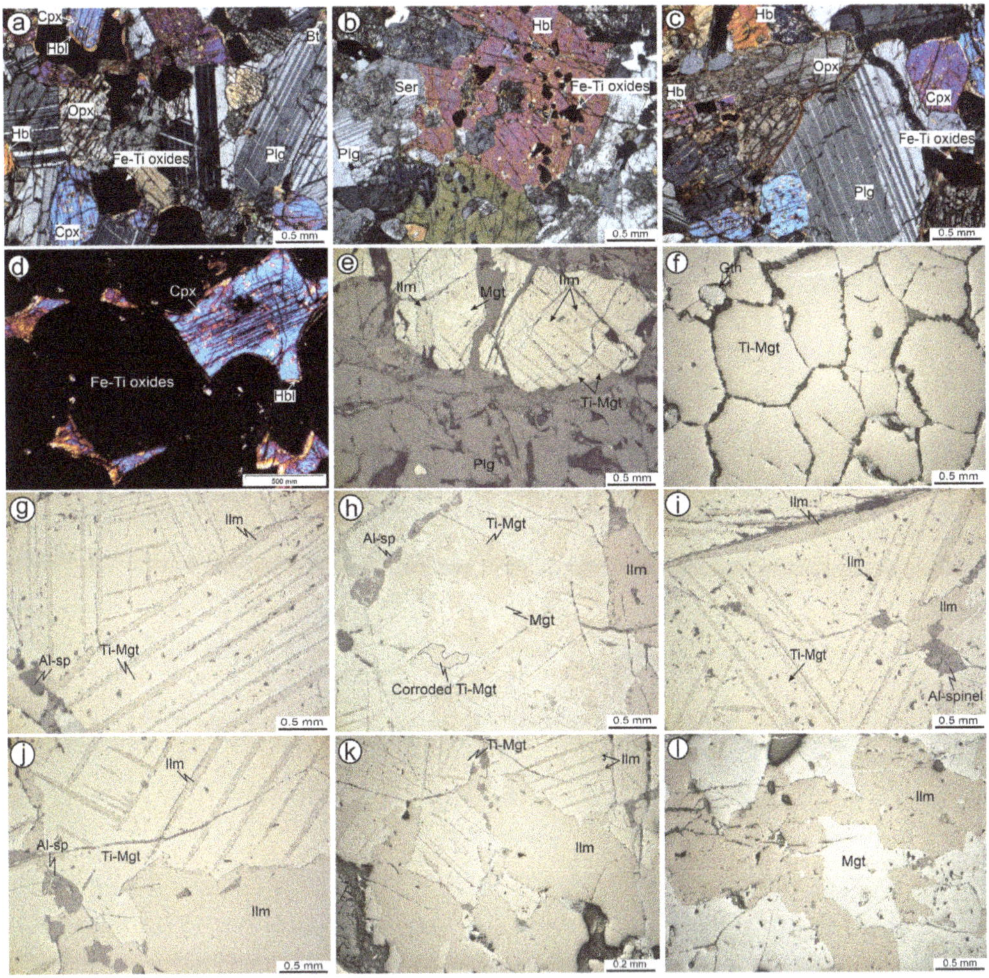

Figure 3. Photomicrographs of Fe–Ti oxide ores and hosting pyroxene gabbro in the El-Baroud intrusion. All photos were taken under reflected light, except a–d taken under crossed polarized light. (**a**) Interstitial Fe-Ti oxides between Cpx, Opx, and plagioclase in pyroxene gabbro. (**b**) Hornblende with Fe-Ti oxide inclusions and altered plagioclase to sericite in the host pyroxene gabbro. (**c**) Prismatic Opx and Cpx granules with Fe-Ti oxides and plagioclase in gabbronorite. (**d**) Interstitial prismatic Cpx in titaniferous iron ores. (**e**) Titanomagnetite–ilmenite intergrowth existence as interstitial phases between silicate minerals. (**f**) Polygonal titanomagnetite with boundaries converging at ~120° triple junctions within Fe-Ti oxide ore. (**g**) Titanomagnetite showing distinct thick sandwich-type ilmenite lamellae, followed by a fine-grained, second-generation trellis-type lamellae. (**h**) Corroded titanomagnetite and converted to interstitial magnetite, with exsolved ilmenite and Al-spinel. (**i**) Triangular-shaped thick lamellae of ilmenite exsolved in the host titanomagnetite. (**j**) Cross-hatching-like texture and coarse grain of ilmenite within the host titanomagnetite, with inclusions of Al-spinel. (**k**) The interstitial ilmenite between titanomagnetite that enclosed ilmenite lamellae. (**l**) Interstitial ilmenite between the altered magnetite matrix, showing sharp contact. Abbreviations, Cpx = clinopyroxene, Opx = orthopyroxene, Hbl = hornblende, Plg = plagioclase, Bt = biotite, Ser = sericite, Ilm = ilmenite, Mgt = magnetite, Ti-Mgt = titanomagnetite, Gth = geothite, Al-sp = Al-spinel.

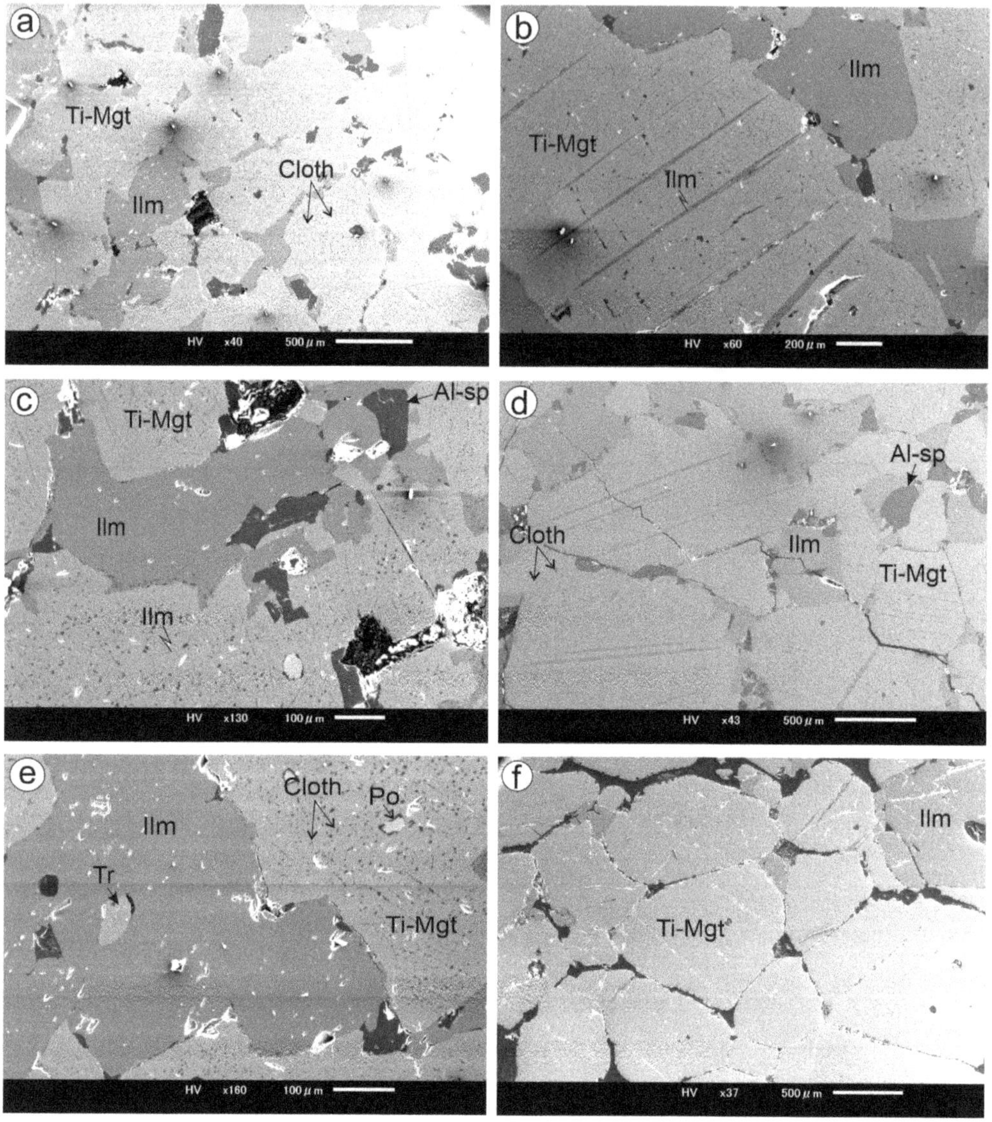

Figure 4. Scanning electron microscopy (SEM) images of Fe-Ti oxide ores in the studied mafic intrusion showing the textures resulting from the exsolution of ilmenite in titanomagnetite. (**a**) Composite-type texture, where granules of ilmenite are hosted by titanomagnetite. (**b**) Sandwich-type texture, as thick primarily oriented ilmenite lamellae in titanomagnetite. (**c**) Trellis-type textures showing thin ilmenite lamellae distributed in the host titanomagnetite; Al-spinel is an accessory mineral. (**d**) Cloth texture, as ultrafine and discontinuous lamellae of ilmenite in titanomagnetite. (**e**) Troilite and pyrrhotite grains hosted by ilmenite and titanomagnetite, respectively, in addition to cloth texture. (**f**) Polygonal titanomagnetite and ilmenite grains with boundaries that meet at ~120° triple junctions. Abbreviations, Ilm = ilmenite, Ti-Mgt = titanomagnetite, Al-sp = Al-spinel, Tr = troilite, Po = pyrrhotite.

Trace element concentrations of the interstitial Cpx in the titaniferous iron ores (Table 3) were in situ determined using laser-ablation (193 nm ArF excimer: MicroLas GeoLas Qplus)-inductively coupled plasma mass spectrometry (Agilent 7500 S) (LA-ICP-MS) at Kanazawa University, Japan. For the analysis, spots with a diameter of 60 μm were ablated at a frequency of 6 Hz, and the energy density was 8 J/cm^2 per pulse. The precision or reproducibility is better than 5% for most elements, except for Sc, Cr, and Ni, where it is better than 12%. Calibration was carried out using international standards; NIST 612 glass was used as an external standard, and ^{29}Si was used as an internal standard based on the SiO_2 value obtained by the EPMA [31]. NIST 614 glass (secondary standard) was measured for quality control in each analysis.

4. Petrography and Oxide Micro-Textures

Pyroxene gabbro is the main host of titaniferous iron ores and comprises plagioclase (50–55 vol.%), clinopyroxene (Cpx: 15–25 vol.%), orthopyroxene (Opx: 3–5 vol.%), and amphibole (5–10 vol.%); it is rich with iron oxides, mainly Ti-magnetite and subordinate ilmenite and magnetite, reaching up to 15 vol.% (Figure 3a,b). Plagioclase is found in the form of columnar hypidiomorphic crystals with a grain size of 0.8–1 mm, showing a granoblastic texture. It exhibits lamellar and percline twinning. An appreciable amount of Cpx and Opx are dominant, and both occur as intercumulus anhedral to prismatic crystals between plagioclase grains. There is a sharp contact with the triple junction between pyroxene and amphibole, suggesting the primary origin. The Cpx and amphibole totally and partially enclose plagioclase minerals, forming ophitic and sub-ophitic textures, which are particularly characteristic of gabbroic rocks (Figure 3a,b). In a few samples, there is amphibole replacement after pyroxene, and an amphibole rim is found around the core of pyroxene, forming a corona texture (Figure 3a).

In mineralogy, gabbronorite resembles pyroxene gabbros, but it is characterized by high modal proportions of pyroxene (up to 35 vol.%); it comprises plagioclase (45 vol.%), Opx (~20 vol.%), Cpx (~15 vol.%), Fe-Ti oxides (~15 vol.%), and traces of amphibole (Figure 3c). Opx is found in the form of long prismatic crystals, but Cpx exists as granules or short prismatic crystals. Exsolution lamellae are noted in both types of pyroxenes. The gabbronorite shows orthocumulate and mesocumulate textures. Pyroxene forms intercumulus anhedral crystals between plagioclase grains. Cpx, or amphibole crystals, enclose laths of plagioclase crystals, forming ophitic and sub-ophitic textures. The Fe-Ti oxides (10 vol.%) are interstitially found between silicate crystals or as inclusions in minerals (Figure 3c).

Some titaniferous iron ore samples contain a significant amount of Cpx, which occurs either in the form of isolated grains or aggregates (Figure 3d); this Cpx is considered a silicate matrix between titanomagnetite and ilmenite crystals. In addition, the Fe-Ti oxides between silicate minerals (15 vol.%) in the host gabbroic rocks consist of titanomagnetite, ilmenite, and magnetite with the same morphology and compositions (Figure 3e).

The Fe-Ti oxide ores hosted by the pyroxene gabbro are identified as titanomagnetite (80–87 vol.%), with subordinate ilmenite (10–15 vol.%) and magnetite (3–5 vol.%) (Figures 3 and 4). The massive and granular-textured ores consist of polygonal grains (0.8–1.2 mm across), predominantly composed of titanomagnetite (Figure 3f). The majority of boundaries separating titanomagnetite grains are generally straight to slightly curved, intersecting at clear triple junctions with $\approx 120^\circ$ interfacial angles, suggesting the magmatic origin (Figures 3f and 4f).

4.1. Textures of Fe-Ti Oxide Ores

The exsolved lamellae of ilmenite in titanomagnetite crystals indicate the true solid solution between both of them, showing exsolution or intergrowth textures (Figure 3g–k) (e.g., [49]). Textural analysis of the studied Fe-Ti oxides indicates that 75% of the oxides exhibit exsolution or intergrowth textures (heterogeneous or multiphase oxides; Figure 3g–k), while 25% of the oxides are free of exsolution or intergrowth lamellae, forming homogeneous or monophase oxides (Figure 3f). The monophase oxides are predominantly titanomagnetite (>95%) with minor ilmenite (<5%) (Figure 3f). The multiphase oxides are dominated by ilmenite exsolution or intergrowth lamellae (10–20%) in the host titanomagnetite (60–80%) (Figure 3g,i). The titanomagnetite occurs as cumulus or polygonal crystals, showing magmatic origin (Figure 3f); it is altered to magnetite that partially grows at the expense of titanomagnetite (Figure 3h). Sometimes, with high alteration and severe interaction with hydrothermal fluids, titanomagnetite was completely converted to magnetite as a matrix around ilmenite (Figure 3l). The Fe-Ti oxides exhibit heterogeneous characteristics with different textures, as follows.

4.1.1. Composite-Type Texture

In this texture type, the ilmenite intergrowth, exsolved as blebs or worm-like crystals, is interstitially found between the titanomagnetite grains. These ilmenite grains of various sizes typically exhibit sharp contacts with the host titanomagnetite (Figures 3k and 4a). Notably, they do not exhibit any preferred orientation within the studied ores.

4.1.2. Sandwich-Type Texture

In this particular texture type, the ilmenite exsolution lamellae are mainly characterized by thick laths within the titanomagnetite. These laths are irregular and oriented, with sharp terminations (Figure 4b).

4.1.3. Trellis-Type Texture

The ilmenite exsolution or intergrowth lamellae appear as 1–3 directional laths within the host titanomagnetite, forming a trellis-like structure. These ilmenite laths exhibit parallel sides and tapered terminations (Figures 3g,j and 4c). In the majority of the oxides exhibiting the trellis-type texture, the ilmenite exsolution lamellae are observed along three sides of the titanomagnetite octahedral planes (three directional).

4.1.4. Cross-Hatching-Like Texture

The exsolution or intergrowth lamellae of ilmenite show two directions perpendicular to each other, creating a pattern similar to the cross-hatching texture (Figures 3j and 4c).

4.1.5. Cloth-Type Texture

In this texture type, extremely ultrafine, discontinuous exsolution or intergrowth lamellae of ilmenite (<1 μm in width) occur as a network within the host titanomagnetite (Figure 4d,e). This cloth texture of fine needle-like ilmenite in the host titanomagnetite is dominant over other textural types (Figure 4).

Some accessory minerals in the titaniferous iron ores include geothite, Al-spinel, and sulfide minerals (chalcopyrite, pyrite, pyrrhotite, pentlandite, troilite, and hormanite) (Figure 5). Sulfides are a few grains disseminated in Fe-Ti oxide minerals. Chalcopyrite and pyrite are rimmed by secondary geothite (Figure 5a,b). Chalcopyrite and hormanite are exsolved as bands within troilite, and all are hosted by magmatic titanomagnetite (Figure 5c). In addition to the intergrowth between pentlandite and pyrrhotite (Figure 5d), chalcopyrite is converted to goethite, which grows at the expense of the former (Figure 5e). The numerous inclusions of Al-spinel are noted within titanomagnetite and between Fe-Ti oxide minerals (Figure 5f).

Figure 5. Photomicrographs of accessory minerals (sulfides, goethite, and Al-spinel) in titaniferous iron ores under reflected light. (**a**) Goethite with colloform texture and chalcopyrite in the El-Baroud pyroxene gabbros. (**b**) Pyrite within goethite and ilmenite in pyroxene gabbros. (**c**) Bands of chalcopyrite and hormanite within troilite crystal embedded in titanomagnetite. (**d**) Pyrrhotite–pentlandite exsolution within titanomagnetite, where one grain of pyrrhotite–pentlandite is embedded in titanomagnetite, taken as a scanning electron microscopy (SEM) image. (**e**) Converted chalcopyrite to goethite growing on the expenses of the former. (**f**) Exsolved Al-spinel inclusions in titanomagnetite. Abbreviations, Gth = goethite, Cpy = chalcopyrite, Py = pyrite, Ilm = ilmenite, Ti-Mgt = titanomagnetite, Ho = hormanite, Tr = troilite, Po = pyrrhotite, Pn = pentlandite, Al-sp = Al-spinel.

5. Ore Bulk Chemistry

5.1. Major Oxides

Based on the TAS nomenclature diagram of Cox et al. [50], the studied samples were classified into gabbro (Figure 6a). For the major elements of the host pyroxene gabbro and gabbronorite, SiO_2 contents vary between 34.0 and 53.33 wt.%, and Al_2O_3 contents range from 9.92 to 18 wt.% (Table 1). Al_2O_3 is positively correlated with Sr (395–768 ppm) (Figure 6b), suggesting that plagioclase is the main host of Sr [51]. CaO and Na_2O range from 8.21 to 13.05 wt.% and 2.34 to 4.74 wt.%, respectively, while K_2O displays a relatively limited range between 0.25 and 0.89 wt.%. There is a notable increase in the proportions of Fe_2O_3 and TiO_2, which range from 8.8 to 16.5 wt.% (except sample Mt. 93) and 0.9 to 7.19 wt.%, respectively. This is due to the different amounts of disseminated Fe-Ti oxides (up to 15 vol.% in sample Mt. 93). The studied gabbros are enriched with P_2O_5 up to 2.48 wt.%.

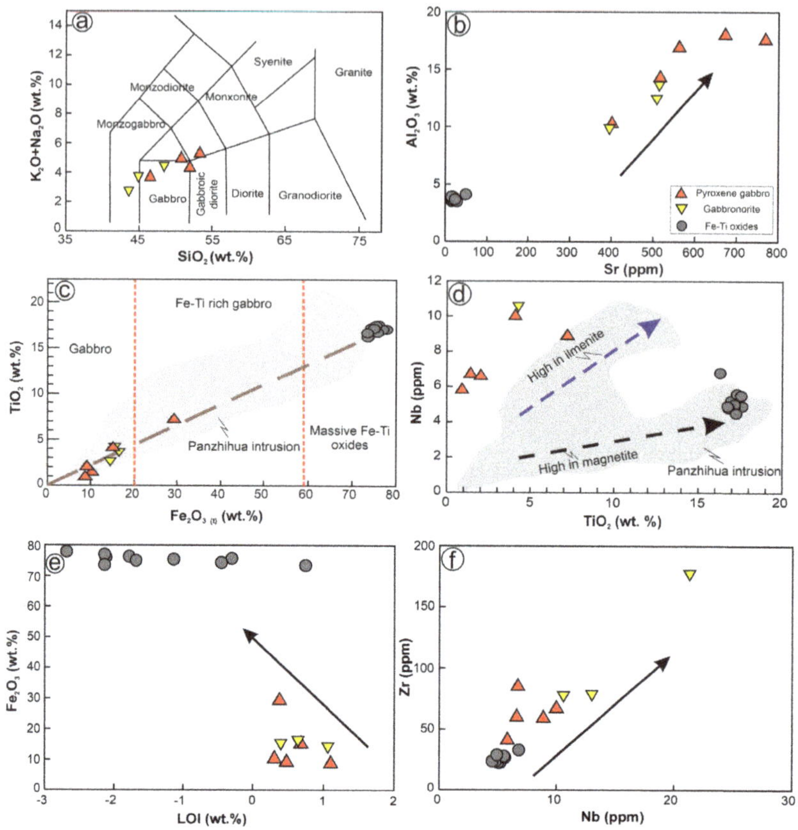

Figure 6. Binary variation diagrams of whole-rock geochemical analyses of the El-Baroud mafic intrusion. (**a**) Nomenclature TAS diagram for the studied samples [50]. (**b**) Sr versus Al_2O_3. (**c**) Fe2O3(t) versus TiO2. (**d**) TiO2 versus Nb. (**e**) LOI versus Fe_2O_3. (**f**) Nb versus Zr. The fields of the Panzhihua intrusion [52] are used for comparison.

The massive titaniferous iron ores are characterized by their high Fe_2O_3 content, which ranges from 73.48 to 77.84 wt.% (75.66 wt.% on average). The SiO_2 content is relatively low, ranging from 0.58 to 3.59 wt.%. A relatively high TiO_2 content ranges from 16.30 to 17.60 wt.% due to the subordinate ilmenite and TiO_2 content of titanomagnetite. The studied ores have significant amounts of Al_2O_3 (3.5–4.1 wt.%). The CaO and V_2O_5 contents are low, with an average of 0.36 wt.% and 0.44 wt.%, respectively (Table 1). The concentration of TiO_2 shows a well-positive correlation with $Fe_2O_{3(t)}$, and TiO_2 is also positively correlated with Nb; this resembles those of the Panzhihua intrusion (Figure 6c,d) [52].

The titaniferous iron ores exhibit very low loss of ignition (LOI) values, from −2.68 to 0.74 wt.%, but the host gabbros are relatively moderate, falling within the range of 0.31 to 1.1 wt.%, reflecting that these samples are relatively fresh (Table 1). Negative LOI of titanomagnetite samples means that the iron is in the form of bivalent iron; the bivalent iron will absorb oxygen and be oxidized while burning the powders of ores to calculate the LOI; then, it gains weight rather than losing it. This leads to a negative LOI when the mass increase surpasses the mass loss from burning. This usually occurs when the sample is in a reduced state (e.g., FeO) and is simple to oxidize while burning. Therefore, the higher the iron content in the sample, the lower the LOI (Figure 6e).

Table 1. Major (wt.%) and trace element (ppm) abundances of El-Baroud gabbros and their titaniferous iron ores.

Rock	Titaniferous Iron Ores										Pyroxene Gabbro					Gabbronorite		
Sam.#	**Mt.62**	**Mt.64**	**Mt.94**	**Mt.102**	**Mt.105**	**Mt.118**	**Mt.120**	**Mt.121**	**Mt.125**	**Mt.128**	**Mt.28**	**Mt.90**	**Mt.93**	**Mt.119**	**Mt.124**	**Mt.18**	**Mt.37**	**Mt.97**
SiO_2	1.26	1.28	0.58	0.91	0.75	3.59	1.14	0.70	1.93	1.71	51.98	53.33	34.07	50.82	46.55	44.88	48.49	43.54
TiO_2	16.95	17.60	17.15	17.15	16.80	16.30	17.30	17.55	17.25	16.75	1.40	0.90	7.19	2.01	4.07	3.71	2.68	4.26
$Al2O_3$	3.82	3.91	3.50	3.70	3.48	4.12	3.65	3.66	3.90	3.66	16.92	17.58	10.27	18.00	14.30	12.51	13.74	9.92
Fe_2O_{3t}	76.27	76.04	77.84	76.92	75.74	73.59	74.38	75.40	74.99	73.48	10.30	8.80	29.31	9.19	15.09	16.51	14.48	15.50
MnO	0.37	0.38	0.38	0.38	0.37	0.39	0.35	0.38	0.38	0.28	0.16	0.20	0.25	0.13	0.18	0.24	0.22	0.25
MgO	2.27	2.81	2.43	2.33	2.38	2.92	3.05	2.57	3.09	2.66	5.45	3.63	5.48	3.72	5.49	5.48	4.32	7.24
CaO	0.17	0.18	0.09	0.10	0.25	0.62	0.22	0.16	0.37	0.31	9.18	8.29	8.21	9.52	9.74	8.99	8.24	13.05
K_2O	<0.01	<0.01	<0.01	<0.01	<0.01	0.01	<0.01	<0.01	<0.01	<0.01	0.32	0.53	0.25	0.42	0.32	0.75	0.89	0.29
Na_2O	<0.01	<0.01	<0.01	<0.01	<0.01	0.18	<0.01	<0.01	0.01	<0.01	3.98	4.74	2.34	4.50	3.34	3.00	3.57	2.49
P_2O_5	<0.01	0.03	<0.01	<0.01	<0.01	0.12	0.01	0.05	<0.01	0.01	0.35	0.40	1.60	0.82	0.40	2.48	1.58	2.48
LOI	−1.79	−2.12	−2.68	−2.14	−0.31	−2.14	−0.46	−1.14	−1.69	0.74	0.31	1.10	0.38	0.48	0.68	0.64	1.06	0.40
Sum	101.11	102.23	101.97	101.49	99.77	101.84	100.10	100.47	101.92	98.86	100.04	98.40	98.97	99.13	99.48	99.19	99.27	99.42
Mg#	0.050	0.062	0.053	0.051	0.053	0.066	0.068	0.057	0.068	0.061	0.485	0.424	0.250	0.419	0.393	0.372	0.347	0.454
Trace and REE elements (ppm)																		
Li	1.0	2.5	0.6	0.8	4.2	1.0	0.4	0.5	0.7	5.9	5.3	4.6	2.1	4.7	1.7	6.9	7.2	3.5
Cs	<0.01	<0.01	<0.01	<0.01	<0.01	<0.01	<0.01	<0.01	<0.01	<0.01	0.04	0.25	0.01	0.05	0.03	0.2	0.2	0.1
Rb	<0.2	<0.2	<0.2	<0.2	<0.2	0.4	<0.2	<0.2	0.2	0.3	1.7	6.7	2.0	3.7	3.0	14.2	13.6	3.1
Ba	18.2	9.2	8.7	27.3	9.9	68.5	17.4	8.8	20.3	13.4	220	428	117.0	184.5	135.5	213.0	292.0	113.0
Th	0.02	0.03	0.01	0.02	0.01	0.09	0.02	0.02	0.01	0.02	1.52	0.18	0.32	0.45	0.27	2.1	1.5	0.4
U	0.07	0.08	0.09	0.11	0.06	0.24	0.06	0.10	0.05	0.13	0.54	0.06	0.11	0.15	0.08	0.57	0.55	0.13
Nb	5.2	4.9	4.9	5.0	4.8	6.8	5.6	5.5	4.5	4.9	6.7	5.8	8.9	6.6	10.0	13.0	21.3	10.6
Ta	0.4	0.3	0.3	0.4	0.3	0.5	0.4	0.4	0.4	0.3	0.3	0.3	0.6	0.4	0.7	0.8	1.1	0.6
Pb	<0.5	0.9	<0.5	0.8	<0.5	<0.5	<0.5	<0.5	<0.5	<0.5	3.5	4.2	0.8	1.6	0.6	3.0	4.7	0.7
Sr	20.7	17.0	15.8	18.2	27.8	49.2	16.2	17.7	14.8	23.7	562	768	401	672	516	508	514	395
Zr	23	28	24	23	24	33	27	28	24	29	85	41	59	60	67	79	178	78
Hf	0.7	0.8	0.8	0.8	0.8	1.0	0.9	0.9	0.8	0.9	2.5	1.1	1.6	1.6	1.8	2.1	4.3	2.1
Y	0.7	1.5	0.4	0.4	0.7	5.3	1.4	1.5	1.5	1.5	36.3	21.0	26.9	19.8	17.5	40.5	48.0	43.0
La	0.4	0.8	0.4	0.3	0.4	4.0	0.6	0.8	0.6	0.8	14.7	10.4	12.0	10.4	6.8	27.4	31.8	18.5
Ce	1.1	2.0	0.8	1.1	1.2	8.8	2.2	1.9	1.5	1.9	37.0	25.1	33.0	26.1	18.0	74.6	79.8	49.1
Pr	0.16	0.31	0.12	0.10	0.17	1.22	0.25	0.30	0.22	0.28	5.10	3.70	5.40	4.02	2.87	10.85	11.70	8.08
Nd	0.7	1.5	0.5	0.5	0.9	5.5	1.4	1.5	1.3	1.4	23.0	17.8	28.0	19.8	14.9	52.4	54.1	43.6
Sm	0.17	0.35	0.10	0.11	0.21	1.13	0.34	0.38	0.34	0.32	5.88	4.29	7.03	4.72	3.76	12.80	11.45	10.70

Table 1. *Cont.*

Rock	Titaniferous Iron Ores										Pyroxene Gabbro					Gabbronorite		
Sam.#	Mt.62	Mt.64	Mt.94	Mt.102	Mt.105	Mt.118	Mt.120	Mt.121	Mt.125	Mt.128	Mt.28	Mt.90	Mt.93	Mt.119	Mt.124	Mt.18	Mt.37	Mt.97
Eu	0.09	0.14	0.04	0.03	0.11	0.36	0.10	0.12	0.12	0.13	1.79	3.38	2.82	2.44	1.91	3.99	3.90	4.18
Gd	0.21	0.44	0.12	0.08	0.22	1.31	0.39	0.43	0.38	0.38	6.29	4.85	7.94	5.27	4.43	12.05	12.45	12.55
Tb	0.02	0.06	0.01	0.02	0.03	0.17	0.05	0.05	0.06	0.05	0.98	0.73	1.07	0.75	0.64	1.63	1.77	1.68
Dy	0.13	0.29	0.07	0.06	0.17	0.95	0.29	0.31	0.31	0.28	5.99	4.15	5.68	4.12	3.57	8.69	9.65	8.80
Ho	0.02	0.05	0.02	0.01	0.03	0.18	0.05	0.06	0.06	0.05	1.32	0.80	1.02	0.73	0.67	1.62	1.78	1.60
Er	0.05	0.10	<0.03	<0.03	0.05	0.42	0.10	0.13	0.11	0.09	3.62	2.14	2.36	1.81	1.66	3.97	4.57	3.78
Tm	0.01	0.02	0.01	0.01	0.02	0.06	0.02	0.02	0.05	0.02	0.52	0.31	0.32	0.26	0.24	0.50	0.64	0.51
Yb	0.05	0.10	0.03	0.04	0.07	0.31	0.10	0.12	0.11	0.10	3.28	1.73	1.56	1.33	1.30	2.84	3.39	2.54
Lu	0.01	0.02	0.01	0.01	0.01	0.06	0.02	0.02	0.02	0.02	0.50	0.29	0.25	0.23	0.21	0.42	0.57	0.41
Sc	18.8	18.6	16.0	15.2	15.3	16.6	15.6	15.4	16.8	15.0	27.3	24.8	30.8	21.9	30.9	26.1	24.8	44.6
V	2030	1970	2000	2130	1840	1795	2050	1900	2100	2140	230	58	698	129	252	244	240	248
Cr	920	320	890	930	850	360	420	370	450	500	60	50	180	30	60	30	80	50
Co	211	210	188.0	182.0	164.0	190.5	226	241	216	185.5	37.3	17.8	53.5	27.0	40.1	43.5	36.9	46.8
Ni	306	247	282	244	190.0	162.5	290	241	262	249	62.4	9.4	43.4	9.7	11.2	8.6	24.8	7.1
Cu	18.4	45.3	1.1	8.3	4.2	4.8	8.8	3.8	12.6	13.4	21.8	10.4	27.6	21.2	76.4	23.9	33.3	22.8
Zn	439	458	420	448	435	479	428	487	438	292	124	106	161	69	103	156	162	113
Ga	46.7	45.2	48.6	45.7	43.0	46.2	44.5	46.0	43.8	41.8	20.7	21.7	25.9	21.3	18.85	20.6	21.4	16.10
Be	<0.05	<0.05	<0.05	<0.05	<0.05	<0.05	<0.05	<0.05	<0.05	<0.05	0.99	0.80	0.26	0.58	0.38	0.78	1.03	0.42
Ti*	101,588	105,484	102,787	102,787	100,689	97,692	103,686	105,184	103,386	100,389	8390	5394	43,092	12,046	24,393	22,236	16,062.34	25531.9
Mn	2870	2750	2810	2850	2830	3000	2570	2850	2780	2090	1215	1450	1800	987	1385	1790	1590	1835
Ge	0.35	0.34	0.62	0.46	0.33	0.58	0.35	0.34	0.40	0.32	0.16	0.17	0.17	0.14	0.14	0.21	0.21	0.18
Mo	4.42	2.44	3.83	4.55	12.20	3.18	10.25	6.30	2.15	12.60	2.48	4.12	2.09	2.55	4.26	3.04	4.47	3.67
Cd	0.03	0.05	0.03	0.03	0.04	0.03	0.04	0.04	0.04	0.04	0.15	0.19	0.11	0.09	0.11	0.12	0.17	0.14
Sn	1.8	1.6	1.7	1.6	1.7	1.8	1.6	1.6	1.5	1.6	1.0	0.4	1.0	0.6	0.6	0.2	1.3	<0.2
Sb	<0.05	<0.05	<0.05	<0.05	<0.05	<0.05	<0.05	<0.05	<0.05	<0.05	<0.05	0.11	<0.05	<0.05	<0.05	<0.05	<0.05	<0.05
Cu/Ni	0.06	0.18	0.00	0.03	0.02	0.03	0.03	0.02	0.05	0.05	0.35	1.11	0.64	2.19	6.82	2.78	1.34	3.21
SREE	3.12	6.18	2.25	2.39	3.59	24.47	5.91	6.14	5.18	5.82	109.97	79.67	108.45	81.98	60.96	213.76	227.57	166.03
SLREE	2.83	5.54	2.08	2.22	3.21	22.32	5.28	5.43	4.46	5.21	93.76	69.52	96.19	72.75	52.67	194.1	205.2	146.7
SHREE	0.29	0.64	0.17	0.17	0.38	2.15	0.63	0.71	0.72	0.61	16.21	10.15	12.26	9.23	8.29	19.67	22.37	19.32

Fe_2O_{3t}: the total iron; Ti*: calculated Ti ppm from TiO_2 wt.% by XRF.

5.2. Trace and REE

The titaniferous iron ores have concentrations of Mn (up to 3000 ppm), V (1795–2140 ppm), Cr (601 ppm), Zn (432.4 ppm), Ni (247.4 ppm), and Co (201.4 ppm on average) (Table 1). Vanadium has moderate and very high concentrations in the host gabbro samples, ranging from 58–698 ppm, while Ni and Zr concentrations show considerable diversity, ranging from 7.1 to 62.4 ppm and 41 to 178, respectively. There is a sharp positive correlation between Zr and Nb (5.8–21.3 ppm) (Figure 6f).

Scandium displays limited variation, ranging from 21.9 to 44.6 ppm. Although Sc contents are typically associated with Cpx crystallization (e.g., [53]). There is no observable correlation between Sc values and the Cpx modal abundance in the gabbroic rocks. The gabbroic rocks contain relatively high values of Ba (up to 428 ppm), and the control of Ba contents is commonly attributed to plagioclase crystallization [54].

The titaniferous iron ores are poor in rare earth elements (REE: 2.25–24.5 ppm) relative to their host gabbro (ΣREE: 61–110 ppm), where their light rare earth elements (ΣLREE: 2–22.3 ppm) have higher values than the heavy rare earth elements (ΣHREE: 0.17–2.1 ppm) (Table 1). The chondrite-normalized REE of the studied titaniferous iron ores shows nearly flat patterns, similar to the REE patterns of Fe-Ti oxide ores in the Panzhihua intrusion. Their REE and spider patterns coincide with those of the Panzhihua intrusion (Figure 7a,b) [5,52].

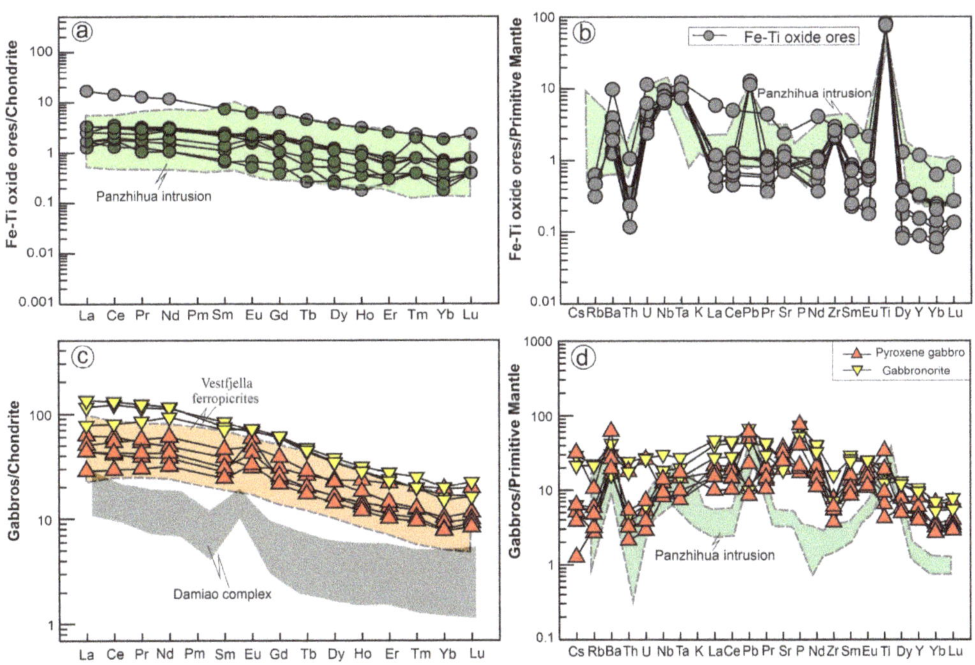

Figure 7. Whole-rock chemistry of the El-Baroud intrusion. (**a**) The chondrite-normalized REE patterns of the El-Baroud Fe-Ti oxide ores compared with those from the Panzhihua intrusion [5]. (**b**) The primitive mantle-normalized trace element patterns of the El-Baroud Fe-Ti oxide ores compared with the Panzhihua intrusion [52]. (**c**) The chondrite-normalized REE patterns of the El-Baroud pyroxene gabbro compared with those from the Damiao complex [28] and Vestfjella ferropicrites [55]. (**d**) The primitive mantle-normalized trace element patterns of the studied pyroxene gabbro compared with the Panzhihua intrusion [52]. Trace element concentrations of chondrites and primitive mantle are from Sun and McDonough [56].

Chondrite-normalized REE of the studied gabbros shows an inclined pattern, reflecting the enrichments of LREE relative to HREE contents with small positive Eu anomalies, similar to those of the Vestfjella ferropicrite and Damiao complex [28,55]. The REE patterns normalized to the primitive mantle also coincide with those of the Panzhihua intrusion (Figure 7c,d) [5,52].

6. Mineral Chemistry

6.1. Major Oxides

The Fe-Ti oxides in the El-Baroud area include titanomagnetite, ilmenite, and magnetite with few amounts of goethite; this is consistent with the compositions of the Fe-Ti oxide ores identified in the Korab Kansi intrusion [29,31] (Figure 8a). The El-Baroud titanomagnetite has TiO_2, Fe_2O_3, and FeO contents ranging from 0.18–16.62 wt.%, 56.4–70.14 wt.%, and 21.88–31.33 wt.%, respectively (Table 2). The Al_2O_3 content of this titanomagnetite is relatively high, ranging from 0.19 to 4.48 wt.% (2.33 wt.% on average). The Cr_2O_3 content is notably low, ranging from 0.01 to 0.15 wt.%, relative to Cr_2O_3 in the Rahaba ores (0.28–0.71 wt.%) [32].

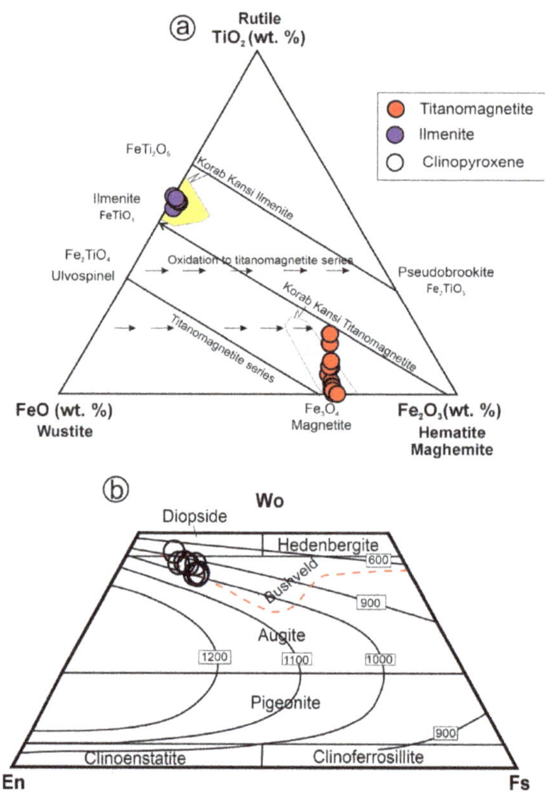

Figure 8. Tri-elemental variation diagrams of titanomagnetite, ilmenite, and Cpx from the El-Baroud intrusion. (**a**) Triangular diagram (FeO-TiO_2-Fe_2O_3) of Fe-Ti oxides exhibiting oxides, mainly titanomagnetite, ilmenite, and magnetite (fields of Korab Kansi titanomagnetite and ilmenite are from Khedr et al. [29]. (**b**) Wo-En-Fs Cpx nomenclature diagram [57]. The isothermal lines are from Lindsley Stromer Jr. [58], and the dashed red line indicates the Bushveld trend [59].

Table 2. Representative electron microprobe analyses of ilmenite, magnetite, and clinopyroxene in the El-Baroud gabbroic intrusion, Egypt.

Rock	G.Nr	Fe-Ti Ores							G.Nr	Fe-Ti Ores							Px.Gb	Fe-Ti Ores		Px.Gb
					Px.Gb		Fe-Ti Ores													
Sample#	Mt.18	Mt.63	Mt. 64		Mt. 93		Mt. 94		Mt.97	Mt. 107				Mt. 118			Mt. 119	Mt. 121		Mt. 124
Mineral	Ilm	Mgt	Ilm	Mgt	Ilm	Mgt	Ilm	Mgt	Mgt	Ilm	Mgt	Cpx	Cpx	Mgt	Cpx	Cpx	Ilm	Ilm	Mgt	Ilm
SiO_2	0.01	0.00	0.03	0.03	0.03	0.02	0.00	0.03	0.02	0.03	0.02	52.93	52.18	0.02	53.05	52.59	0.00	0.00	0.03	0.02
TiO_2	48.49	3.56	54.27	1.05	53.26	0.36	54.75	2.09	0.16	52.74	5.61	0.66	1.02	14.41	0.79	0.87	49.97	54.01	16.62	53.39
Al_2O_3	0.01	0.74	0.02	0.47	0.00	0.37	0.00	0.33	0.22	0.01	1.26	2.52	2.78	2.21	2.58	2.60	0.02	0.02	4.48	0.02
Cr_2O_3	0.00	0.09	0.00	0.06	0.00	0.06	0.04	0.12	0.00	0.01	0.13	0.00	0.01	0.04	0.00	0.02	0.01	0.02	0.11	0.03
FeO*	50.67	90.59	42.48	93.32	46.35	93.87	40.32	93.08	94.09	44.52	84.63	8.81	9.69	77.52	7.72	5.46	49.11	42.45	72.63	46.41
MnO	1.27	0.08	1.03	0.01	0.97	0.01	1.17	0.03	0.00	1.23	0.12	0.32	0.34	0.43	0.27	0.16	1.13	1.08	0.36	0.91
MgO	0.10	0.48	3.87	0.40	0.45	0.03	4.26	0.38	0.01	1.74	0.46	14.29	14.49	1.94	14.48	15.26	0.08	3.88	2.96	0.55
CaO	0.00	0.00	0.00	0.00	0.00	0.00	0.00	0.00	0.00	0.00	0.00	20.73	19.58	0.00	20.27	21.94	0.00	0.00	0.00	0.00
Na_2O	0.03	0.00	0.01	0.01	0.00	0.00	0.00	0.02	0.03	0.00	0.00	0.82	0.69	0.00	1.34	1.00	0.00	0.00	0.00	0.00
K_2O	0.01	0.00	0.00	0.00	0.00	0.00	0.00	0.00	0.00	0.01	0.00	0.00	0.00	0.00	0.00	0.03	0.01	0.00	0.00	0.00
NiO	0.01	0.05	0.01	0.00	0.00	0.00	0.03	0.02	0.00	0.00	0.00	0.00	0.03	0.01	0.00	0.00	0.00	0.02	0.06	0.00
Total	100.6	95.6	101.7	95.3	101.0	94.7	100.6	96.1	94.5	100.3	92.2	101.1	100.8	96.6	100.5	99.9	100.3	101.5	97.2	101.3
Mg#	2.09	0.0	1.86	0.02	1.97	0.00	1.82	0.02	0.00	1.93	0.03	0.74	0.73	0.13	0.77	0.83	2.0	1.86	0.19	1.96
Fe#	1.0	1.0	1.1	1.0	1.0	1.0	1.2	1.0	1.0	1.1	1.0	0.3	0.3	0.9	0.2	0.2	1.0	1.1	0.8	1.0
Ti	0.94	0.13	0.99	0.04	1.00	0.01	1.00	0.08	0.01	0.99	0.20	0.02	0.03	0.46	0.02	0.02	0.96	0.99	0.51	1.00
Al	0.00	0.04	0.00	0.03	0.00	0.02	0.00	0.02	0.01	0.00	0.07	0.11	0.12	0.11	0.11	0.11	0.00	0.00	0.21	0.00
Cr	0.00	0.00	0.00	0.00	0.00	0.00	0.00	0.00	0.00	0.00	0.01	0.00	0.00	0.00	0.00	0.00	0.00	0.00	0.00	0.00
Mn	0.03	0.00	0.02	0.00	0.02	0.00	0.02	0.00	0.00	0.03	0.01	0.01	0.01	0.02	0.01	0.00	0.02	0.02	0.01	0.02
Mg	0.00	0.03	0.14	0.03	0.02	0.00	0.15	0.03	0.00	0.06	0.03	0.78	0.80	0.12	0.79	0.84	0.00	0.14	0.18	0.02
Fe^{3+}	0.15	1.87	0.03	1.95	0.00	1.97	0.00	1.92	1.98	0.02	1.78	0.03	0.03	1.54	0.06	0.06	0.10	0.03	1.42	0.00
Fe^{2+}	0.91	0.92	0.83	0.96	0.96	0.99	0.82	0.95	1.00	0.90	0.89	0.24	0.27	0.69	0.18	0.11	0.93	0.82	0.61	0.96
Xilm	0.8	-	0.8	-	1.0	-	0.8	-	-	0.9	-	-	-	-	-	-	0.9	0.8	-	1.0
Xusp	-	0.100	-	0.0	-	0.0	-	0.1	0.0	-	0.2	-	-	0.4	-	-	-	-	0.4	-
Wo	-	-	-	-	-	-	-	-	-	-	-	43.4	41.2	-	43.4	46.1	-	-	-	-
En	-	-	-	-	-	-	-	-	-	-	-	41.6	42.4	-	43.2	44.6	-	-	-	-
Fs	-	-	-	-	-	-	-	-	-	-	-	14.9	16.5	-	13.4	9.2	-	-	-	-
F1	-	-	-	-	-	-	-	-	-	-	-	−0.9	−0.9	-	−0.9	−0.9	-	-	-	-
F2	-	-	-	-	-	-	-	-	-	-	-	−2.5	−2.5	-	−2.5	−2.5	-	-	-	-

Mg# = (Mg/Mg + Fe); Fe# = (Fe/Fe + Mg); Xilm = $[Fe^{2+}/(Fe^{2+} + Fe^{3+} + Mg + Mn)]$; Xusp = $[Ti^{+4}/(Ti^{+4} + Fe^{+3}/2)]$; Wo = Ca/(Ca + Mg + Fe + Mn); En = Mg/(Mg + Ca + Fe + Mn); Fs = (Fe + Mn)/(Mg + Fe + Mn + Ca); F1 = −0.012 × SiO_2 − 0.0807 × TiO_2 + 0.0026 × Al_2O_3 − 0.0012 × FeO* − 0.0026 × MnO + 0.0087 × MgO − 0.0128 × CaO − 0.0419 × Na_2O; F2 = −0.0469 × SiO_2 − 0.0818 × TiO_2 − 0.0212 × Al_2O_3 − 0.0041 × FeO* − 0.1435 × MnO − 0.0029 × MgO + 0.0085 × CaO + 0.0160 × Na_2O; Abbreviations, G.Nr = gabbronorite, Px.Gb = pyroxene gabbro.

The investigated ilmenite has significant amounts of TiO_2, FeO, and Fe_2O_3 contents within the ranges of 52.74–54.75 wt.%, 40.13–43.12 wt.%, and 0.13–2.19 wt.%, respectively (Table 2). Magnesium and manganese are incompatible within magnetite and are notably enriched in the ilmenite phase (MgO, up to 4.26 wt.%; MnO, 1.27 wt.%) compared to the coexisting magnetite (MgO, up to 2.96 wt.%; MnO, 0.43 wt.%).

The interstitial Cpx (Mg#: 0.72–0.83) in the titaniferous iron ores is characterized by endmembers ranging from $En_{41.6-44.9}$, $Fs_{9.2-16.5}$, and $Wo_{41.2-46.1}$, suggesting a diopside composition (Figure 8b). It is highly depleted in Cr_2O_3 (<0.1 wt.%). The concentrations of Al_2O_3, TiO_2, and MgO in the Cpx range between 1.82–2.9 wt.%, 0.66–1.02 wt.%, and 14.29–15.41wt.%, respectively (Table 2).

6.2. In Situ Trace and REE

REEs of the interstitial Cpx in the titaniferous iron ores were normalized to chondrite, while trace elements were normalized to the primitive mantle [56] (Figure 9). The chondrite-normalized REE patterns of this Cpx (Figure 9a) display a nearly convex upward shape, indicating the depletion of LREE (from La to Nd) relative to HREE (Tb–Lu) (Table 3). These REE patterns resemble the Cpx patterns in abyssal peridotites [60] and Lalibela ferropicrites [61] (Figure 9a). The primitive mantle-normalized spider patterns of this Cpx display negative anomalies for Rb, Ba, Nb, Ta, Pb, Ni, Sr, and Zr (Figure 9b), similar to those of Lalibela ferropicrites. The depletion of highly incompatible elements, such as high field strength elements (HFSE: Nb, Ta, Ti and Zr), along with Pb and Cr, in the studied Cpx suggests the coexistence of accessory phases, particularly ilmenite and magnetite.

Table 3. Representative LA-ICP-MS (ppm) of interstitial clinopyroxene in Fe-Ti oxides of the El-Baroud intrusion, Egypt.

Rock Name	Titaniferous Iron Ores					
Sample#	Mt. 107			Mt. 118		
Mineral	Cpx					
Li	0.90	1.02	0.65	0.12	0.63	0.09
B	26.80	27.30	24.51	25.29	24.04	23.74
Sc	93.96	98.37	92.41	105.98	87.46	96.18
V	184.83	182.61	172.05	177.75	163.68	192.86
Cr	35.96	35.68	38.47	17.09	15.75	18.99
Co	45.09	73.18	38.94	29.23	28.12	32.09
Ni	49.45	78.95	34.31	17.20	17.35	19.42
Rb	0.11	0.94	0.00	0.00	1.41	0.23
Sr	49.12	60.35	44.25	67.15	73.92	67.83
Y	31.48	32.10	28.92	34.26	32.65	31.43
Zr	30.96	36.87	30.22	38.00	40.89	30.43
Nb	0.19	0.76	0.12	0.15	0.45	0.47
Ba	3.12	36.36	2.60	5.39	7.32	5.93
Hf	1.22	1.49	1.17	1.74	1.50	1.29
Ta	0.00	0.09	0.00	0.01	0.06	0.01
Pb	0.26	0.31	0.15	0.46	0.72	0.50
Th	0.08	0.06	0.04	0.10	0.37	0.13
La	2.76	2.87	2.43	3.64	4.44	3.82
Ce	11.48	11.88	10.19	14.33	16.24	13.32
Pr	2.39	2.55	2.13	2.87	3.13	2.61
Nd	15.34	16.39	13.90	17.48	18.26	15.67
Sm	5.78	6.00	5.21	6.17	6.20	5.50
Eu	2.18	2.34	2.08	2.43	2.35	2.29
Gd	7.11	7.35	6.40	7.57	7.21	6.66
Tb	1.06	1.12	0.98	1.17	1.12	1.05
Dy	6.70	7.10	6.17	7.44	7.06	6.79
Ho	1.31	1.35	1.18	1.37	1.34	1.28
Er	3.33	3.50	3.04	3.53	3.41	3.25

Table 3. *Cont.*

Rock Name	Titaniferous Iron Ores					
Sample#	Mt. 107			Mt. 118		
Mineral	Cpx					
Tm	0.46	0.46	0.41	0.47	0.46	0.45
Yb	2.75	2.88	2.59	2.89	3.02	2.87
Lu	0.40	0.41	0.36	0.43	0.41	0.37
∑REE	63.04	66.20	57.09	71.80	74.66	65.93
∑HREE	16.00	16.82	14.74	17.30	16.83	16.05
∑LREE	47.04	49.38	42.35	54.50	57.83	49.88
∑LREE	74.62	74.60	74.17	75.91	77.46	75.65
$(La/Yb)_N$	0.68	0.68	0.64	0.86	0.99	0.90
$(La/Gd)_N$	0.33	0.33	0.32	0.40	0.52	0.48
$(Gd/Yb)_N$	2.10	2.10	1.99	2.10	1.93	1.88

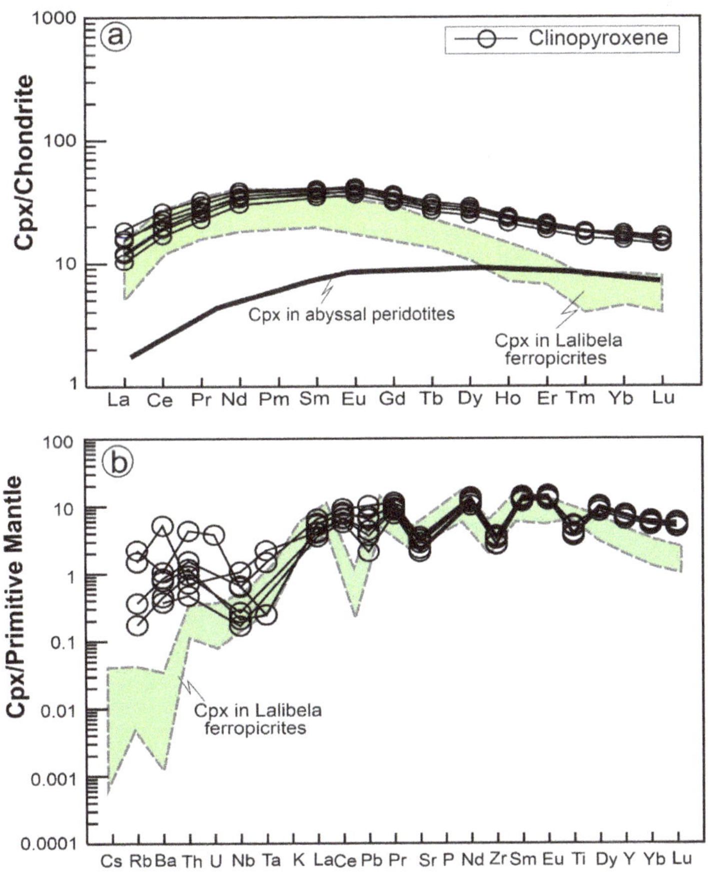

Figure 9. REE and trace element patterns of Cpx in magnetite from Wadi El-Baroud intrusion. (**a**) Chondrite-normalized REE pattern compared with the Cpx in abyssal peridotite [60] and Lalibela ferropicrites [61]. (**b**) Primitive mantle-normalized REE pattern compared with Lalibela ferropicrites. Normalized chondrite and primitive mantle values are from Sun and McDonough [56].

7. Discussion

7.1. The Physico-Chemical Conditions (P, T, fO2) of El-Baroud Gabbroic Intrusion

Titanomagnetite reaches saturation before ilmenite in high oxygen fugacity (fO_2) conditions, but this crystallization order is reversed under low oxygen fugacity in basaltic systems [62,63]. The predominance of titanomagnetite (80–87 vol.%) over ilmenite (10–15 vol.%) in El-Baroud titaniferous iron ores (Figure 3f–l) can be attributed to the high oxygen fugacity (fO_2) conditions (Figure 10a) and the low Ti-Fe ratio of the parent melts (e.g., [31]). The abundance of hydrous minerals (Figure 3a–d) and high LREE relative to HREE (Figure 7) in both Fe-Ti oxide ores and their host gabbros indicate the hydrous condition of parent melts due to the addition of H_2O from the slab-derived fluids, forming metasomatized sub-arc lithospheric mantle; this huge H_2O is a significant source of free oxygen for buffering the system, and this highly oxidizing condition is suitable for the precipitation of magnetite from the system with more Fe^{+3} than Fe^{+2} [64]. In contrast, where melts have high TiO_2 content, ilmenite reaches saturation prior to magnetite under high fO_2 conditions [65]. In our specific case, the parent melts are richer in iron than titanium (Figures 3 and 4; Table 2), leading to the formation of titanomagnetite over ilmenite at high fO_2.

Melting experiment studies on ferrobasalts reveal that the partitioning of V in magnetite is dependent on oxygen fugacity, which plays a crucial role in determining the ratios of V^{3+}, V^{4+}, and V^{5+} in the magma. The partition coefficient for V^{3+} in magnetite-melt is significantly higher than that for V^{4+} and V^{5+} [66]. In simpler terms, magnetite formed under reducing conditions contains higher concentrations of V compared to magnetite formed under oxidizing conditions, primarily because of the elevated V^{3+} content in the magma. Vanadium in our titaniferous iron ore samples ranges from 1795 to 2140 ppm, and this is considered lower, compared to ~13,000 ppm V in magnetite of the Bushveld Complex [67] and ~18,500 ppm V in the Fedorivka layered intrusion [10]; it exhibits similarities to that of Suwalki anorthosite (V: 2000–6700 ppm), which crystallized under relatively high oxidized conditions [68], like our case. The relatively low V concentrations of El-Baroud magnetite suggest its crystallization under higher oxygen fugacity (e.g., [69]) relative to the Bushveld Complex.

The pyroxene thermometer [58] indicates that the interstitial pyroxenes between titanomagnetite grains likely crystallized within a temperature range of 800 °C to 1000 °C (Figure 8b). By plotting the Cpx data on the X_{PT}–Y_{PT} diagram [70] of clinopyroxene crystallization PT conditions (Figure 10b), it is notably noted that the low-pressure conditions of this Cpx range from 1.5 to 3.0 kbar. By applying the conversion factor of 1 kbar = 3.7 km for the continental crust [71] and considering a calculated error factor of $\pm$0.5 kbar for pressure, the Cpx possibly crystallized at a depth estimated to be less than 12 km in the lower crust. The formation condition (800–1000 °C, depth < 12 Km, and P < 3.5 Kbar) of the El-Baroud intrusion is similar to that of the Wajilitag layered mafic–ultramafic intrusion (1253 °C, depth < 24 Km, and P < 6.5 Kbar; [72]), the Korab Kansi mafic-ultramafic intrusion (700–1100 °C, depth < 15 Km, and P < 5 Kbar; [29]) and the Rahaba layered intrusion (820–1000 °C, depth < 15 Km, and P < 5 Kbar; [32]). By using the Al_{vi}+2Ti+Cr versus the Al_{iv}+Na diagram of the Cpx oxygen fugacity, the investigated Cpx plots within the field of high oxygen fugacity or oxidized conditions [73] (Figure 10a). The formation condition of the El-Baroud intrusion is consistent with that of the Korab Kansi and Rahaba intrusions in the SED of Egypt (Figure 10). Therefore, we suggest that the Fe-Ti-rich mafic intrusions in the Eastern Desert of Egypt were possibly formed under the same conditions of oxygen fugacity, temperature, and pressure.

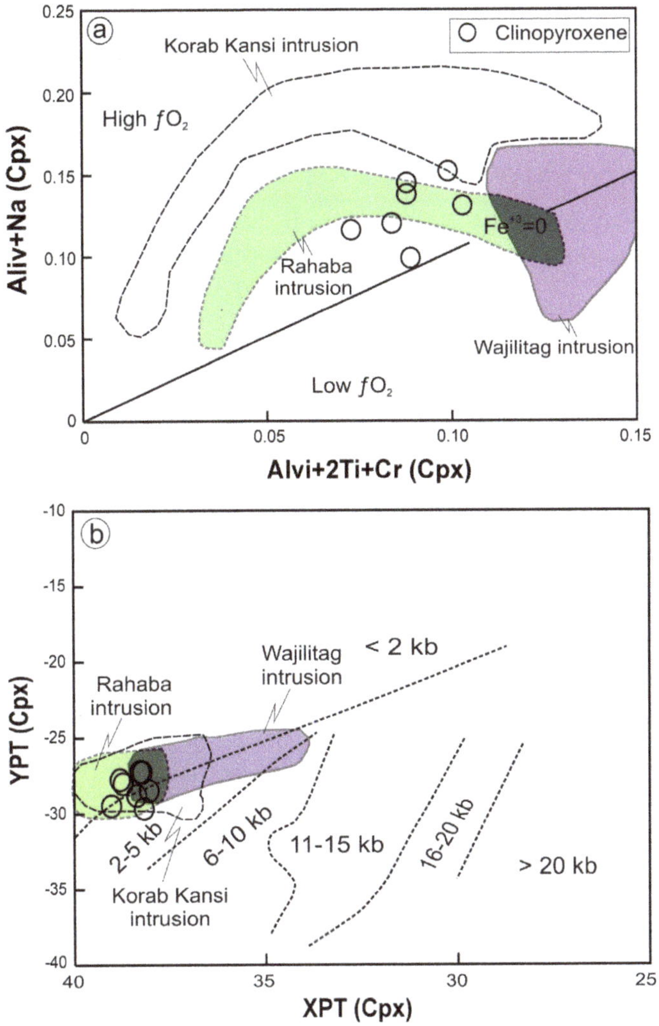

Figure 10. (**a**) Al^{vi}+2Ti+Cr versus Al^{iv}+Na diagram of Cpxs oxygen fugacity, indicating crystallization under relatively high oxygen fugacity or oxidized conditions [73]. (**b**) X_{PT} versus Y_{PT} diagrams for determination of the clinopyroxene crystallization pressure [70]. The data of the Wajilitag intrusion are taken from Li et al. [72], Korab Kansi intrusion from Khedr et al. [29], and Rahaba intrusion from Khedr et al. [32].

$$X_{PT} = 0.446SiO_2 + 0.187TiO_2 - 0.404Al_2O_3 + 0.346FeO_t - 0.052MnO + 0.309MgO + 0.431CaO - 0.446Na_2O,$$

$$Y_{PT} = -0.369SiO_2 + 0.535TiO_2 - 0.317Al_2O_3 + 0.323FeO_t + 0.235MnO - 0.516MgO - 0.167CaO - 0.153Na_2O.$$

7.2. Crystallization Sequence of El-Baroud Fe-Ti Oxide Minerals

The sequence of mineral formation in El-Baroud titaniferous iron ores can be established through the process of intra-oxide re-equilibration (e.g., [74,75]). During the process of intra-oxide re-equilibration, titanium was transferred from the early-formed titanomagnetite to ilmenite as the ores cooled. This often leads to exsolved ilmenite lamella in titanomagnetite (Figure 3g–l), forming oxy-exsolution textures [49,76,77]. It can be described using this equilibrium equation: $6Fe_2TiO_4$ (in magnetite) + O_2 = $6FeTiO_3$ (ilmenite) + $2Fe_3O_4$ (magnetite), where extra oxygen may be supplied from water and other fluids [78].

Many re-equilibration textures are observable in both gabbro and Fe-Ti oxides, reflecting variations in oxidation state and diffusion rates. These textures include: (i) the formation of single-phase homogeneous Al-spinel (Figure 5f); (ii) trellis intergrowths of thin ilmenite lamellae within the host titanomagnetite (Figure 4c); (iii) sandwich intergrowths of thick lamellae of ilmenite (Figure 4b); and (iv) granules or occasional ilmenite lamellae at the external borders of titanomagnetite (Figure 4a,b) [76].

During the cooling process, the intercumulus liquid was expelled, and titanomagnetite within the ores recrystallized, forming a cumulus texture and coarse and polygonal grains with well-annealed junctions (Figures 3f and 4f), suggesting its magmatic origin. Upon cooling, the Fe_2TiO_4 component in titanomagnetite parent melts may transfer in the form of ilmenite exsolutions by sub-solidus oxidation. In many cases, exsolution–oxidation is the primary mechanism driving this transformation and forms the ilmenite exsolved lamellae in titanomagnetite (Figure 3g–k) (e.g., [76,79]). Ilmenite formation occurs through either direct crystallization from parent melts as subhedral to euhedral or polygonal grains with sharp contacts with primary titanomagnetite (Figures 3j,k and 4e,f) or as by-product of anhedral blebs, wormy-like crystals, and/or lamellae of oxidation-exsolution from titanomagnetite (Figures 3e,i,j and 4a,b) (e.g., [80]). Two distinct types of ilmenite lamellae are observed in titanomagnetite grains (Figure 3g,h), reflecting two stages of ilmenite formation. The coarse ilmenite lamellae (Figure 3g) suggest their formation through direct exsolution at high temperatures, while the thin ilmenite lamellae (Figure 3h) could originate from inter-oxide re-equilibration processes (e.g., [49,79]). This is consistent with the Korab Kansi titanomagnetite that has (111) oxidation–exsolution lamellae of ilmenite in the mafic–ultramafic intrusion [31]. Consequently, this oxidation-exsolution texture characterizes the gabbroic layered intrusions.

Further cooling and interaction between titanomagnetite and hydrothermal fluids may lead to the exsolution of Al-spinel in the form of inclusions or small granules (Figures 3g–j, 4c–d, and 5f) and its transformation to magnetite as the matrix around subhedral ilmenite (Figure 3l). The dominant Al-spinel granules and inclusions (up to 2 vol.%) in the El-Baroud Fe-Ti oxide ores indicate the existence of Al in the parent oxide melts; this is confirmed by the significant contents of Al_2O_3 (3.5–4.1 wt.%) in these ores (Table 1). Exsolutions of Al-spinel occur either due to a decrease in the solubility of Fe, Zn, and Al upon cooling or because of local excesses of Al and Zn resulting from the removal of ulvöspinel during the early oxidation [81].

In summary, the titanomagnetite crystallized early before ilmenite, which occurs as composites (Figure 4a), granules with polygonal shapes (Figures 3f and 4f), and exsolved lamellae (Figures 3g and 4b). This is due to the composition of primitive melts that were richer in Fe than Ti and had high oxygen fugacity. During cooling, the excess Ti in the titanomagnetite may be exsolved as ilmenite granules and lamellae, generating composite-, sandwich-, trellis-, cloth-, and cross-hatching-like textures (Figures 3 and 4) (e.g., [82]). Some of the investigated titanomagnetite grains suffered from the post-magmatic process and were converted into a matrix of magnetite (Figure 3l).

7.3. Source and Mechanism of Unusual Fe-Ti Enriched Parental Magma

The El-Baroud gabbroic rocks are rich in Fe ($Fe_2O_{3\ total}$, 8.8–16.5 wt.%; Table 1), and this is similar to those of the Lalibela ferropicrite (14 wt.%, [61]) and the Vestfjella ferropicrite (14.53 wt.%, [55]), which were generated due to the mantle plume. The concept of ferropicrite formation being linked to mantle plumes has been widely suggested by several studies [83–86]. The exceptionally Fe-rich source, such as ferropicritic melts, is attributed to the partial melting of the metasomatized (high Rb, Ba, Sr, Pb, and LREE relative to HREE, Figure 7) lithosphere mantle by the mantle plume or during asthenosphere upwelling [29,31,32,55,87]. This lithosphere mantle-derived melt may mix with traces of melts from the head of mantle plumes or asthenosphere melts as a source of high Ti and Fe. This is confirmed by the high Ti, Fe, Nb, and Ta (HFSE) in the gabbroic sample and Fe-Ti oxide ores (Figure 7). Furthermore, the chondrite-normalized REE patterns of the studied Cpx resemble those of Cpxs in the abyssal peridotite, indicating that our samples have the same magma source (primitive mantle) as the abyssal peridotite. We cannot exclude the partial melting of the subducted slab as a source of Ti and slab-derive fluids as a source of Fe for the metasomatized lithosphere mantle beneath the Eastern Desert (ED) of Egypt as part of the ANS [29,31,32]. Analysis of a worldwide ferropicrite dataset indicates the existence of the recycled Fe-Ti gabbro component in numerous ferropicritic parent melts [55]. The mixture of Fe-Ti-rich gabbroic (and/or basalts) melts and the mantle peridotite melts is the possible Fe-Ti-rich source of ferropicrites.

The ratios of $(La/Sm)_N$ and $(Sm/Yb)_N$ of the studied gabbros are 1.0–1.73 and 1.95–4.9, respectively, similar to those ratios (La_N/Sm_N: 1.2–1.3 and Sm_N/Yb_N: 4.5) of Vestfjella ferropicrites. The studied gabbro samples are similar in the $(V/Lu)_N$ ratio (0.087–1.23; Table 1) to that observed in the ferropicrites (0.5–1.13) as the source melts of the oceanic Fe-Ti gabbros [55]. By using the Nb/Y versus Zr/Ti diagram of Pearce [88] (Figure 11a), we found that the studied gabbroic samples typically show relatively low Nb/Y values (0.2–0.6), similar to those of Vestfjella ferropicrites (0.2–1.1) and Lalibela ferropicrites (0.68–0.85). The REE pattern of El-Baroud gabbros is typically similar to that of the Vestfjella ferropicrites (Figure 7c).

The composition of the melt equilibrated (Supplementary Table S2) with the El-Baroud Cpx was calculated through the Cpx/melt partition coefficients [89,90]. The REE of this equilibrated melt normalized to chondrite [56] exhibits inclined patterns, with decreasing REE concentrations from LREE to HREE (Figure 11b). This REE trend resembles the patterns observed in Greenstone ferropicritic melts [91], tholeiitic basaltic melts derived from the mantle [92], and melts of the ocean–island basalts [56]. The chemical characteristics of the Cpx suggest ferrobasaltic parent melts for El-Baroud Fe-Ti ores.

Moreover, samples of gabbroic rocks were plotted on the Nb/Yb vs. Th/Yb diagram [93], and these samples fall toward the Oceanic Island basalt (OIB) trend (Figure 11c). The mantle melting beneath intra-plate volcanic islands is related to mantle plumes and/or dynamic upwelling and decompression processes [94]. The OIB that erupted through a thick lithosphere is richer in Ti and Fe compared to that that erupted through a thin lithosphere. Thick-lithosphere-related OIB typically shows high contents of Fe, Ti, Mg, and P (Table 1), along with high Sm/Yb ratios and poor Si and Al content, like our case (Figure 11d) (e.g., [95–97]). Consequently, we suggest that the El-Baroud mafic intrusion was possibly generated and emplaced in the thick part of the lithosphere, where the enrichment of Fe and Ti can form Fe-Ti oxide ores.

The interstitial Cpx in El-Baroud titaniferous iron ores has Mg# of 72.4–83.27, TiO_2 (0.66–1.02 wt.%), and Al_2O_3 (1.82–2.9 wt.%), and this is consistent with the Cpx phenocrysts in Lalibela ferropicrites, i.e., Mg#: 74–88; TiO_2: 0.84–1.82 wt.%; and Al_2O_3: 1.2–3.2 wt.%. The REE and trace element patterns of our Cpx are similar to the Cpx REE patterns of ferropicrite and picritic ferrobasalt rocks from the Lalibela area (Figure 9b).

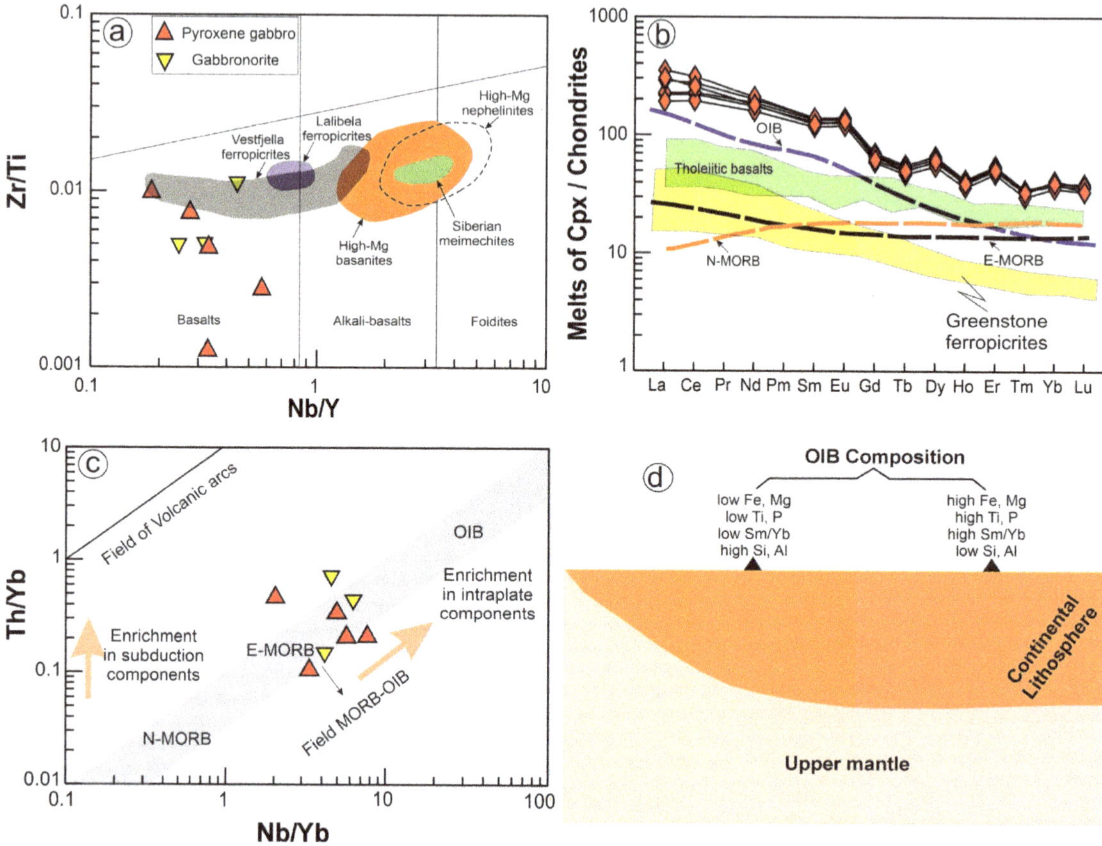

Figure 11. (**a**) Variations of Zr/Ti vs. Nb/Y [88] of El-Baroud gabbros. High-Mg nephelinites and basanites data compiled from GEOROC: http://georoc.mpch-mainz.gwdg.de/georoc/. Siberian meimechites and related alkaline picrites and Vestfjella and Lalibela ferropicrites are from Arndt et al. [98]; Fedorenko and Czamanske [99]; Kogarko and Ryabchikov [100]; Heinonen and Luttinen [55]; and Desta et al. [61] and are shown for comparison. (**b**) Chondrites-normalized [56] REE patterns for calculated melts in equilibrium with Cpx from El-Baroud intrusion. Cpx/melt partition coefficients are from Hart and Dunn [88] and Stosch [101]. Normal mid-ocean ridge basalt (N-MORB), Ocean Island basalt (OIB), and enriched mid-ocean ridge basalt (E-MORB) compositions are from Sun and McDonough [56]. For comparison, REE patterns of tholeiitic basalts from mantle-derived melts [92] and Greenstone ferropicrites [91] are used for comparison. (**c**) Nb/Yb versus Th/Yb tectonic discrimination diagram [93]. (**d**) Geochemical characteristics of OIB-like melts within the thick and thin parts of the continental lithosphere [94].

The Nb/Ta and Zr/Hf ratios may be indexes for the nature of magma, where the primitive mantle has a higher Nb/Ta ratio (17.8) and Zr/Hf ratio (37), from McDonough and Sun [102], than the crust (Nb/Ta: 11 and Zr/Hf: 33) [103]. The studied gabbro samples have the Nb/Ta ratio of 14.29–22.33 (17.57 on average) and the Zr/Hf ratio of 34–41.39 (37.37 on average); these ratios align closely with those of the primitive mantle. The studied gabbro samples show Th/Ta ratios ranging from 0.2 to 2.59 (1.24 on average), similar to that (Th/Ta = 2.1) of the primitive mantle [56]. Therefore, the El-Baroud gabbros were possibly generated from the primitive mantle-derived melts.

The thick Fe-Ti oxide layer in the Emeishan Large Igneous Province (ELIP) is explained by the gravitational accumulation of these oxides during the process of fractionation from the ferropicritic melts [104–106]. This is similar to the thick Fe-Ti oxide layers in the El-Baroud gabbroic intrusion (Figure 2a–f), suggesting their ferropicritic parent melts. The studied Fe-Ti oxides occur as cumulus phases (Figure 3c), similar to those of the upper part of the Bushveld Complex and the central regions of the Skaergaard intrusion, indicating that their magma was derived from the primitive mantle [18,21,67]. In summary, El-Baroud gabbroic rocks and their associated Fe-Ti oxide ore crystallized from the ferropicritic melts generated by the mantle plume.

7.4. Genesis of El-Baroud Fe-Ti Oxide Ores

Parent melts of Fe-Ti oxide ores are enriched in Ti and Fe due to one of the following: (1) mantle-derived melt already enriched in Fe and Ti; (2) mantle-derived magma that acquired enrichment in Fe and Ti through differentiation before its ultimate emplacement; or (3) a combination of both processes (e.g., [29,31,32,107]).

As we mentioned above, titanomagnetite first crystallized, then ilmenite in the oxidizing environment. Therefore, oxygen fugacity is the primary factor influencing the stability boundaries of magnetite and ilmenite in parental magmas with relatively low FeO_t and TiO_2 concentrations [108]. The elevation in oxygen fugacity can lead to the early crystallization of thick layers of titanomagnetite [109]. This high oxygen fugacity is related to the dominant H_2O and hydrous magmas of El-Baroud gabbros and their Fe-Ti oxide ores.

The appreciable amount of biotite (10–15 vol.%.; Khedr et al., under preparation) and amphibole (20–25 vol.%) was observed in the El-Baroud gabbroic intrusion (Figure 3a–d). The dominant hydrous minerals and enrichments of LREE relative to HREE with high Cs, Rb, Ba, Pb, and Sr (fluid mobile elements) of Fe-Ti oxides and host gabbros (Figure 7) suggest that the parent magma (hydrous nature) was supplied with water and volatiles from the subducted slab during mantle metasomatism [29,110]. The primitive magma is ferropicrite melts that were derived from the metasomatized mantle by hydrous fluids, which produce melts by decompressing. These mantle-derived melts are rich in H_2O and fluid mobile elements (e.g., LREE, Rb, Th, U, Pb, and Sr; Figure 7). Based on Nb/Zr versus Th/Zr and Th/Yb against Sr/Nd diagrams from Zafar et al. [111] and Woodhead et al. [112], respectively (Figure 12a,b), the studied gabbro samples fall in the field of slab dehydration and fluid metasomatism trends. This will facilitate the liquid immiscibility processes as a mechanism of El-Baroud Fe-Ti oxide formation (Figure 12c,d).

Based on the section discussed above, the El-Baroud titaniferous iron ores were generated from the ferropicritic parent melts derived from the metasomatized lithosphere mantle during interaction with the mantle plume heads. During this interaction, the lithosphere mantle fertilized with HFSE (high Ti, Nb, Ta) from the trace melts of the plume heads, forming Fe-Ti-rich melts derived from this mantle part; therefore, gabbro samples plot close to E-MORB (Figure 11c). The magma type of El-Baroud gabbros is low-K tholeiitic magma (Figure 13a,b), similar to Panzhihua, Korab Kansi, and Rahaba intrusions, as well as Lalibela and Vestfjella ferropicrites [5,29,32,55,61].

Based on textures (Figures 3 and 4) and chemistry (Figures 7 and 9), the studied titaniferous iron ores are magmatic in origin. For example, all magnetite compositions fall within the magmatic magnetite field based on the Ni/Cr versus Ti diagram (Figure 13c) [113]; they also plot close to the field of magnetite in the Korab Kansi and Abu Ghalaga intrusions [29,31]. The trace elements are used to indicate the origin of magnetite [113]. According to the Ga + Co versus Ge diagram (Figure 13d) [114], the analyzed Fe-Ti oxide ore samples exist in the field of the layered intrusion and close to the Panzhihua intrusion, reflecting the same origin by the liquid immiscibility for both of them [115]. There are several factors that control the immiscible oxide melt formation from the silicate magma; one of them is the abrupt change in oxygen fugacity, in addition to magma mixing, mineral fractionation, and/or the injection of fluids into the system [5].

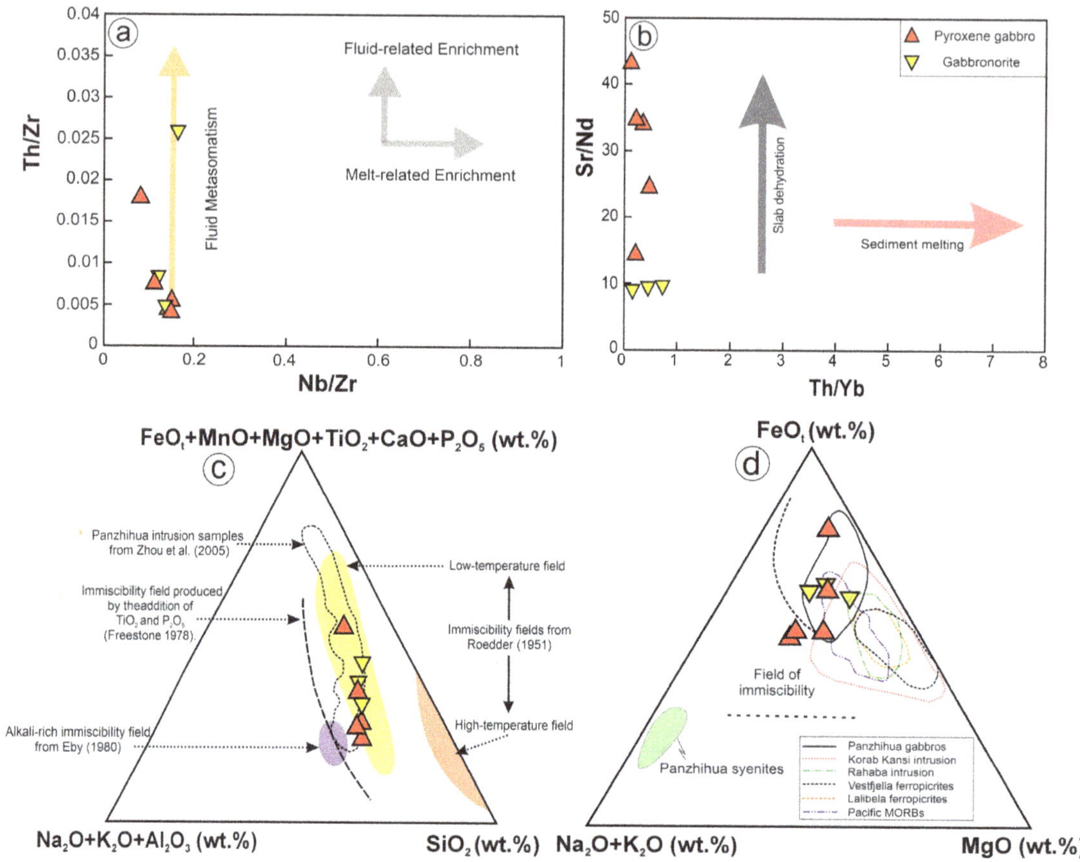

Figure 12. Binary variation diagrams as evidence of fluid metasomatism and liquid immiscibility for the origin of pyroxene gabbros and their Fe-Ti oxide ores. (**a**) Nb/Zr versus Th/Zr [111]. (**b**) Th/Yb versus Sr/Nb [112]. (c) Triangular liquid immiscibility diagram from Roedder [116] showing the two fields of potential liquid immiscibility. The dashed line represents the immiscibility field created by the addition of TiO_2 and P_2O_5 [117]. The alkali-rich immiscibility field is represented by the small elliptical area and is from Eby [118]. Panzhihua intrusion data are from Zhou et al. [5]. (**d**) Na_2O + K_2O-FeOt-MgO triangular diagram. The field of immiscibility is from Philpotts [119]. The data of gabbros and syenites of Panzhihua, Korab Kansi, Rahaba, Vestfjella, and Lalibela ferropicrites are used for comparison [13,29,32,55,61]. The Pacific MORBs are downloaded from PetDB: http://www.earthchem.org/petdb (1 January 2018).

The El-Baroud gabbros and their disseminated oxides (Figure 3a,b) crystallized from primitive ferropicritic melts by in situ crystallization as one process of fractional crystallization (Figure 14). These disseminated oxide ores and some Fe-Ti-rich gabbro samples are settled at the bottom of the El-Baroud mafic intrusion (Figure 14). Then, the remains of this primitive magma suffered from immiscibility due to the evolution of magma and a decrease in temperature, followed by the segregation of liquids by density differences into Fe-rich and silica-rich liquids, including (1) silicate melts as a source of the more evolved gabbros and (2) oxide melts as a source of massive Fe-Ti oxide ores as lenses at the middle or top of the El-Baroud mafic intrusion (Figure 2a,e,f).

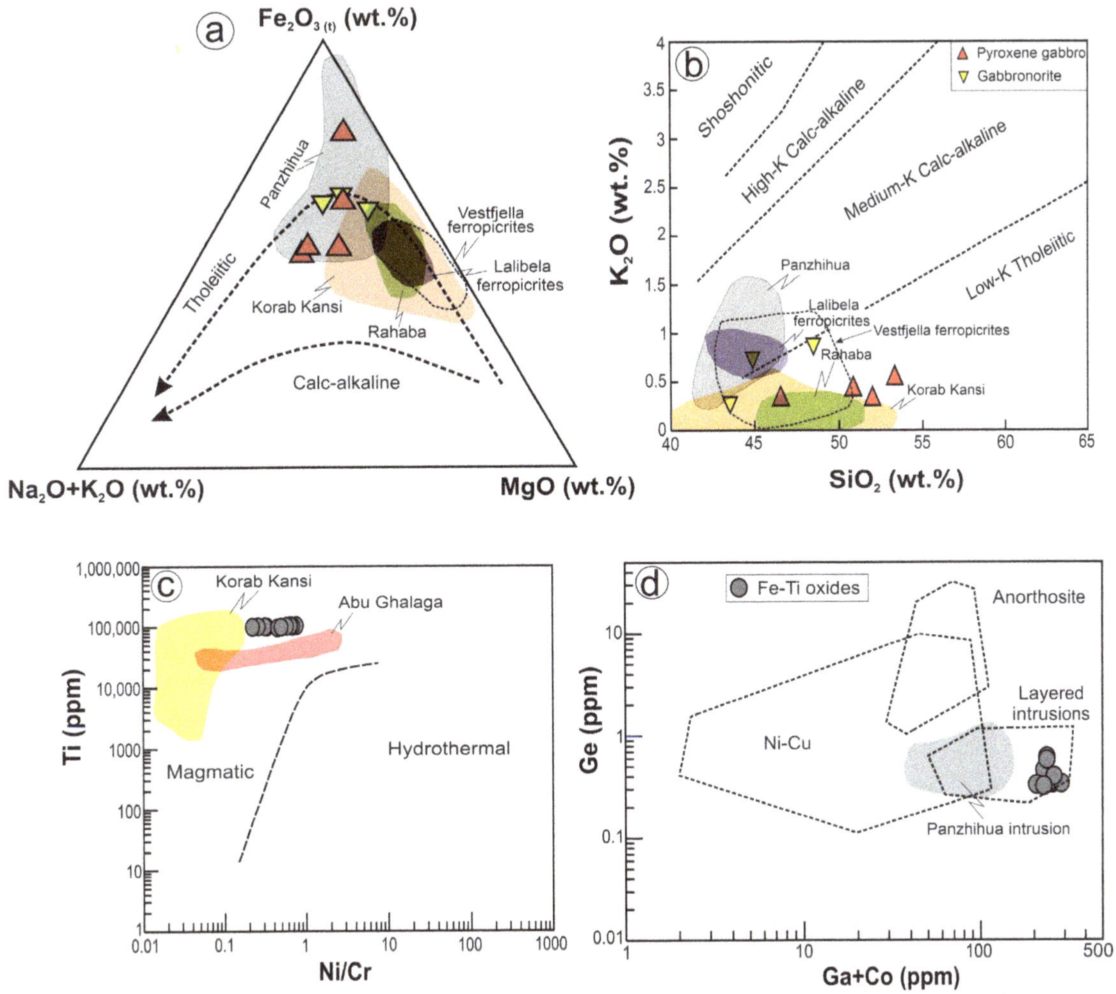

Figure 13. Different discrimination diagrams for magnetite and pyroxene gabbros. (**a**) AFM diagram showing geochemical variations in the El-Baroud intrusion. The tholeiitic and calc-alkaline trends are from Wilson [120]. (**b**) SiO_2 vs. K_2O diagram [121]. (**c**) Ni/Cr vs. Ti diagram [113]. (**d**) Ga + Co vs. Ge diagram [114]. Data of Panzhihua are from Liu et al. [115]. Fields of Korab Kansi, Abu Ghalaga, and Rahaba [29,31,32], Panzhihua [5], Vestfjella [55], and Lalibela ferropicrites [61] are used for comparison.

To explain the genesis of the Fe-Ti oxide ores in the mafic intrusions, a number of theories have been proposed, including: (1) segregation of magma by liquid immiscibility [6,16,31,122–125]; (2) magma addition and/or mixing [17,126–128]; (3) mineral settling and sorting [18,52,129]; (4) fractional crystallization by in situ crystallization [8,130,131]; (5) polybaric crystallization [68,132]; and (6) solid-state remobilization [133,134].

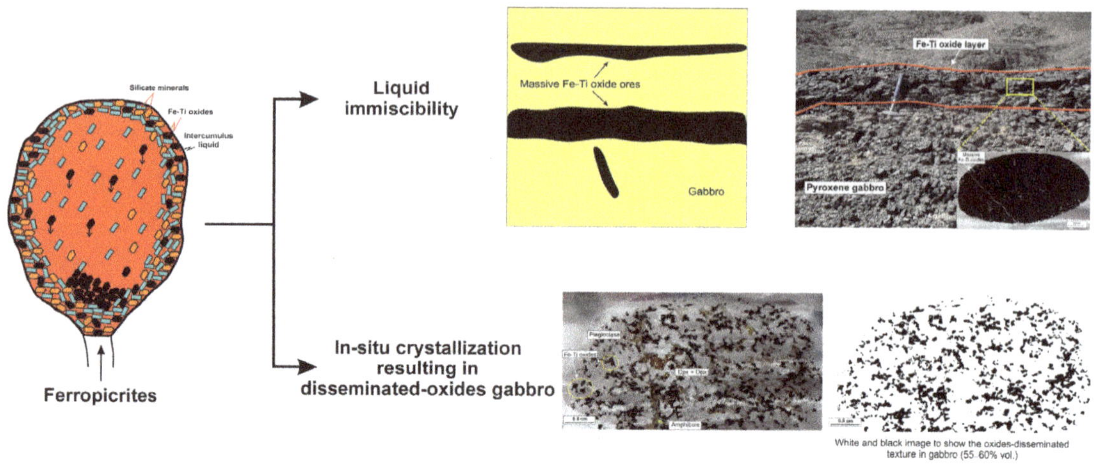

Figure 14. A schematic model for magma evolution and the genesis of Fe-Ti oxide ore deposits in El-Baroud intrusion.

The primitive magma derived from the mantle by plume interaction or decompression melting during asthenosphere upwelling interacted with the continental lithosphere, leading to the in situ crystallization of the layered gabbro and disseminated oxides (Figure 3a–c). Then, the remains of this primitive magma suffered from immiscibility due to the evolution of magma and a decrease in temperature, followed by the segregation of liquids by density differences into Fe-rich and silica-rich liquids (Figure 14), including: (1) silicate melts as a source of the host gabbros and (2) oxide melts as a source of disseminated and massive Fe-Ti oxide ores (Figure 2).

The liquid immiscibility may occur at temperatures ~1100 °C through the experimental study within the K_2O–FeO–Al_2O_3–SiO_2 system [116,135]. In basaltic magma systems, the separation of magma into Si-rich and Fe-rich melts typically occurs in the later stages of the magma evolution, generally above 1000 °C [136,137]. These temperatures and formation conditions resemble those of ferropicritic/ferrobasaltic melts as a source of the studied titaniferous iron ores, which are similar in texture, mineralogy, and chemistry to the Panzhihua Fe-Ti oxide deposits that were originated by the liquid immiscibility model [5].

The liquid immiscibility phenomenon primarily occurs below temperatures ranging from 1000 to 1020 °C, representing a relatively advanced stage in the evolution of basaltic magma. Notably, the immiscible melts exhibit significantly lower MgO contents compared to the starting materials [137]. The results (MgO = 3.63–7.24 wt.% in Table 1; T = 800–1000 °C) obtained for the studied gabbroic rocks are consistent with these conditions. During liquid immiscibility, Ca and P are typically distributed into Fe-rich melts [123,138,139]. Increasing H_2O and fO_2 expands the two-liquid field, which facilitates the separation of the Fe–Ca–P melt from the silicic magma, leading to the formation of iron oxide ores (e.g., [140]). Ca and P in our Fe-Ti oxide ores range from 0.04 to 0.41 wt.% and 80–770 ppm, respectively, suggesting that liquid immiscibility is the primary process for the ore formation in the El-Baroud area.

In addition, the chemical evidence of liquid immiscibility as a model for the origin of El-Baroud titaniferous iron ores is confirmed by using the major oxide triangular diagram after Roedder [116] (Figure 12c). By plotting the studied gabbro samples as the host rock of the titaniferous iron ores, they are found in the low-temperature immiscibility field (Figure 12c). By using the Na_2O+K_2O–FeO_t–MgO triangular diagram [119] (Figure 12d), the El-Baroud gabbro samples plot in the field of immiscibility and are also comparable with Panzhihua gabbro samples [13], Korab Kansi [29], Rahaba [32], as well as Lalibela and Vestfjella ferropicrites [55,61]. Therefore, we suggest that the liquid immiscibility process plays an important and effective role in the formation of the massive Fe-Ti oxide layers in the El-Baroud intrusion.

The Fe-rich liquids crystallized as Fe-Ti oxide lenses at the top of the mafic intrusion (Figure 2a,e,f), and some intercumulus liquid may be expelled due to pressure changes (e.g., [20]). We propose that there was the replenishment of the primitive ferropicritic magma, and pulses of new magma were added to the primitive magma chamber; this process was continuously repeated and formed interlayered gabbros with the Fe-Ti oxide lenses or layers (Figure 2b–f). This suggestion is supported by the numerous layers (Figure 2d), dykes (Figure 2c), veins, and lenses of Fe-Ti oxide ores (Figure 2f). According to the field relationships, the cap of titanomagnetite lies above the pyroxene gabbro (Figure 2e), indicating the liquid immiscibility origin of the El-Baroud intrusion.

The Fe-Ti oxide ores in the El-Baroud intrusion can be found either between the silicate minerals as disseminated in the host Fe-Ti-rich gabbros (Figure 3e) or as accumulated grains, forming massive ores (Figure 3f–l). Some of the oxide inclusions hosted by hornblende in the pyroxene gabbro are euhedral (Figure 3b) and could be trapped during the growth of hornblende crystals. This suggests that the crystallization of El-Baroud Fe-Ti oxide minerals usually occurs in the late magmatic stages, where Fe and Ti increase by fractionation (e.g., [29,63,141]). This is confirmed by the enrichment of LREE relative to HREE (Figure 7a) and enrichment in fluid mobile elements (e.g., Rb, Th, U, Pb, and Sr) of the studied Fe-Ti oxide ores (Figure 7b). This is also observed in the host gabbroic rocks (Figure 7c,d).

In conclusion, the Fe-Ti oxide layers in the El-Baroud area were formed through two distinct processes: (1) in situ crystallization and (2) liquid immiscibility (Figure 14). The mantle-derived Fe-Ti enriched primitive magma is the primary source of Fe and Ti for the formation of titaniferous iron ores. The formation of these ores can be attributed to: (1) the composition and differentiation of the parental magma; (2) high oxygen fugacity; (3) the existence of volatiles in the magma; and (4) the structures (mainly NW-SE and NE-SW trends) controlling the foliation of the host pyroxene gabbros and the distribution of titaniferous iron ores.

7.5. *Tectono-Metallogenetic Setting of Titaniferous Iron Ores*

The sporadic mountains (Figure 1b) of El-Baroud gabbroic intrusions that were found in the tonalite–granodiorite [45] suggest that the El-Baroud pyroxene gabbro and its Fe-Ti oxide ores formed after the El-Baroud syn-collisional granites. The El-Baroud gabbroic rocks may have formed in an extensional tectonic setting (for example, arc rifting) after the syn-collisional tonalite and granodiorite (Figure 1b). In addition, the abundance of dykes with bimodal compositions in the studied area (Figure 1b) may indicate an extensional regime.

The Najd fault system or shear zone with the NW-SE structural trend (Figure 1b) controls mineralization such as rare-metal granites in the Central Eastern Desert (CED) [142–144], Fe-Ti-rich gabbros, and mineralized granites in the SED of Egypt [31,32]. The Wadi El-Baroud area includes various small-offset strike-slip faults with both NW-SE and NE-SW striking trends and various dykes like rhyolite, andesite, and basalt with a NW trend (Figure 1b) (e.g., [48]). The Fe-Ti oxide layers and foliation of their host gabbros follow the NW-SE trend in the studied area (Figure 2b,d,f). Therefore, the NW-SE Najd fault system and its conjugate NE-SW strike-slip fault trend (Figure 1b) may control the distribution of El-Baroud titaniferous iron ores in the NED.

According to the Zrx3–Nbx50–Ce/P_2O_5 triangular diagram proposed by Muller and Groves [145], it is evident that all samples fall in the field of a post-collisional arc setting (Figure 15a). On the La/10–Y/15–Nb/8 tectonic diagram (Figure 15b) [146], the composition of the studied gabbros aligns with the compositional field of intraplate continental basalt, suggesting the continental extensional (rifting one) setting. The chemistry of Cpx is useful for determining the tectonic setting of the mafic intrusion. The El-Baroud Cpx data plot in the fields of within-plate tholeiitic basalt or ocean floor basalt; this coincides with the post-collisional Rahaba intrusion [32] but differs slightly with the island-arc Korab Kansi and Abu Ghalaga tectonic settings [29,31] (Figure 15c). By using the Zr versus Ti diagram, most of the studied gabbro samples lie within the field of within-plate basalts (Figure 15d), indicating the rift-related tectonic setting, i.e., arc rifting setting.

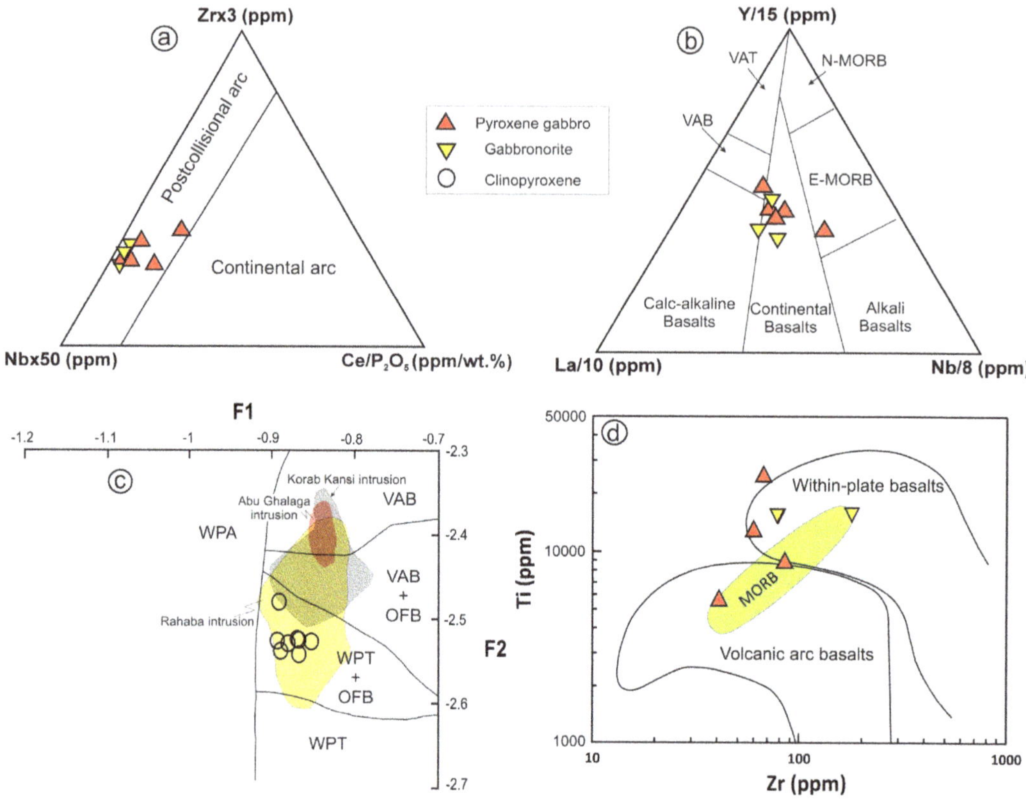

Figure 15. Tectonic setting discrimination diagrams. (**a**) Zrx3–Nbx50–Ce/P_2O_5 triangular diagram [145]. (**b**) La/10–Y/15–Nb/8 diagram [146]. (**c**) F1 vs. F2 diagram of Cpx [147]. (**d**) Zr vs. Ti diagram [148]. The fields of Rahaba intrusion [32], Korab Kansi [29], and Abu Ghalaga [31] are used for comparison.

Intraplate tectonic and magmatic activities related to continental or arc rifting are associated with deep mantle plumes and asthenosphere upwellings [29,31,32,149,150]. Many intraplate and anorogenic ore systems are either directly or indirectly influenced by the magmas found in this arc-related rifting, which includes layered intrusions, continental and oceanic flood lavas, and anorogenic A-type granitic rocks [149]. So, we adapt the concept that the mantle plume is responsible for the generation of El-Baroud mafic intrusion (Figure 16). The evidence was deduced from whole chemistry (Figures 6 and 7) and Cpx chemistry (Figure 9), whereas the mantle plume can generate intra-plate OIB-like magma that is rich in Fe and Ti [94]. The studied Fe-Ti oxide ores and their host gabbros are rich in HFSE (Nb, Ta, and Ti) and Fe (Figure 7b, d) and plot close to the E-MORB field (Figure 11c), indicating their parent melts originated due to the mantle plume interaction.

Partial melting in the plume head occurs by adiabatic decompression, producing lower temperatures and low-Mg melts (tholeiitic basalts). In contrast, melting in the high-temperature tail generates high-temperature and high-Mg melts (picrites, komatiites) [151] (Figure 16a). Then, plume–lithosphere interaction occurs at the base of the lithosphere, and the surface expression of mantle plumes typically appears as crustal doming, forming extensions in the crustal dome shape. This extensional stress will lead to the formation of continental rift or drift during the extensional setting (Figure 16b) (e.g., [152]). This geological phenomenon is often associated with the formation of rift basins and intraplate volcanism [153,154] (Figure 16b). The titaniferous iron ores increased in abundance from the NED to the SED of Egypt (Figure 1a) and are not haphazard. This highly distributed Fe-Ti oxide ores in the SED suggest the dominant mantle plumes and/or shear zones in the SED relative to the NED (Figure 1a). This observation needs more future studies to emphasize if the abundance of Fe-Ti ores and host gabbro have been emplaced by the shear zones or structures or if they should first be formed by mantle plume mechanisms before emplacement along the fractures.

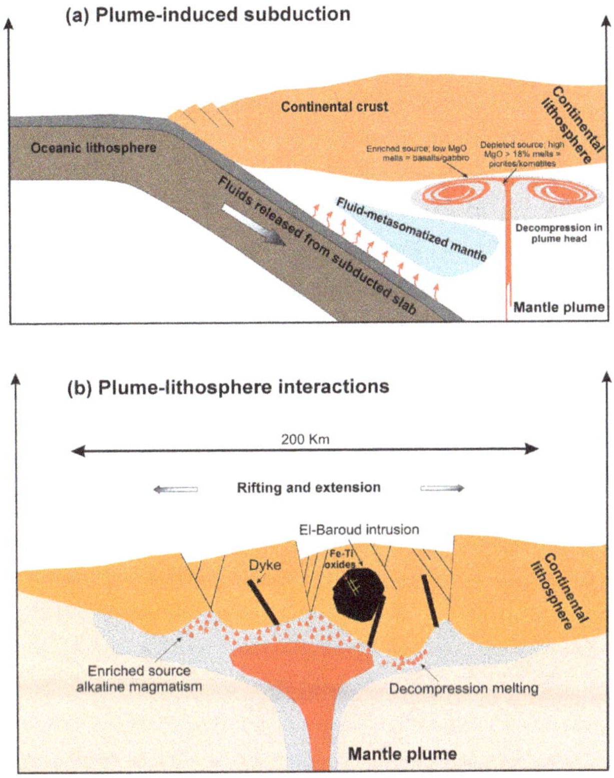

Figure 16. A schematic sketch showing the tectonic setting of the El-Baroud intrusion. (**a**) Plume-induced subduction as the first stage after subduction and formation may be due to slab window. (**b**) Plume–lithosphere interactions, where traces of melts from the plume head mix with lithosphere melts (modified from Campbell and Griffiths [155]; Griffiths and Campbell [156]).

8. Concluding Remarks

1. The El-Baroud mafic intrusion hosts Fe-Ti oxide ores, which occur as dykes, horizontal layers, caps, and lenses and comprise titanomagnetite with subordinate ilmenite and magnetite.
2. Titanomagnetite formed first from the oxide-rich melt and is associated with some polygonal ilmenite grains (composite grains). As it cooled, some of the extra Ti may have been expelled as ilmenite exsolution lamellae in the host titanomagnetite. More cooling changed the titanomagnetite into a magnetite matrix during the post-magmatic process.

3. The ilmenite exsolution in magnetite occurs in several forms, including composite-type, sandwich-type, and trellis-type textures, reflecting the magmatic origin of both minerals; this exsolution texture indicates the formation of most ilmenite crystals as a result of the oxidation–exsolution of titanomagnetite, and this oxidation–exsolution texture characterizes the gabbroic layered intrusions in Egypt.
4. The El-Baroud mafic intrusion was formed from ferropicritic parent melts generated from the metasomatized (high Rb, Ba, Sr, Pb, and LREE relative to HREE) lithosphere mantle by the mantle plume interaction or during asthenosphere upwelling. This lithosphere mantle-derived melt may mix with traces of melts from the head of mantle plumes or asthenosphere melts as a source of high Ti and Fe. This is confirmed by the high Ti, Fe, Nb, and Ta (HFSE) in the gabbroic sample and Fe-Ti oxide ores
5. The El-Baroud gabbroic rocks and their Fe-Ti oxide ores crystallized under high oxygen fugacity at 800–1000 °C and a pressure of 1.5–3 kbar for a depth of 6–12 Km and were emplaced in the extensional environment, i.e., the arc-rifting setting.
6. The El-Baroud titaniferous iron ores and their host gabbros were formed by two different processes, in situ crystallization and liquid immiscibility, forming Fe-Ti oxide layers and disseminated oxides in gabbros. They were possibly generated in the rift-related tectonic setting, i.e., the arc rifting setting.
7. The formation of thick layers or lenses of Fe-Ti oxide ores in the El-Baroud mafic intrusion can be attributed to: (1) the composition (rich in Ti and Fe) and differentiation of the ferropicritic parental magma; (2) the high oxygen fugacity condition; (3) hydrous- and volatiles-enriched parent magma; and (4) the structurally controlled NW-SE trend of the Najd fault system and its conjugated NE-SE trend.
8. The titaniferous iron ores increased in abundance from the NED to the SED of Egypt (Figure 1a), reflecting the dominant mantle plumes and/or shear zones in the SED relative to the NED.

Supplementary Materials: The following supporting information can be downloaded at: https://www.mdpi.com/article/10.3390/min14070679/s1, Supplementary Table S1: Representative microprobe analyses of ilmenite, magnetite, and clinopyroxene in gabbros and titaniferous iron ores in El-Baroud intrusion. Supplementary Table S2: Calculated melt of Cpx in titaniferous iron ores.

Author Contributions: Conceptualization, M.Z.K.; data curation, M.Z.K. and A.M.; formal analysis, M.Z.K., A.T., W.T., Y.I. and E.T.; funding acquisition, A.Y.K. and K.A.; investigation, M.Z.K., A.T., W.T., Y.I. and E.T.; methodology, M.Z.K. and A.M., project administration, M.Z.K., A.M. and N.H.E.-S.; resources, A.Y.K. and K.A.; supervision, M.Z.K. and N.H.E.-S.; validation and visualization, M.Z.K. and A.M.; writing—original draft preparation, M.Z.K. and A.M.; writing—review and editing, M.Z.K. and N.H.E.-S. All authors have read and agreed to the published version of the manuscript.

Funding: This research was funded by the Researchers Supporting Project number (RSPD2024R546) at King Saud University in Riyadh, Saudi Arabia.

Data Availability Statement: All data derived from this research are presented in the enclosed figures and tables and Supplementary Materials.

Acknowledgments: The first author is grateful to all colleagues in the Geology Department at both Niigata and Chiba Universities, Japan, for their support during his stay in August 2022 and December 2023 to conduct polish thin sections, petrography, EDS-SEM, and mineral chemistry (EPMA). We are grateful to the anonymous reviewers for their careful reading of our manuscript and their many insightful comments. The authors extend their sincere appreciation to the Researchers Supporting Project number (RSPD2024R546) at King Saud University in Riyadh, Saudi Arabia.

Conflicts of Interest: The authors declare no conflicts of interest.

References

1. Lister, G.F. The composition and origin of selected iron-titanium deposits. *Econ. Geol.* **1966**, *61*, 275–310. [CrossRef]
2. Shirazy, A.; Ziaii, M.; Hezarkhani, A.; Timkin, T. Geostatistical and Remote Sensing Studies to Identify High Metallogenic Potential Regions in the Kivi Area of Iran. *Minerals* **2020**, *10*, 869. [CrossRef]

3. Eales, H.V.; Cawthorn, R.G. The Bushveld Complex. In *Layered Intrusions. Developments of Petrology, 15*; Cawthorn, R.G., Ed.; Elsevier: Amsterdam, The Netherlands, 1996; pp. 181–230.
4. McBirney, A.R. Differentiation of the Skaergaard Intrusion. *Nature* **1975**, *253*, 691–694. [CrossRef]
5. Zhou, M.F.; Robinson, P.T.; Lesher, C.M.; Keays, R.R.; Zhang, C.J.; Malpas, J. Geochemistry, Petrogenesis and Metallogenesis of the Panzhihua Gabbroic Layered Intrusion and Associated Fe–Ti–V Oxide Deposits, Sichuan Province, SW China. *J. Petrol.* **2005**, *46*, 2253–2280. [CrossRef]
6. Ripley, E.M.; Serverson, M.J.; Hauck, S.A. Evidence for sulfide and Fe-Ti-P-rich liquid immiscibility in the Duluth Complex, Minnesota. *Econ. Geol.* **1998**, *93*, 1052–1062. [CrossRef]
7. Duchesne, J.C.; Charlier, B. Geochemistry of cumulates from the Bjerkreim-Sokndal layered intrusion (S. Norway). Part I: Constraints from major elements on the mechanism of cumulate formation and on the jotunite liquid line of descent. *Lithos* **2005**, *83*, 229–254. [CrossRef]
8. Namur, O.; Charlier, B.; Holness, M.B. Dual origin of Fe–Ti–P gabbros by immiscibility and fractional crystallization of evolved tholeiitic basalts in the Sept Iles layered intrusion. *Lithos* **2012**, *154*, 100–114. [CrossRef]
9. Duchesne, J.-C. Fe-Ti deposits in Rogaland anorthosites (South Norway): Geochemical characteristics and problems of interpretation. *Miner. Depos.* **1999**, *34*, 182–198. [CrossRef]
10. Duchesne, J.C.; Shumlyanskyy, L.; Charlier, B. The Fedorivka layered intrusion (Korosten Pluton, Ukraine): An example of highly differentiated ferrobasaltic evolution. *Lithos* **2006**, *89*, 353–376. [CrossRef]
11. Charlier, B.; Namur, O.; Bolle, O.; Latypov, R.; Chesne, J.-C. Fe-Ti-V-P ore deposits associated with Proterozoic massif-type anorthosites and related rocks. *Earth Sci. Rev.* **2015**, *141*, 56–81. [CrossRef]
12. Khalil, K.; Summers, P.; El Shazly, A. Origin of the post collisional younger gabbroic rocks and the associated Fe–Ti oxide ores, Abu Ghalaga area, Southern Eastern Desert, Egypt: Mineralogical and geochemical constraints. *Arab. J. Geosci.* **2023**, *16*, 160. [CrossRef]
13. Zhou, M.F.; Chen, W.T.; Wang, C.Y.; Prevec, S.A.; Liu, P.P.; Howarth, G.H. Two stages of immiscible liquid separation in the formation of Panzhihua-type Fe-Ti-V oxide deposits, SW China. *Geosci. Front.* **2013**, *4*, 481–502. [CrossRef]
14. Pang, K.N.; Zhou, M.F.; Qi, L.; Shellnutt, J.G.; Wang, C.Y.; Zhao, D.G. Flood basalt-related Fe-Ti oxide deposits in the Emeishan large igneous province, SW China. *Lithos* **2010**, *119*, 123–136. [CrossRef]
15. Cawthorn, R.G.; Ashwal, L.D. Origin of anorthosite and magnetitite layers in the Bushveld Complex, constrained by major element compositions of plagioclase. *J. Petrol.* **2009**, *50*, 1607–1637. [CrossRef]
16. Reynolds, I.M. The nature and origin of titaniferous magnetite rich layers in the Upper Zone of the Bushveld Complex: A review and synthesis. *Econ. Geol.* **1985**, *80*, 1089–1108. [CrossRef]
17. Harney, D.M.W.; Merkle, R.K.W.; von Gruenewaldt, G. Platinum-group element behaviour in the lower part of the Upper Zone, Eastern Bushveld Complex. Implications for the formation of the main magnetite layer. *Econ. Geol.* **1990**, *85*, 1777–1789. [CrossRef]
18. Wager, L.R.; Brown, G.M. *Layered Igneous Rocks*; Freeman: San Francisco, CA, USA, 1967.
19. Klemm, D.D.; Henckel, J.; Dehm, R.; von Gruenewaldt, G. The geochemistry of titanomagnetite in magnetite layers and their host rocks of the Eastern Bushveld Complex. *Econ. Geol.* **1985**, *80*, 1075–1088. [CrossRef]
20. Cawthorn, R.G.; McCarthy, T.S. Incompatible trace element behavior in the Bushveld Complex. *Econ. Geol.* **1985**, *80*, 1016–1026. [CrossRef]
21. Tegner, C.; Cawthorn, R.G.; Kruger, F.J. Cyclicity in the main and upper zones of the Bushveld Complex, South Africa: Crystallization from a zoned magma sheet. *J. Petrol.* **2006**, *47*, 2257–2279. [CrossRef]
22. Morse, S.A. Kiglapait mineralogy III: Olivine compositions and Rayleigh fractionation models. *J. Petrol.* **1996**, *37*, 1037–1061. [CrossRef]
23. Park, Y.R.; Ripley, E.M.; Miller, J.D.; Li, C.; Mariga, J.; Shafer, P. Stable isotopic constraints on fluid-rock interaction and Cu-PGE-S redistribution in the Sonju Lake Intrusion. *Minn. Econ. Geol.* **2004**, *99*, 325–338. [CrossRef]
24. Pintoa, V.M.; Koesterb, E.; Debruynea, D.; Chemale, F.; Marquesb, J.C.; Porcherb, C.C.; Passosd, L.H.; Lenze, C. Petrogenesis of the mafic-ultramafic Canindé layered intrusion, Sergipano Belt, Brazil: Constraints on the metallogenesis of the associated Fe–Ti oxide ores. *Ore Geol. Rev.* **2020**, *122*, 503–535. [CrossRef]
25. Wang, K.; Dong, H.; Ou, Q.; Liu, R. Large-scale liquid immiscibility in the Hongge layered intrusion hosting a giant Fe-Ti oxide deposit in SW China. *Ore Geol. Rev.* **2021**, *136*, 104268. [CrossRef]
26. Bai, Z.J.; Zhong, H.; Zhu, W.G.; Hu, W.J.; Chen, C.J. The genesis of the newly discovered giant Wuben magmatic Fe–Ti oxide deposit in the Emeishan Large Igneous Province: A product of the late-stage redistribution and sorting of crystal slurries. *Miner. Depos.* **2019**, *54*, 31–46. [CrossRef]
27. Eldougdoug, A.; El-Rahman, Y.; Harbi, H. The Ediacaran postcollisional Khamal gabbro-anorthosite complex from the Arabian Shield and its Fe-Ti-P ore: An analogy to Proterozoic massif-type anorthosites. *Lithos* **2020**, *105674*, 372–373.
28. He, H.; Song, Y.; Song, X.; Du, Z.; Dai, Z.; Zhou, T.; Xie, W. Origin of nelsonite and Fe–Ti oxides ore of the Damiao anorthosite complex, NE China: Evidence from trace element geochemistry of apatite, plagioclase, magnetite and ilmenite. *Ore Geol. Rev.* **2016**, *79*, 367–381. [CrossRef]
29. Khedr, M.Z.; El-Awady, A.; Arai, S.; Hauzenberger, C.; Tamura, A.; Stern, R.J.; Morishita, T. Petrogenesis of the ~740 Korab Kansi mafic-ultramafic intrusion, South Eastern Desert of Egypt: Evidence of Ti-rich ferropicritic magmatism. *Gondwana Res.* **2020**, *82*, 48–72. [CrossRef]

30. Basta, E.Z.; Takla, M.A. Petrological studies on Abu Ghalaga ilmenite occurrence, Eastern Desert. *J. Geol.* **1968**, *12*, 43–72.
31. Khedr, M.Z.; Takazawa, E.; Arai, S.; Stern, R.J.; Morishita, T.; El-Awady, A. Styles of Fe–Ti-V ore deposits in the Neoproterozoic layered mafic-ultramafic intrusions, South Eastern Desert of Egypt: Evidence for fractional crystallization of V-rich melts. *J. Afr. Earth Sci.* **2022**, *194*, 104620. [CrossRef]
32. Khedr, M.Z.; Hauzenberger, C.; Morishita, T.; Tamura, A.; Bamousa, A.O.; Whattam, S.A.; Ali, S. The layered gabbroic intrusion in the Southern Eastern Desert of Egypt: Implications for Fe-Ti-rich magmatism at the end of the East African Orogeny. *Geochemistry* **2024**, *84*, 126077. [CrossRef]
33. Kharbish, S.; Eldosouky, A.M.; Amer, O. Integrating mineralogy, geochemistry and aeromagnetic data for detecting Fe–Ti ore deposits bearing layered mafic intrusion, Akab El Negum, Eastern Desert, Egypt. *Sci. Rep.* **2022**, *12*, 15474. [CrossRef]
34. Basta, E.Z.; Girgis, M.H. Petrological, mineralogical and geochemical studies on the magnetite-ilmenite-apatite ore (Nelsonite) from Kolmnab, South Eastern Desert, U.A.R. *Proceed Egypt Acad. Sci.* **1969**, *22*, 147–157.
35. Sabet, A.H.; Khalaf, I.M. Petrological, mineralogical and geochemical studies on magnetite-ilmenite-apatite ore (Nelsonite) from Kolminab, South Eastern Desert UAR. *Proceed Egypt Acad. Sci.* **1989**, *22*, 235–440.
36. Basta, E.Z.; Girgis, M.H. The mineralogy of titaniferous iron ores of Um Effin, South Eastern Desert, Egypt. *Proceed Egypt Acad. Sci.* **1968**, *21*, 127–144.
37. El-Ramly, M.F. A new geological map for the basement rocks in the Eastern and Southwestern Deserts of Egypt. *Ann. Geol. Surv. Egypt Cairo* **1972**, *2*, 1–18.
38. El-Gaby, S.; List, F.K.; Tahrani, R. The basement complex of the Eastern Desert and Sinai. In *The Geology of Egypt*; Said, R., Ed.; Balkema: Kotterdam, The Netherlands; Brookfield, VT, USA, 1990; pp. 175–184.
39. Mohamed, F.H.; Hassanen, M.A. Geochemical evolution of arc-related mafic plutonism in the Umm Naggat district, Eastern Desert of Egypt. *J. Afr. Earth Sci.* **1996**, *22*, 269–283. [CrossRef]
40. Azer, M.K.; El-Gharbawy, R.I. Contribution to the Neoproterozoic layered mafic-ultramafic intrusion of Gabal Imleih, south Sinai, Egypt: Implication of post-collisional magmatism in the north Arabian-Nubian Shield. *J. Afr. Earth Sci.* **2011**, *60*, 253–272. [CrossRef]
41. Abdel Halim, A.H.; Helmy, H.M.; Abd El-Rahman, Y.M.; Shibata, T.; El Mahallawi, M.M.; Yoshikawa, M.; Arai, S. Petrology of the Motaghairat mafic-ultramafic complex, Eastern Desert, Egypt: A high-Mg post-collisional extension-related layered intrusion. *J. Asian Earth Sci.* **2016**, *116*, 164–180. [CrossRef]
42. Surour, A.A.; Ahmed, A.H.; Harbi, H.M. Mineral chemistry as a tool for understanding the petrogenesis of Cryogenian (arc-related) Ediacaran (post-collisional) gabbros in the western Arabian Shield of Saudia Arabia. *Int. J. Earth Sci.* **2017**, *106*, 1597–1617. [CrossRef]
43. El-Bialy, M.Z.; Khedr, M.Z.; El-Bialy, B.M.; Hassan, H.F. Continental arc plutonism in a juvenile crust: The Neoproterozoic metagabbro-diorite complexes of Sinai, northern Arabian-Nubian Shield. *Minerals* **2024**, *14*, 145. [CrossRef]
44. Abd El-Rahman, Y.; Polat, A.; Dilek, Y.; Fryer, B.J.; El-Sharkawy, M.; Sakran, S. Geochemistry and tectonic evolution of the Neoproterozoic incipient arc–forearc crust in the Fawakhir area, Central Eastern Desert of Egypt. *Precambr. Res.* **2009**, *175*, 116–134. [CrossRef]
45. El-Bialy, M.Z.; Omar, M.M. Spatial association of Neoproterozoic continental arc I-type and post-collision A-type granitoids in the Arabian-Nubian Shield: The Wadi Al-Baroud Older and Younger Granites, North Eastern Desert, Egypt. *J. Afr. Earth Sci.* **2014**, *103*, 1–29. [CrossRef]
46. Hamimi, Z.; Abd El-Wahed, M.A. Suture(s) and Major Shear Zones in the Neoproterozoic Basement of Egypt. In *The Geology of Egypt, Regional Geology Reviews*; Hamimi, Z., El-Barkooky, A., Martínez Frías, J., Fritz, H., Abd El-Rahman, Y., Eds.; Springer Nature: Cham, Switzerland, 2020; Chapter 5; pp. 153–189.
47. Hamimi, Z.; Fowler, A. *Najd Shear System in the Arabian-Nubian Shield*; Springer Geology: Cham, Switzerland, 2021; pp. 359–392.
48. Omran, A. Geology, mineralogy and radioelements potentiality of microgranite dikes to the south of Abu Hadieda area, Northern Eastern Desert, Egypt. *Al Azhar Bull. Sci.* **2014**, *25*, 47–62. [CrossRef]
49. Haggerty, S.E. Oxide textures, a mini-atlas. *Rev. Miner. Geochem.* **1991**, *25*, 129–219.
50. Cox, K.G.; Bell, J.D.; Pankhurst, R.J. *The Interpretation of Igneous Rocks*; George, Allen and Unwin: London, UK, 1979.
51. Prowatke, S.; Klemme, S. Trace element partitioning between apatite and silicate melts. *Geochim. Cosmochim. Acta* **2006**, *70*, 4513–4527. [CrossRef]
52. Song, X.Y.; Qi, H.W.; Hu, R.Z.; Chen, L.M.; Yu, S.Y.; Zhang, J.F. Formation of thick stratiform Fe-Ti oxide layers in layered intrusion and frequent replenishment of fractionated mafic magma: Evidence from the Panzhihua intrusion, SW China. *Geochem. Geophys. Geosyst.* **2013**, *14*, 712–732. [CrossRef]
53. Rollison, H.R. *Using Geochemical Data: Evaluation, Presentation, Interpretation*; Longman: Singapore, 1993; p. 352.
54. Zellmer, G.F.; Sparks, R.S.J.; Hawkesworth, C.J.; Wiedenbeck, M. Magma emplacement and remobilization timescales beneath Montserrat: Insights from Sr and Ba zonation in plagioclase phenocryst. *J. Petrol.* **2003**, *44*, 1413–1431. [CrossRef]
55. Heinonen, J.S.; Luttinen, A.V. Jurassic dikes of Vestfjella, western Dronning Maud Land, Antarctica: Geochemical tracing of ferropicrite sources. *Lithos* **2008**, *105*, 347–364. [CrossRef]
56. Sun, S.S.; McDonough, W.F. Chemical and isotopic systematics of oceanic basalts: Implications for mantle composition and processes. In *Magmatism in the Ocean Basins*; Saunders, A.D., Norry, M.J., Eds.; Geological Society, London, Special Publication: London, UK, 1989; Volume 42, pp. 313–345.

57. Morimoto, N.; Fabries, J.; Ferguson, A.K.; Ginzburg, I.V.; Ross, M.; Seifert, F.A.; Zussman, J.; Gottardi, D. Nomenclature of pyroxenes. *Amer. Miner.* **1988**, *62*, 53–62.
58. Stromer, J.C., Jr. The effects of recalculation on estimates of temperatures and oxygen fugacity from analyses of multicomponent iron-titanium oxides. *Amer. Miner.* **1983**, *68*, 586–594.
59. Atkin, F.B. Pyroxenes of the Bushveld intrusion, South Africa. *J. Petrol.* **1969**, *10*, 222–249. [CrossRef]
60. Zou, H.; Reid, M.R. Quantitative modeling of trace element fractionation during incongruent dynamic melting. *Geochim. Cosmochim. Acta* **2001**, *65*, 153–162. [CrossRef]
61. Desta, M.T.; Ayalew, D.; Ishiwatari, A.; Arai, S.; Tamura, A. Ferropicrite from the Lalibela area in the Ethiopian large igneous province. *J. Miner. Petrol. Sci.* **2014**, *109*, 191–207. [CrossRef]
62. Snyder, D.; Carmichael, I.S.E.; Wiebe, R.A. Experimental study of liquid evolution in an Fe-rich, layered mafic intrusion: Constraints of Fe-Ti oxide precipitation on the T-fO2, and T-p paths of tholeiitic magmas. *Contrib. Mineral. Petrol.* **1993**, *113*, 73–86. [CrossRef]
63. Toplis, M.J.; Carroll, M.R. An experimental study of the influence of oxygen fugacity on Fe-Ti oxide stability, phase relations, and mineral-melt equilibria in ferro basaltic systems. *J. Petrol.* **1995**, *36*, 1137–1170. [CrossRef]
64. Carmichael, I.S.E.; Ghiorso, M.S. Oxidation-reduction relations in basic magma: A case for homogenous equilibria. *Earth Planet. Sci. Lett.* **1986**, *78*, 200–210. [CrossRef]
65. Charlier, B.; Sakoma, E.; Sauvé, M.; Stanaway, K.; Vander Auwera, J.; Duchesne, J.-C. The Grader layered intrusion (Havre-Saint-Pierre Anorthosite, Quebec) and genesis of nelsonite and other Fe–Ti–P ores. *Lithos* **2008**, *101*, 359–378. [CrossRef]
66. Toplis, M.J.; Corgne, A. An experimental study of element partitioning between magnetite, clinopyroxene and iron-bearing silicate liquids with particular emphasis on vanadium. *Contrib. Mineral. Petrol.* **2002**, *144*, 22–37. [CrossRef]
67. Cawthorn, R.G.; Molyneux, T.G. Vanadiferous Magnetite Deposits of the Bushveld Complex. In *Mineral Deposits of Southern Africa*; Geol. Society South Africa: Johannesburg, South Africa, 1986; pp. 1251–1266.
68. Charlier, B.; Namur, O.; Duchesne, J.C.; Wiszniewska, J.B.; Parecki, A.; Auwera, J.V. Cumulate Origin and Polybaric Crystallization of Fe-Ti Oxide Ores in the Suwalki Anorthosite, Northeastern Poland. *Econ. Geol.* **2009**, *104*, 205–221. [CrossRef]
69. Shellnutt, J.G.; Wang, K.-L.; Zellmer, G.F.; Iizuka, Y.; Jahn, B.-M.; Pang, K.-N.; Qi, L.; Zhou, M.-F. Three Fe-Ti oxide ore-bearing gabbro-granitoid complexes in the Panxi region of the Emeishan large igneous province, SW China. *Am. J. Sci.* **2011**, *311*, 773–812. [CrossRef]
70. Soesoo, A. A multivariate statistical analysis of clinopyroxene composition: Empirical coordinates for the crystallization PT-estimations. *GFF* **1997**, *119*, 55–60. [CrossRef]
71. Tulloch, A.J.; Challis, G.A. Emplacement depths of Paleozoic-Mesozoic plutons from western New Zealand estimated by Hornblende-Al Geobarometry. *N. Z. J. Geol. Geophys.* **2000**, *43*, 555–567. [CrossRef]
72. Li, Y.-Q.; Li, Z.-L.; Chen, H.-L.; Yang, S.-F.; Yu, X. Mineral characteristics and metallogenesis of the Wajilitag layered mafic–ultramafic intrusion and associated Fe–Ti–V oxide deposit in the Tarim large igneous province, northwest China. *J. Asian Earth Sci.* **2012**, *49*, 161–174. [CrossRef]
73. Schweitzer, E.L.; Papike, J.J.; Bence, A.E. Statitical analysis of clinopyroxenes from deep sea basalts. *Am. Mineral.* **1979**, *64*, 501–513.
74. Frost, B.R.; Lindsley, D.H.; Andersen, D.J. Fe-Ti oxide-silicate equilibria: Assemblages with fayalitic olivine. *Amer. Miner.* **1988**, *73*, 727–740.
75. Frost, B.R. Magnetic petrology: Factors that control the occurrence of magnetite in crustal rocks. *Rev. Mineral. Geochem.* **1991**, *25*, 489–509.
76. Buddington, A.F.; Lindsley, D.H. Iron–titanium oxide minerals and synthetic equivalents. *J. Petrol.* **1964**, *5*, 310–357. [CrossRef]
77. Haggerty, S.E. *Opaque Mineral Oxides in Terrestrial Igneous Rocks*; Mineralogical Society of America; Short Course Notes 3; Washington, DC, USA, 1976; pp. 101–300.
78. Verhoogen, J. Oxidation of iron-titanium oxides in igneous rocks. *J. Geol.* **1962**, *70*, 168–181. [CrossRef]
79. Tan, W.; Liu, P.; He, H.; Wang, C.Y.; Liang, X. Mineralogy and origin of exsolution in Ti-rich magnetite from different magmatic Fe-Ti oxide-bearing intrusions. *Canad. Mineral.* **2016**, *54*, 539–553. [CrossRef]
80. Frost, B.R.; Lindsley, D.H. Occurrence of iron-titanium oxides in igneous rocks. In *Oxide Minerals: Petrologic and Magnetic Significance*; Lindsley, D.H., Ed.; Reviews in Mineralogy; Mineralogical Society of America: Washington, DC, USA, 1991; Volume 25, pp. 433–468.
81. Arguin, J.-P.; Pagé, P.; Barnes, S.-J.; Girard, R.; Duran, C. An Integrated Model for Ilmenite, Al-Spinel, and Corundum Exsolutions in Titanomagnetite from Oxide-Rich Layers of the Lac Doré Complex (Québec, Canada). *Minerals* **2018**, *8*, 476. [CrossRef]
82. Krause, J.; Brugmann, G.E.; Pushkarev, E.V. Chemical composition of spinel from Uralian-Alaskan-type Mafic–Ultramafic complexes and its petrogenetic significance. *Contrib. Mineral. Petrol.* **2011**, *161*, 255–273. [CrossRef]
83. Gibson, S.A.; Thompson, R.N.; Dickin, A.P. Ferropicrites; geochemical evidence for Fe-rich streaks in upwelling mantle plumes. *Earth Planet. Sci. Lett.* **2000**, *174*, 355–374. [CrossRef]
84. Gibson, S.A. Major element heterogeneity in Archean to Recent mantle plume starting heads. *Earth Planet. Sci. Lett.* **2002**, *195*, 59–74. [CrossRef]
85. Riley, T.R.; Leat, P.T.; Curtis, M.L.; Millar, I.L.; Duncan, R.A.; Fazel, A. Early-Middle Jurassic Dolerite Dykes from Western Dronning Maud Land (Antarctica): Identifying Mantle Sources in the Karoo Large Igneous Province. *J. Petrol.* **2005**, *46*, 1489–1524. [CrossRef]

86. Ichiyama, Y.; Ishiwatari, A.; Hirahara, Y.; Shuto, K. Geochemical and isotopic constraints on the genesis of the Permian ferropicritic rocks from the Mino-Tamba belt, SW Japan. *Lithos* **2006**, *89*, 47–65. [CrossRef]
87. Koppers, A.A.P.; Becker, T.W.; Jackson, M.G.; Konrad, K.; Müller, R.D.; Romanowicz, B.; Steinberger, B.; Whittaker, J.M. Mantle plumes and their role in Earth processes. *Nat. Rev. Earth Environ.* **2021**, *2*, 382–401. [CrossRef]
88. Pearce, J.A. A user's guide to basalt discrimination diagrams. In *Trace Element Geochemistry of Volcanic Rocks: Applications for Massive Sulphide Exploration*; Wyman, D.A., Ed.; Short Course Notes; Geological Association of Canada: St. Johns, NL, Canada, 1996; Volume 12, pp. 79–113.
89. Hart, S.R.; Dunn, T. Experimental Cpx/melt partitioning of 24 trace elements. *Contrib. Mineral. Petrol.* **1993**, *113*, 1–8. [CrossRef]
90. Johnson, K.T.M. Experimental determination of partition coefficients for rare earth and high field strength elements between clinopyroxene, garnet and basaltic melt. *Contrib. Mineral. Petrol.* **1998**, *133*, 60–68. [CrossRef]
91. Thurston, P.C. Igneous rock associations 19. Greenstone belts and granite Greenstone Terranes: Constraints on the nature of the Archean world. *Geosci. Can.* **2015**, *42*, 437–484. [CrossRef]
92. Clift, P.D.; Fitton, J.G. Trace and rare earth element chemistry of volcanic ashes from sites 918 and 919: Implications for Icelandic volcanism. *Proc. Ocean Drill. Progr. Sci. Results* **1998**, *152*, 67–84.
93. Pearce, J.A. Geochemical fingerprinting of oceanic basalts with applications to ophiolite classification and the search for Archean oceanic crust. *Lithos* **2008**, *100*, 14–48. [CrossRef]
94. Niu, Y.; Wilson, M.; Humphreys, E.R.; O'Hara, M.J. The Origin of Intra-plate Ocean Island Basalts (OIB): The Lid Effect and its Geodynamic Implications. *J. Petrol.* **2011**, *52*, 1443–1468. [CrossRef]
95. Niu, Y.L.; Bideau, D.; Hèkinian, R.; Batiza, R. Mantle compositional control on the extent of mantle melting, crust production, gravity anomaly, ridge morphology, and ridge segmentation: A case study at the Mid-Atlantic Ridge 33-358N. *Earth Planet. Sci. Lett.* **2001**, *286*, 383–399. [CrossRef]
96. Niu, Y.L.; O'Hara, M.J. Global correlations of ocean ridge basalt chemistry with axial depth: A new perspective. *J. Petrol.* **2008**, *49*, 633–664. [CrossRef]
97. Humphreys, E.R.; Niu, Y.L. On the composition of ocean island basalts (OIB): The effects of lithospheric thickness variation and mantle metasomatism. *Lithos* **2009**, *112*, 118–136. [CrossRef]
98. Arndt, N.; Lehnert, K.; Vasil'ev, Y. Meimechites: Highly magnesian lithospherecontaminated alkaline magmas from deep subcontinental mantle. *Lithos* **1995**, *34*, 41–59. [CrossRef]
99. Fedorenko, V.; Czamanske, G.K. Results of new field and geochemical studies of the volcanic and intrusive rocks of the Maymecha-Kotuy area, Siberian flood-basalt province, Russia. *Int. Geol. Rev.* **1997**, *39*, 479–531. [CrossRef]
100. Kogarko, L.N.; Ryabchikov, I.D. Geochemical evidence for meimechite magma generation in the subcontinental lithosphere of polar Siberia. *J. Asian Earth Sci.* **2000**, *18*, 195–203. [CrossRef]
101. Stosch, H.G. Rare earth partitioning between minerals from anhydrous spinel peridotite xenoliths. *Geochem. Cosmochim. Acta* **1982**, *46*, 793–811. [CrossRef]
102. McDonough, W.F.; Sun, S.S. The composition of the earth. *Chem. Geol.* **1995**, *120*, 223–253. [CrossRef]
103. Taylor, S.R.; McLennan, S.M. *The Continental Crust: Its Composition and Evolution*; Blackwell Scientific Publications: London, UK, 1985; p. 312.
104. Pang, K.-N.; Li, C.; Zhou, M.-F.; Ripley, E.M. Abundant Fe-Ti oxide inclusions in olivine from the Panzhihua and Hongge layered intrusions, SW China: Evidence for early saturation of Fe-Ti oxides in ferrobasaltic magma. *Contrib. Mineral. Petrol.* **2008**, *156*, 307–321. [CrossRef]
105. Pang, K.-N.; Zhou, M.-F.; Lindsley, D.H.; Zhao, D.; Malpas, J. Origin of Fe-Ti oxide ores in mafic intrusions: Evidence from the Panzhihua intrusion. *J. Petrol.* **2008**, *49*, 295–313. [CrossRef]
106. Zhang, Z.; Mahoney, J.J.; Mao, J.; Wang, F. Geochemistry of picritic and associated basalt flows of the western Emeishan flood basalt province, China. *J. Petrol.* **2006**, *47*, 1997–2019. [CrossRef]
107. Pang, K.-N. Origin of the Permian Panzhihua Layered Gabbroic Intrusion and the Hosted Fe-Ti-V Oxide Deposits, Sichuan Province, SW China. Ph.D. Thesis, The University of Hong Kong, Hong Kong, China, 2008.
108. Ryabchikov, I.D.; Kogarko, L.N. Magnetite compositions and oxygen fugacities of the Khibina magmatic system. *Lithos* **2006**, *91*, 35–45. [CrossRef]
109. Ganino, C.; Arndt, N.T.; Zhou, M.F.; Gaillard, F.; Chauvel, C. Interaction of magma with sedimentary wall rock and magnetite ore genesis in the Panzhihua mafic intrusion, SW China. *Miner. Depos.* **2008**, *43*, 677–694. [CrossRef]
110. Ghoneim, M.; Lebda, E.; Nasr, B.; Khedr, M. Geology and tectonic evolution of the area around Wadi Arais, Southeastern Desert, Egypt. In Proceedings of the 6th International Conference on Geology of the Arab World, Cairo University, Cairo, Egypt, February 2002; Volume 1, pp. 45–66.
111. Zafar, T.; Leng, C.B.; Mahar, M.A.; Alam, M.; Zhang, X.C.; Chen, W.T.; Rehman, H.U.; Rehman, S.U. Petrogenesis, platinum-group element geochemistry and geodynamic evolution of the Cretaceous Chilas gabbros, Kohistan Island arc, NE Pakistan. *Lithos* **2020**, *372–373*, 105691. [CrossRef]
112. Woodhead, J.D.; Eggins, S.M.; Johnson, R.W. Magma genesis in the New Britain Island Arc: Further insights into melting and mass transfer processes. *J. Petrol.* **1998**, *39*, 1641–1668. [CrossRef]
113. Dare, S.A.S.; Barnes, S.-J.; Beaudoin, G.; Méric, J.; Boutroy, E.; Potvin-Doucet, C. Trace elements in magnetite as petrogenetic indicators. *Miner. Depos.* **2014**, *49*, 785–796. [CrossRef]

114. Liu, P.P.; Zhou, M.F.; Chen, W.T.; Gao, J.F.; Huang, X.W. In-situ LA-ICPMS trace elemental analyses of magnetite: Fe-Ti-(V) oxide-bearing mafic-ultramafic layered intrusions of the Emeishan Large igneous Province, SW China. *Ore Geol. Rev.* **2015**, *65*, 853–871. [CrossRef]
115. Liu, P.-P.; Zhou, M.-F.; Chen, W.T.; Boone, M.; Cnudde, V. Using multiphase solid inclusions to constrain the origin of the baima Fe-Ti-(V) oxide deposit, SW China. *J. Petrol.* **2014**, *55*, 951–976. [CrossRef]
116. Roedder, E. Low temperature liquid immiscibility in the system K_2O-FeO-Al_2O_3-SiO_2. *Am. Miner.* **1951**, *36*, 282–286.
117. Freestone, I.L. Liquid immiscibility in alkali-rich magmas. *Chem. Geol.* **1978**, *23*, 115–123. [CrossRef]
118. Eby, G.N. Minor and trace element partitioning between immiscible ocelli-matrix pairs from lamprophyre dykes and sills, Monteregian hills petrographic province, Quebec. *Contrib. Mineral. Petrol.* **1980**, *75*, 269–278. [CrossRef]
119. Philpotts, A.R. Silicate liquid immiscibility in tholeiitic basalts. *J. Petrol.* **1979**, *20*, 99–118. [CrossRef]
120. Wilson, M. *Igneous Petrogenesis;* Unwin Hyman: London, UK, 1989; p. 494.
121. Peccerillo, A.; Taylor, S.R. Geochemistry of Eocene Calc-Alkaline Volcanic Rocks from the Kastamonu Area, Northern Turkey. *Contrib. Mineral. Petrol.* **1976**, *58*, 63–81. [CrossRef]
122. Holness, M.; Stripp, G.; Humphreys, M.; Veksler, I.; Nielsen, T.; Tegner, C. Silicate liquid immiscibility within the crystal mush: Late-stage magmatic microstructures in the Skaergaard intrusion, East Greenland. *J. Petrol.* **2011**, *52*, 175–222. [CrossRef]
123. Veksler, I.; Dorfman, A.; Danyushevsky, L.; Jakobsen, J.; Dingwell, D. Immiscible silicate liquid partition coefficients: Implications for crystal-melt element partitioning and basalt petrogenesis. *Contrib. Mineral. Petrol.* **2006**, *152*, 685–702. [CrossRef]
124. Veksler, I.V.; Dorfman, A.M.; Borisov, A.A.; Wirth, R.; Dingwell, D.B. Liquid immiscibility and the evolution of basaltic magma. *J. Petrol.* **2007**, *48*, 2187–2210. [CrossRef]
125. Kamenetsky, V.S.; Charlier, B.; Zhitova, L.; Sharygin, V.; Davidson, P.; Feig, S. Magma chamber–scale liquid immiscibility in the Siberian Traps represented by melt pools in native iron. *Geology* **2013**, *41*, 1091–1094. [CrossRef]
126. Morisset, C.E.; Scoates, J.S.; Weis, D.; Sauv'e, M.; Stanaway, K.J. Rutile-bearing ilmenite deposits associated with the Proterozoic saint-Urbain and lac allard anorthosite massifs, Grenville province, Quebec. *Can. Miner.* **2010**, *48*, 821–849. [CrossRef]
127. Clark, A.H.; Kontak, D.J. Fe-Ti-P Oxide Melts Generated through Magma Mixing in the Antauta Subvolcanic Center, Peru: Implications for the Origin of Nelsonite and Iron Oxide-Dominated Hydrothermal Deposits. *Econ. Geol.* **2004**, *99*, 377–395. [CrossRef]
128. Spera, F.J.; Schmidt, J.S.; Bohrson, W.A.; Brown, G.A. Dynamics and thermodynamics of magma mixing: Insights from a simple exploratory model. *Am. Miner.* **2016**, *101*, 627–643. [CrossRef]
129. Wang, J.; Wang, M.; Wei, B. In situ crystallization of Fe-Ti oxides in the Panzhihua layered intrusion, SW China: Constraints from crystal size distribution and high-resolution X-ray computed tomography analyses. *J. Asian Earth Sci.* **2023**, *257*, 105858. [CrossRef]
130. Latypov, R.M.; Chistyakova, S.Y.; Page, A.; Hornsey, R. Field evidence for the in-situ crystallization of the Merensky Reef. *J. Petrol.* **2016**, *56*, 2341–2372. [CrossRef]
131. Latypov, R.M.; Chistyakova, S.Y.; Namur, O.; Barnes, S. Dynamics of evolving basaltic magma chambers: Textural and chemical evolution of cumulates at the arrival of new liquidus phases. *Earth Sci. Rev.* **2020**, *210*, 103388. [CrossRef]
132. Charlier, B.; Duchesne, J.-C.; Auwera, J.V.; Storme, J.-Y.; Maquil, R.; Longhi, J. Polybaric fractional crystallization of high-alumina basalt parental magmas in the Egersund-Ogna massif-type anorthosite (Rogaland, SW Norway) constrained by plagioclase and high-alumina orthopyroxene megacrysts. *J. Petrol.* **2010**, *51*, 2547–2570. [CrossRef]
133. Marshall, B.; Gilligan, L.B. An introduction to remobilization: Information from ore-body geometry and experimental considerations. *Ore Geol. Rev.* **1987**, *2*, 87–131. [CrossRef]
134. Hobbs, B.E. Principles involved in mobilization and remobilization. *Ore Geol. Rev.* **1987**, *2*, 37–45. [CrossRef]
135. Pernet-Fisher, J.F.; Howarth, G.H.; Liu, Y.; Chen, Y.; Taylor, L.A. Estimating the lunar mantle water budget from phosphates: Complications associated with silicate-liquid immiscibility. *Geochim. Cosmochim. Acta* **2014**, *144*, 326–341. [CrossRef]
136. Philpotts, A.R. Comments on: Liquid immiscibility and the evolution of basaltic magma. *J. Petrol.* **2008**, *49*, 2171–2175. [CrossRef]
137. Charlier, B.; Grove, T. Experiments on liquid immiscibility along tholeiitic liquid lines of descent. *Contrib. Mineral. Petrol.* **2012**, *164*, 27–44. [CrossRef]
138. Philpotts, A.R. Compositions of immiscible liquids in volcanic rocks. *Contrib. Mineral. Petrol.* **1982**, *80*, 201–218. [CrossRef]
139. Jakobsen, J.K.; Veksler, I.V.; Tegner, C.; Brooks, C.K. Crystallization of the Skaergaard intrusion froman emulsion of immiscible iron- and silica-rich liquids: Evidence from melt inclusions in plagioclase. *J. Petrol.* **2011**, *52*, 345–373. [CrossRef]
140. Hou, T.; Charlier, B.; Holtz, F.; Veksler, I.; Zhang, Z.; Thomas, R.; Namur, O. Immiscible hydrous Fe–Ca–P melt and the origin of iron oxide-apatite ore deposits. *Nat. Commun.* **2018**, *9*, 1415. [CrossRef] [PubMed]
141. Thy, P.; Lesher, C.E.; Nielsen, T.F.D.; Brooks, C.K. Experimental constraints on the Skaergaard liquid line of descent. *Lithos* **2006**, *92*, 154–180. [CrossRef]
142. Heikal, M.T.S.; Khedr, M.Z.; El-Monesf, M.A.; Gomaa, S.R. Petrogenesis and geodynamic evolution of Neoproterozoic Abu Dabbab Albite Granite, Central Eastern Desert of Egypt: Petrological and geochemical constraints. *J. Afr. Earth Sci.* **2019**, *158*, 103518. [CrossRef]
143. Khedr, M.Z.; Khashaba, S.M.A.; El-Shibiny, N.H.; Takazawa, E.; Hassan, S.M.; Azer, M.K.; Whattam, S.A.; El-Arafy, R.A.; Ichiyama, Y. Integration of remote sensing and geochemical data to characterize mineralized A-type granites, Egypt: Implications for origin and concentration of rare metals. *Int. J. Earth Sci.* **2023**, *112*, 1717–1745. [CrossRef]

144. Abuamarah, B.A.; Azer, M.K.; Asimow, P.D.; Ghrefat, H.; Mubarak, H.S. Geochemistry and Petrogenesis of Late Ediacaran Rare-metal Albite Granites of the Arabian-Nubian Shield. *Acta Geol. Sin.* **2021**, *95*, 459–480. [CrossRef]
145. Muller, D.; Groves, D.I. *Potassic Igneous Rocks and Associated Gold-Copper Mineralization*; Springer: Berlin/Heidelberg, Germany, 1997; p. 238.
146. Cabanis, B.; Lécolle, M. Le diagramme La/10-Y/15-Nb/8: Un outil pour la discrimination des séries volcaniques et la mise en évidence des processus de mélange et/ou de contamination crustale. *Comptes Rendus L'académie Sci. Paris* **1989**, *309*, 2023–2029.
147. Nisbet, E.G.; Pearce, J.A. Clinopyroxene composition in mafic lavas from different tectonic settings. *Contrib. Mineral. Petrol.* **1977**, *63*, 149–160. [CrossRef]
148. Pearce, J.A. Trace element characteristics of lavas from destructive plate boundaries. In *Andesites, Orogenic Andesites and Related Rocks*; Thorpe, R.S., Ed.; Wiley and Sons: Chichester, UK, 1982; pp. 525–548.
149. Pirajno, F. Mantle plumes, associated intraplate tectono-magmatic processes and ore systems. *Episodes* **2007**, *30*, 1. [CrossRef] [PubMed]
150. Pirajno, F.; Santosh, M. Rifting, intraplate magmatism, mineral systems and mantle dynamics in central-east Eurasia: An overview. *Ore Geol. Rev.* **2014**, *63*, 265–295. [CrossRef]
151. Campbell, I.H.; Griffiths, R.W.; Hill, R.I. Melting in an Archaean mantle plume: Heads it's basalts, tails it's komatiites. *Nature* **1989**, *339*, 697–699. [CrossRef]
152. Griffiths, R.W.; Campbell, I.H. Interaction of mantle plume heads with the Earth's surface and onset of small-scale convection. *J. Geophys. Res.* **1991**, *96*, 18295–18310. [CrossRef]
153. Sengör, A.M.C. *Elevation as Indicator of Mantle Plume Activity*; Special Paper; Geological Society of America: Washington, DC, USA, 2001; Volume 352, pp. 183–225.
154. Ernst, R.E.; Buchan, K. Recognising mantle plumes in the geological record: Annual Review. *Earth Planet. Sci.* **2003**, *31*, 469–523. [CrossRef]
155. Campbell, I.H.; Griffiths, R.W. Implications of mantle plume structure for the evolution of flood basalts. *Earth Planet. Sci. Lett.* **1990**, *99*, 79–93. [CrossRef]
156. Griffiths, R.W.; Campbell, I.H. Stirring and structure in mantle plumes. *Earth Planet. Sci. Lett.* **1990**, *99*, 66–78. [CrossRef]

minerals

MDPI

Article

In Situ Carbonate U-Pb Dating of Gold and Mercury Deposits in the Youjiang Metallogenic Province, SW China, and Implications for Multistage Mineralization

Jinwei Li [1], Yuzhou Zhuo [2], Yitong Guo [1], Xinyue Lu [1] and Xinlu Hu [3,*]

[1] Research Center of Ancient Ceramic, Jingdezhen Ceramic University, Jingdezhen 333403, China
[2] Mineral Medicine Research Center, Guizhou University of Traditional Chinese Medicine, Guiyang 550025, China
[3] Faculty of Earth Resources, China University of Geosciences, Wuhan 430074, China
* Correspondence: huxinlu00@foxmail.com

Abstract: The Youjiang metallogenic province (YMP) is a famous ore-concentrating area in South China, known for its substantial Carlin-type gold deposits, antimony deposits, and mercury deposits. Previous studies have yielded conflicting views regarding the ages of mineralization in this area, particularly regarding the occurrence of Yanshanian versus Indosinian ore-forming events during the Mesozoic era. To resolve these discrepancies, this study utilized in situ LA-ICP-MS U-Pb dating on carbonate minerals from the Lannigou Carlin-type Au deposit, the Lanmuchang Hg-(Tl) deposit, and the Sixiangchang Hg deposit to accurately determine their mineralization ages. Our results indicate that the three deposits formed at 137 ± 9 Ma, ~97 Ma, and 454 ± 21 Ma, respectively. By integrating previously reported geochronological data, we propose that the low-temperature Au-As-Sb-Hg-Tl deposits in the YMP were formed during two major periods, Late Triassic and Late Jurassic to Cretaceous, with the latter being more prevalent. Additionally, there was a Paleozoic hydrothermal mercury mineralization event at the northeastern edge of this region. These newly acquired data significantly enhance our understanding of multistage, low-temperature mineralization events in the YMP. Our study also demonstrates that in situ carbonate U-Pb dating is an excellent method for investigating the timing of low-temperature mineralization events.

Keywords: carbonate; LA-ICP-MS U-Pb dating; Carlin-type gold deposit; mercury deposit; Youjiang metallogenic province

1. Introduction

The Youjiang Basin is located at the tri-junction of the Guizhou, Yunnan, and Guangxi provinces in southwest China. This region represents a major Au-As-Sb-Hg-Tl metallogenic province within the South China low-temperature metallogenic domain, covering an extensive area of 90,000 km^2 [1]. It hosts a variety of Carlin-type Au deposits, in addition to low-temperature hydrothermal Sb, Hg, and As deposits [2–4]. Recent isotopic dating results have indicated that Carlin-type Au mineralization in the Youjiang metallogenic province (YMP) occurred during two distinct periods, estimated at 225 to 195 Ma and 145 to 125 Ma [5–11]. This has led to considerable debate regarding the age of low-temperature mineralization, with the primary point of contention being whether there were one or two mineralization events during the Mesozoic era. Additionally, along the margins of the Youjiang Basin, older ages ranging from 470 to 370 Ma [12–14] have been reported, indicating the potential for a Paleozoic low-temperature mineralization event.

In recent years, there has been rapid advancement in the technology for dating carbonate minerals in thin sections using in situ laser ablation inductively coupled plasma mass spectrometry (LA-ICP-MS) U-Pb isotopic analysis. This technique has been successfully applied to the dating of calcite cements in fossils [15], calcite veins [16–18], and paleosols [19]. In situ carbonate U-Pb dating presents numerous advantages, including high spatial resolution, which enables the acquisition of data for multiple mineralization events at a microscopic level, as well as efficient and rapid data acquisition with minimal sample consumption. Carbonate minerals are prevalent as ore-forming gangue minerals in low-temperature deposits within the YMP [20]. Despite successful applications of carbonate U-Pb dating in this region, such as the Shuiyindong and the Yata Au deposits [10,11], there is a notable lack of reports on the use of this technique for other types of low-temperature deposits in the area. In this study, the formation sequence and trace element content of carbonate minerals were first determined to confirm their roles as indicators of the hydrothermal ore-forming process. Afterwards, in situ carbonate U-Pb dating was applied to the Lannigou Carlin-type Au deposit, the Lanmuchang Hg-(Tl) deposit, and the Sixiangchang Hg deposit within the YMP. The newly obtained data will advance our understanding of multistage, low-temperature mineralization events in the YMP.

2. Regional Geology

The YMP is situated along the southwestern margin of the Yangtze Block (Figure 1). This triangular region is delineated by the Mile–Shizong Fault to the northwest, the Shuicheng–Ziyun–Bama Fault to the northeast, and the Youjiang Fault to the south. This region is characterized by a Precambrian metamorphic basement overlaid by Devonian-to-Triassic marine sedimentary sequences. The carbonate platform and deep-water basin sedimentary system emerged during the late Early Devonian to the Early Triassic. Lithologically, the platform facies comprise carbonate sediments and breccias, while the basin facies consist of deep-water sediments, such as pelites and silicolites [20,21]. Although magmatic rocks are infrequently exposed, notable occurrences include Late Permian dolerite intrusions (259 Ma; [22]), quartz porphyry dykes (140~130 Ma; [23]), and lamprophyre dykes (88~85 Ma; [24]) in the southeastern and northern sectors of the region. Gravitational and magnetic data suggest that igneous intrusions may be present at depths ranging from 2 to 5 km [25]. This area is abundant in low-temperature, hydrothermal Au-Sb-Hg-As-Tl deposits, with a number of Carlin-type Au deposits, known as the "Dian–Qian–Gui Golden Triangle" [1].

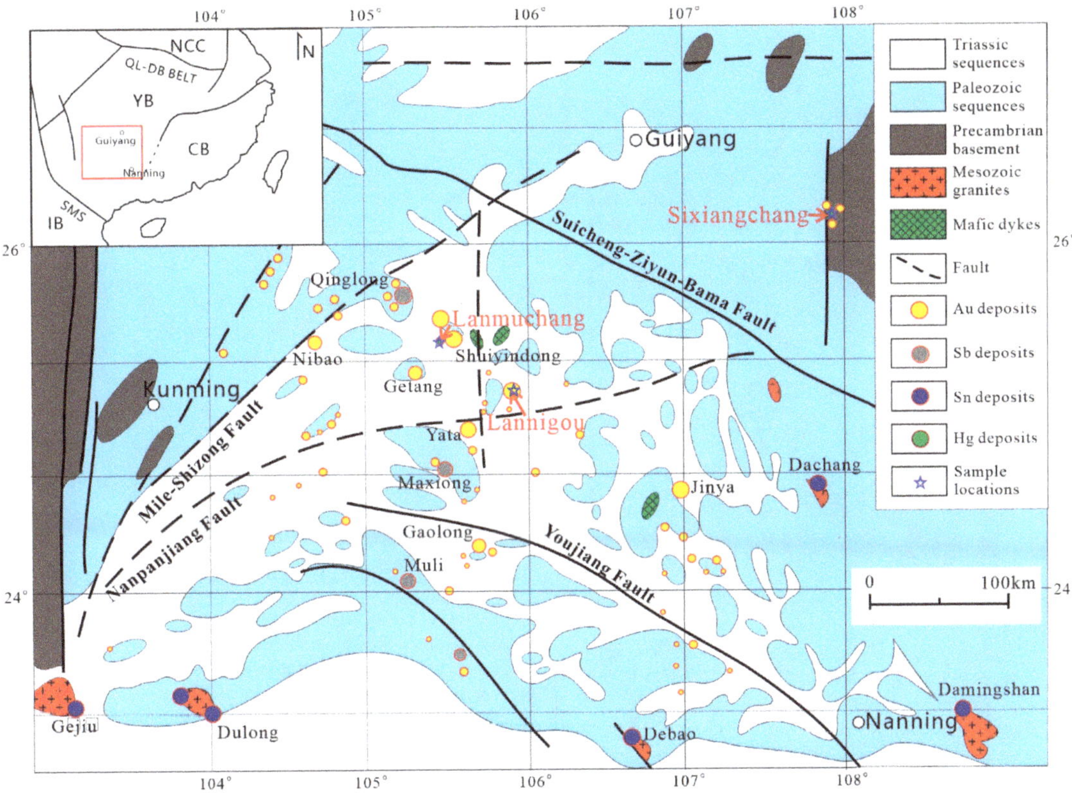

Figure 1. Geologic map of Youjiang metallogenic province (YMP) in southwest China (modified from [20]). The small red square refers to the location of the YMP. Abbreviations: NCC = North China Craton; YB = Yangtze Block; CB = Cathaysia Block; IB = Indochina Block; SMS = Song Ma Suture; QL-DB = Qinling–Dabie.

3. Geology of Ore Deposits

3.1. Lannigou Au Deposit

The Lannigou (also called Jinfeng) Au deposit is situated within the terrestrial sediments of the carbonate platform in the central part of the YMP (Figure 1). The exposed strata in the ore district are dominated by Triassic sequences, with the ore-bearing strata mainly consisting of Middle Triassic Xumeng Formation sandstone and mudstone, Niluo Formation claystone interbedded with siltstone, and Bianyang Formation sandstone interbedded with mudstone. The gold orebody is approximately 1500 m in length and 1200 m in width and varies from 10 to 50 m in thickness. It is primarily hosted in fault zones (Figure 2A). The orebody exhibits a simple and regular shape with a relatively intact structure. Within or near fault zones, there are clusters or vein-like occurrences of calcite and thin layers of bitumen, with Au-bearing pyrite distributed in a fine-disseminated form. As one of the largest Carlin-type Au deposits in the region, the Lannigou deposit has an estimated gold resource of over 5.3 million ounces, with an average grade of 4.5 g/t Au [26]. Associated metallic minerals include pyrite, realgar, orpiment, and cinnabar, while gangue minerals comprise quartz, calcite, dolomite, ankerite, and bitumen [27]. The predominant alteration types are pyritization, illitization, dolomitization, and silicification, with the presence of milky quartz and calcite veins [28,29].

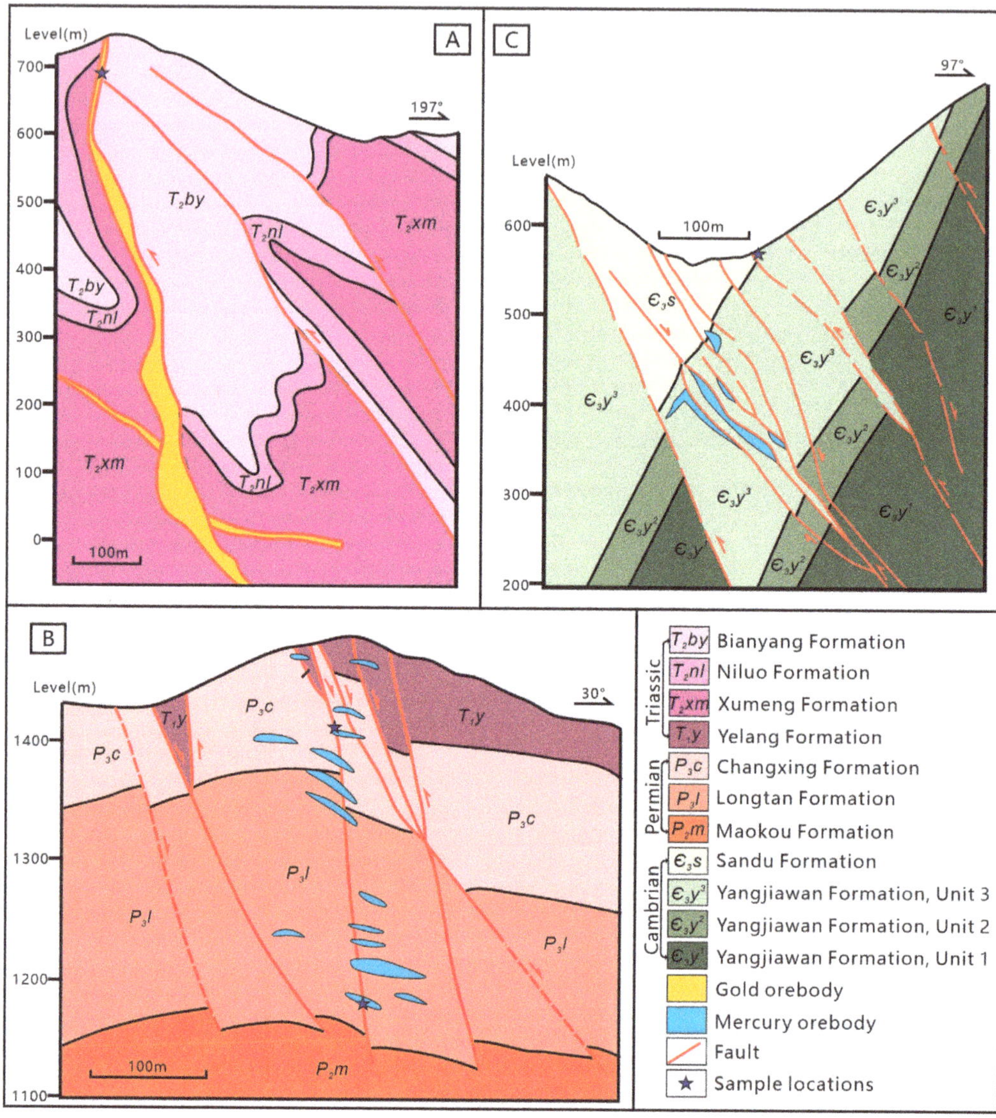

Figure 2. Geological cross sections of ore deposits. (**A**) Lannigou Au deposit (modified from [29]). (**B**) Lanmuchang Hg-(Tl) deposit (modified from [30]). (**C**) Sixiangchang Hg deposit (modified from [31]).

3.2. *Lanmuchang Hg-(Tl) Deposit*

The Lanmuchang Hg-(Tl) deposit lies adjacent to the Shuiyindong Carlin-type Au deposit within the same region, and their geological backgrounds exhibit remarkable similarities. The mineralized layers of the Lanmuchang Hg-(Tl) deposit extend up to 15 layers, primarily distributed in the Permian Longtan Formation and Changxing Formation, followed by the Triassic Yelang Formation (Figure 2B). These orebodies owe their formation and occurrence to a combination of structural and stratigraphic controls, characterized by distinctive layered and lens-shaped features that closely resemble the surrounding sediments [30,32]. Additionally, certain orebodies are associated with fault breccias and zones of alteration. Mineralization predominantly occurs within carbonaceous mudstones, heavily silicified limestones, and their interfacing lithologies. Specifically, independent thallium minerals, primarily composed of lorandite, are predominantly found within the carbonaceous mudstones and sandstones of the Longtan and Changxing Formations [27]. In addition to independent thallium minerals and cinnabar, ore minerals include pyrite enriched in thallium and arsenic. The deposit exhibits diverse alteration types, including silicification, pyritization, argillization, and baritization, with argillization dominated by kaolinization [33,34]. Drilling exploration data reveal that pyritization closely associated with Tl mineralization is well developed in strata ranging from the Longtan Formation to the Yelang Formation [32]. The mercury content in the orebodies typically ranges from 0.08% to 0.3%, with the highest concentration reaching 1.17% and the resource amounting to 4874 tons. Thallium grades generally fall between 0.01% and 0.02%, peaking at 0.035%, with resources exceeding 500 tons [33].

3.3. Sixiangchang Hg Deposit

The Sixiangchang Hg deposit is situated at the northeastern corner outside the YMP. Mercury serves as the primary economic element in this deposit, accompanied by gold as an auxiliary element, with an average gold grade of 7.19 g/t [31]. The ore-bearing stratum is the Upper Cambrian Yangjiawan Formation, which consists of muddy siltstone, thick-bedded detrital limestone interbedded with laminated muddy limestone, limestone, and dolostone (Figure 2C). The principal ore minerals include cinnabar, pyrite, native mercury, and stibnite, while the main gangue minerals are quartz, calcite, dolomite, and ankerite. The deposit exhibits various types of alteration, with silicification, pyritization, calcitization, and dolomitization being predominant [35]. Structural activity within the mining area is intense, characterized by complex and variable faults and folds. Notably, no igneous rock exposures have been observed in the mining area.

4. Sample Description

In the low-temperature deposits of the YMP, multiple stages of carbonate mineral crystallization are observed. The selection of carbonate minerals formed during the metallogenic stage is crucial for the precise determination of the mineralization age. Therefore, we prepared 100 μm thin sections of the samples and performed observations using polarized optical microscopy and cathodoluminescence microscopy. We identified calcite and dolomite, intimately associated with ore minerals in the thin sections, as suitable candidates for U-Pb dating. Most natural calcite and dolomite contain uranium concentrations below 100 ppb, which approaches the resolution limit of LA-ICP-MS technology. Consequently, we conducted preliminary trace element analyses on the calcite and dolomite within the thin sections [33] and selected samples with elevated uranium content (>0.1 ppm) for U-Pb dating.

The dating sample from the Lannigou Au deposit is a sandstone specimen (lng-8) with a notable gold grade of 13.65 g/t. The sample is characterized by the presence of multiple calcite veins (Figure 3A). Notably, the fine calcite veins within the sandstone matrix harbor several realgar particles. Furthermore, the sandstone is interspersed with copious amounts of Au-bearing pyrite (Figure 3B), indicating that these calcite veins crystallized contemporaneously with the ore-forming process. Under cathodoluminescence observation, a close association between quartz and calcite is evident, with the latter exhibiting bright luminescence, in contrast to the subdued appearance of quartz (Figure 3C). Additionally, the fine calcite veins are observed to encapsulate numerous quartz and pyrite particles of varying sizes (Figure 3D).

In the Lanmuchang Hg-(Tl) deposit, dating analyses were conducted on two limestone specimens, each featuring calcite veins, which were extracted from drill hole ZK901 (Figure 4). The first sample, lmc-2, originated from the Permian Longtan Formation at a depth of 67 m. The second, lmc-29, was obtained from a depth of 310 m within a structural alteration zone, also referred to as the SBT. The SBT is a product of regional detachment structures and hydrothermal activities occurring at the interface of the Permian Maokou and Longtan Formations. Lithologically, it includes highly silicified limestone, brecciated silicified limestone, silicified cataclasite, and brecciated claystone, constituting the principal ore-bearing stratum in the YMP [36]. An examination of the microscopic images of these samples shows that calcite is present alongside significant quantities of pyrite and arsenopyrite, suggesting that the calcite deposition coincided with the metallogenic stage (Figure 4C,F).

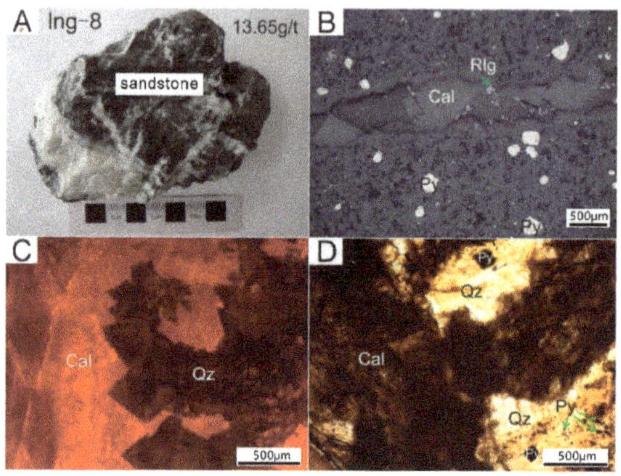

Figure 3. Hand specimen and microphotos of dating sample lng-8 in the Lannigou Au deposit. (**A**) The sandstone sample is interspersed with multiple calcite veins. (**B**) Calcite is associated with realgar and Au-bearing pyrite under reflected light. (**C**) Calcite exhibits bright cathodoluminescence, in contrast to the subdued appearance of quartz. (**D**) Calcite vein contains quartz and pyrite particles under transmission light. Abbreviations: Cal = calcite; Rlg = realgar; Py = pyrite; Qz = quartz.

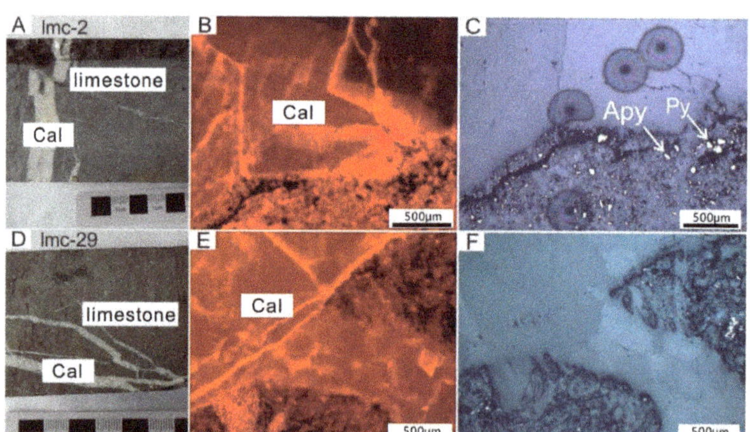

Figure 4. Hand specimen and microphotos of dating samples lmc-2 and lmc-29 in the Lanmuchang Hg-(Tl) deposit. (**A**) The limestone sample lmc-2 is interspersed with calcite veins. (**B**) Cathodoluminescence photo of calcite in the sample lmc-2. (**C**) Calcite is associated with pyrite and arsenopyrite in the sample lmc-2. (**D**) The limestone sample lmc-29 is interspersed with calcite veins. (**E**) Cathodoluminescence photo of calcite in the sample lmc-29. (**F**) The limestone hosting pyrite is intersected by the calcite vein in the sample lmc-29. Abbreviations: Cal = calcite; Py = pyrite; Apy = arsenopyrite.

Regarding the Sixiangchang Hg deposit, the selected sample for dating is a metasedimentary sandstone specimen (sxc-29), derived from the Cambrian strata within the mining district. This sample is rich in dolomite and cinnabar particles (Figure 5A). Upon fresh fracturing, the specimen discloses numerous tiny droplets of silver-white liquid mercury. Cathodoluminescence analysis demonstrates that the dolomite exhibits more intense luminescence compared to cinnabar, and visible cinnabar grows on the dolomite matrix (Figure 5B). This observation is consistent with the microstructures observed under both transmitted light and reflected light (Figure 5C,D), implying a near-simultaneous formation of dolomite and cinnabar. Consequently, the age of the dolomite provides a reliable estimate for the timing of mercury mineralization.

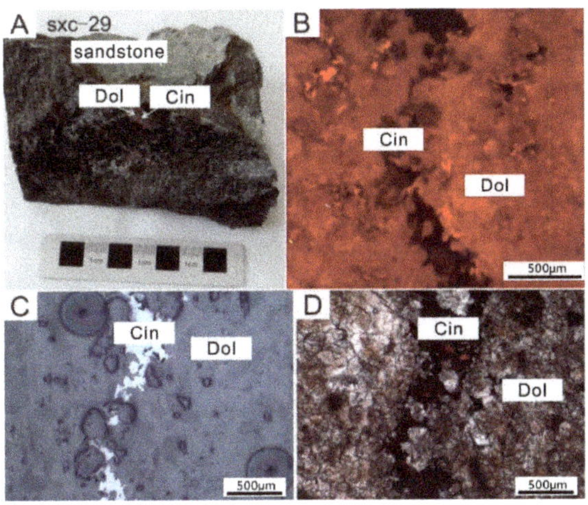

Figure 5. Hand specimen and microphotos of dating sample sxc-29 in the Sixiangchang Hg deposit. (**A**) The sandstone sample sxc-29 is rich in dolomite and cinnabar particles. (**B**) Cathodoluminescence photo of dolomite and cinnabar. (**C**) Dolomite is associated with cinnabar under reflected light. (**D**) Dolomite is associated with cinnabar under transmission light. Abbreviations: Dol = dolomite; Cin = cinnabar.

5. Analytical Methods

In situ carbonate LA-ICP-MS U-Pb dating was performed at the State Key Laboratory of Ore Deposit Geochemistry, Institute of Geochemistry, Chinese Academy of Sciences, using an Agilent 7700× plasma mass spectrometer equipped with a GeoLasPro 193 nm ArF excimer laser system (Agilent Technologies, Santa Clara, CA, USA). The parameters for ablation were as follows: spot size, 120 μm; repetition rate, 10 Hz; and energy density, 8 J/cm^2. Prior to experimentation, it is essential to optimize the performance of ICP-MS using standard reference material Nist 610 in order to achieve optimal sensitivity and ionization efficiency (U/Th $\approx$ 1) while also minimizing oxide yield (ThO/Th < 0.3%) and background values. The signal acquisition of a single sample consists of approximately 20 s of baseline signal, followed by a sampling period lasting 30 s, and concluding with an additional 40-s decay phase until the signal returns to the background level. Nist 610 was used to correct the ^{207}Pb/^{206}Pb ratios, and a matrix-matched carbonate WC-1 was used for the fractionation correction of the ^{238}U/^{206}Pb ratios [37]. The data were processed using ICPMSDataCal [38] to select sample and blank signals, correct for instrument sensitivity drift, and calculate U-Pb isotope ratios. The Tera–Wasserburg diagram was employed in ISOPLOT [39] to derive the lower intersection point age (2σ). The experimental results for WC-1 are listed in Supplementary Table S1. After eliminating any abnormal values, an age of 254 ± 16 Ma was calculated for WC-1 (Supplementary Figure S1), which corresponds to the established standard value [37].

6. Results

The concentrations and isotopic ratios of U, Th, and Pb in the analyzed samples and WC-1 are detailed in Supplementary Table S1. The majority of the analyzed spots exhibit U concentrations above 0.1 ppm, alongside consistent U/Pb ratios and minimal average standard weighted deviations. The $^{207}Pb/^{206}Pb$ and $^{238}U/^{206}Pb$ values demonstrate a pronounced linear correlation, affirming the reliability of the experimental results. Dating based on the U-Pb isotopic ratios resulted in an age of 137 ± 9 Ma (MSWD = 3.0) for calcite within the Lannigou Au deposit (Figure 6A). Two calcite samples from the Lanmuchang Hg-(Tl) deposit yielded ages of 96 ± 2 Ma (MSWD = 3.71) and 97 ± 9 Ma (MSWD = 1.09), respectively (Figure 6B,C). Dolomite from the Sixiangchang Hg deposit was dated at 454 ± 21 Ma (MSWD = 1.17) (Figure 6D). These ages correspond to the formation times of the respective carbonate minerals within each deposit.

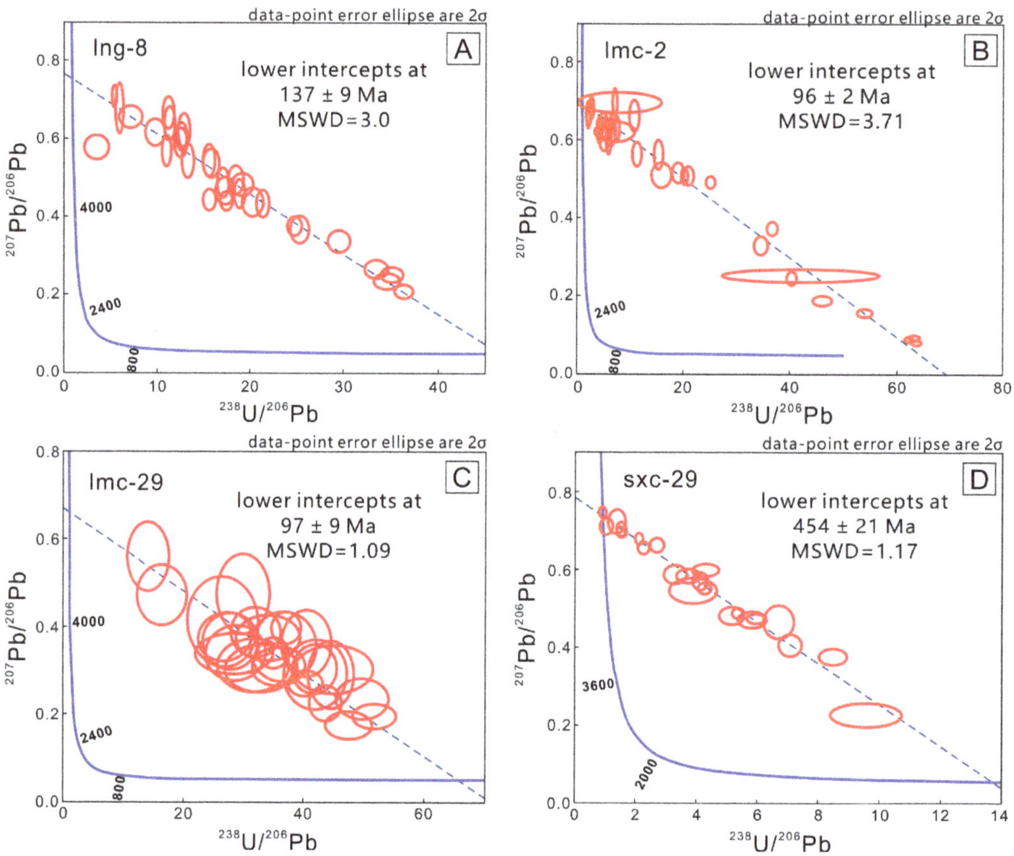

Figure 6. Tera–Wasserburg U-Pb plots for the sample lng-8 (**A**) from the Lannigou Au deposit, the samples lmc-2 (**B**) and lmc-29 (**C**) from the Lanmuchang Hg-(Tl) deposit, and the sample sxc-29 (**D**) from the Sixiangchang Hg deposit. The age errors are quoted at 2 sigma.

7. Discussion

7.1. Interpretation of the Carbonate U-Pb Ages

The U-Pb age of calcite from the Lannigou Au deposit obtained in this study is 137 ± 9 Ma, which is younger than the 148 ± 6 Ma hydrothermal apatite U-Th-Pb age previously reported for this deposit [40]. It is widely accepted that the initial mineralization stage of Carlin-type Au deposits in the YMP is characterized by decarbonation, followed by significant calcite deposition during the late mineralization stage [3,25]. The calcite veins analyzed herein intersect with sandstone ore rich in Au-bearing pyrite (Figure 3B), indicating that the formation of the calcite veins postdates the precipitation of Au-bearing pyrite. Consequently, the calcite veins are indicative of the late mineralization stage. Thus, the U-Pb age of the calcite indicates a minimum age constraint for the mineralization of the Lannigou Au deposit, suggesting that the large-scale precipitation of Au in the deposit did not occur later than 135 Ma. Combining this with the hydrothermal apatite dating results [40], it is inferred that the Lannigou Au deposit formed during the Yanshanian mineralization period.

In the Lanmuchang Hg-(Tl) deposit, two samples yield calcite U-Pb ages of 96 $\pm$ 2 Ma and 97 $\pm$ 9 Ma. Comparing with the neighboring Shuiyindong Au deposit, where calcite U-Pb ages range from 137 to 139 Ma [10], it is evident that the formation of the Lanmuchang Hg-(Tl) deposit occurred at a later stage. Despite their close geographic proximity and similarities in ore-hosting strata and low-temperature ore-forming element association (Au-As-Sb-Hg-Tl), a comparison of the trace element contents in hydrothermal pyrite using LA-ICP-MS reveals distinct differences. The pyrite from the Lanmuchang Hg-(Tl) deposit exhibits enrichment in Tl and W [33], while the pyrite from the Shuiyindong Au deposit is enriched in Au, As, and Cu [41], suggesting variations in ore-forming fluids. The current research proposes that weakly acidic, sulfur-rich, and iron-poor hydrothermal fluids enriched in Au-As-Sb-Hg-Tl-(Cu) interacted with carbonate wall rocks, leading to the decarbonation and subsequent precitation of trace elements (including Au and As) in pyrite [42]. However, the Shuiyindong Au deposit, controlled by the Huaijiapu anticline, lacks well-developed fault structures. Consequently, limited infiltration of rapidly cooling meteoric water and restricted ore-hosting space allowed for residual ore-forming fluids enriched in Tl, W, and Hg to continue migrating along various conduits. Recent studies on the Allchar Au-As-Sb-Tl deposit in North Macedonia, similar to Carlin-type Au deposits, also reveal zonation features from proximal to distal hydrothermal centers, with successive Au, Sb, and As-Tl mineralization [43,44]. Therefore, we posit that both the Lanmuchang and Shuiyindong deposits originate from the same hydrothermal ore-forming system. The mineralization of the Lanmuchang Hg-(Tl) deposit represents the residual enrichment of Hg and Tl fluids after Carlin-type hydrothermal evolution, reacting favorably with geological structures in the surrounding rock. This explanation provides a reasonable account for the delayed occurrence of calcite crystallization in the Lanmuchang Hg-(Tl) deposit compared to the Shuiyindong Au deposit.

The coexistence of dolomite and cinnabar in the samples from the Sixiangchang Hg deposit suggests their simultaneous formation (Figure 4) based on mineral assemblage relationships. In the Sixiangchang mining area, dolomite is widely distributed and serves as the primary gangue mineral associated with hydrothermal activity in this deposit, making its age representative of ore formation. Previous studies have demonstrated that dolomite exhibits higher ablation efficiencies during the laser ablation process compared to calcite and WC-1, and these disparities in ablation rates may result in significant age discrepancies of 4~8% for dolomite [45]. Therefore, despite obtaining a U-Pb age of 454 $\pm$ 21 Ma for dolomite in the Sixiangchang Hg deposit, it is important to acknowledge the potential presence of certain systematic errors. Nevertheless, even when accounting for the error range within 50 Ma, the mineralization age of the Sixiangchang Hg deposit remains notably older than the ages of low-temperature deposits in the YMP. Located on the periphery of the YMP, the Sixiangchang deposit is hosted by Cambrian metasedimentary sandstone strata and exhibits distinct differences in terms of ore-forming geological background compared to Carlin-type Au deposits, vein-type Sb deposits, and the Lanmuchang Hg-(Tl) deposit within the YMP. However, the mineralization age of the Sixiangchang deposit aligns with a series of Paleozoic low-temperature Hg-Pb-Zn-Ba-F deposits located in western Hunan province and eastern Guizhou province, southeast of the YMP [14], supporting the formation of the Sixiangchang Hg deposit during the Late Caledonian mineralization period.

7.2. Implications for Dynamic Setting and Multistage Mineralization

Previous studies on the metallogenic chronology of Carlin-type Au deposits in the YMP have demonstrated that there are two distinct stages of Carlin-type Au mineralization [5–11,40,46]. A comparative analysis of dating results [46] from dissolution-based and in situ data methods reveals that the isotopic ages obtained through single-mineral dissolution samples can be divided into two separate groups, representing Indosinian and Yanshanian mineralization events (Figure 7A). Su et al. further classified Au deposits into two independent metallogenic systems, with those formed during the Yanshanian period located to the north and those formed during the Indosinian period situated to the south along the Youjiang fault [29]. However, based on in situ dating results of hydrothermal accessory minerals, it is evident that Yanshanian mineralization occurred throughout most parts of the region, with only a few deposits (such as the Zhesang deposit) forming during the Indosinian period (Figure 7B). The in situ U-Pb dating of calcite from the Lannigou Au deposit also confirms its formation during the Yanshanian period in this study. Furthermore, previous findings indicate that the mineralization age of Sb deposits in the YMP is attributed to the Yanshanian period, with the Qinglong Sb deposit (the largest Sb deposit in the YMP) was dated to 148~125 Ma [47–49]. In addition, regarding the timing of Hg mineralization, the in situ U-Pb dating of calcite from the Lanmuchang Hg-(Tl) deposit, when compared with the adjacent Shuiyindong Au deposit, reveals that the Hg deposits formed later than the Au deposits but still within the Yanshanian mineralization period.

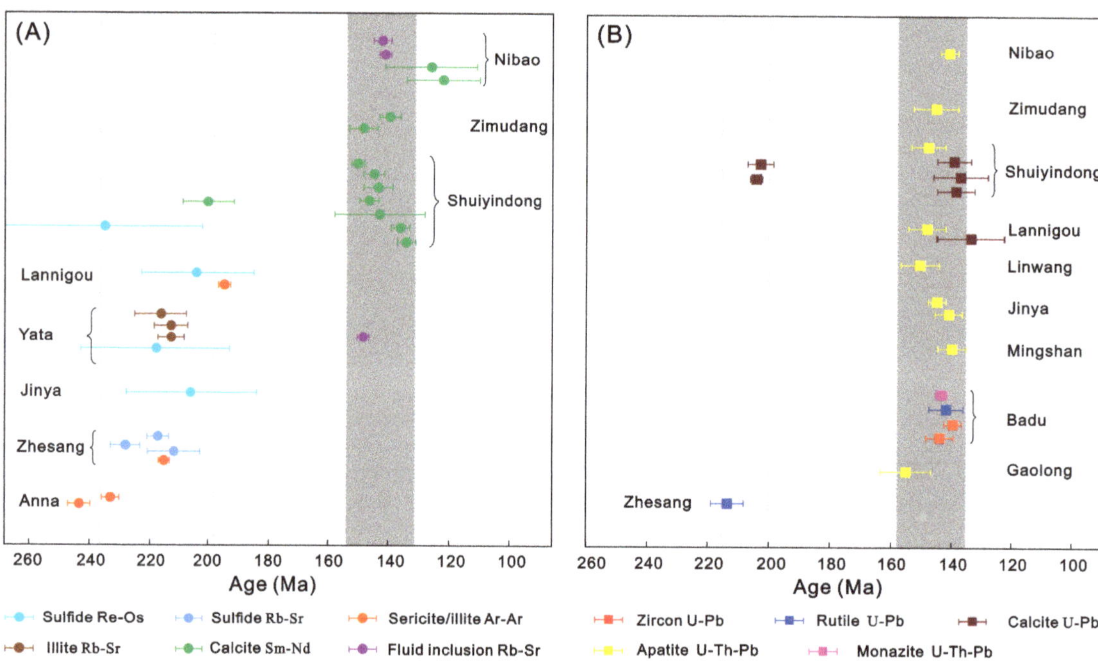

Figure 7. Summary of different isotopic dating results for the Carlin-type Au deposits in the YMP (modified from [46]). (**A**) Dissolution-based methods. (**B**) In situ methods.

Recent studies have indicated that the W-Sn polymetallic deposits in the eastern part of the South China Block formed during three distinct periods: 230~200 Ma, 160~130 Ma, and 100~80 Ma [50,51]. The occurrence of Carlin-type Au deposits, Sb deposits, and Hg deposits in the YMP aligns closely with the extensive W-Sn polymetallic mineralization during the Indosinian and Yanshanian orogenies, suggesting a potential shared tectonic background. Notably, the Late Triassic mineralization event primarily resulted from intracontinental orogenesis following the collision between the South China Block and multiple surrounding blocks [51,52]. In contrast, the Late Jurassic to Cretaceous mineralization event was mainly controlled by processes such as the subduction–retreat of the ancient Pacific plate [53], the formation of slab windows through the tearing of subducted slabs [50], lithospheric delamination-induced asthenospheric upwelling, and subsequent large-scale lithospheric extension [51,54].

The dating results for the Sixiangchang Hg deposit indicate that it arose from a Paleozoic mineralization event, distinctly different from the mineralization timing observed in the Lanmuchang Hg-(Tl) deposit within the YMP triangle area. Sulfur and mercury isotope compositions of cinnabar from both deposits [33,35] show a significant difference between the Sixiangchang and Lanmuchang deposits (Figure 8), suggesting that they are not products of the same mineralization event.

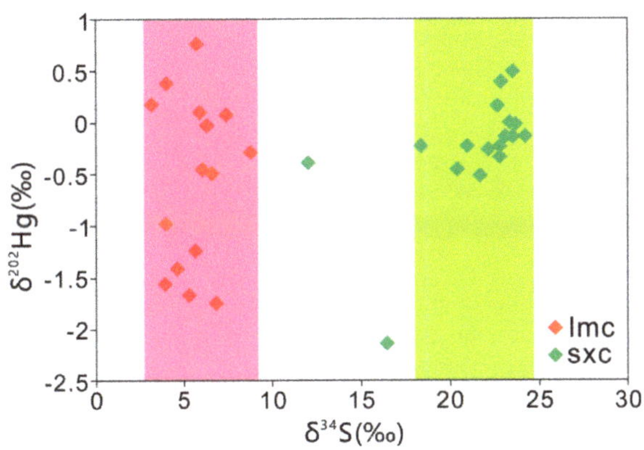

Figure 8. Cinnabar S-Hg isotope diagram of Lanmuchang Hg-(Tl) deposit (lmc) and Sixiangchang Hg deposit (sxc). The data sources are from [33,35].

Considering the timing of mineralization on a larger regional scale, the age of the Sixiangchang Hg deposit is consistent with a series of Hg-Pb-Zn-Ba-F deposits in the western Hunan and eastern Guizhou areas, located in the northeast of the YMP [14]. This similarity indicates that their geodynamic settings may be similar and could have been related to tectono-thermal events during the Late Caledonian orogeny [55]. The Jiangnan Orogenic Belt in South China experienced extensive early Paleozoic granitic magmatism, resulting in a series of gneissic and massive granite intrusions [56]. These granites exhibit zircon U-Pb ages ranging from 480 to 398 Ma, peaking at 456 to 419 Ma, reflecting a transition from an intracontinental orogeny to subsequent crustal extension [57,58]. The movement along NE-SW-trending Caledonian reverse faults led to the inversion of stratigraphic units from the Neoproterozoic sediments to the Early Ordovician sedimentary rocks. The local extensional tectonism during the early Paleozoic orogeny was conducive to the migration of basinal brines along the margins of the Jiangnan Orogenic Belt, leading to the formation of Hg, Pb-Zn, and Ba-F orebodies. The Sixiangchang Hg deposit may represent a distal response to this mineralization event.

8. Conclusions

1. The U-Pb age of hydrothermal calcite from the Lannigou Au deposit is 137 ± 9 Ma, indicating that this mineralization event was associated with Yanshanian tectonic movement.
2. At the Lanmuchang Hg-(Tl) deposit, two samples yielded hydrothermal calcite U-Pb ages of 96 ± 2 Ma and 97 ± 9 Ma, demonstrating that this mineralization event was also related to Yanshanian tectonic movement. However, its ore-forming time was later than that of Carlin-type Au deposits in the YMP.

3. The U-Pb age of hydrothermal dolomite from the Sixiangchang Hg deposit is 454 ± 21 Ma, indicating that this mineralization event was associated with the early Paleozoic Caledonian orogeny.
4. Integrating previous chronological data, the low-temperature Au-As-Sb-Hg-Tl deposits in the YMP primarily underwent two phases of mineralization: one in the Late Triassic and the other spanning from the Late Jurassic to the Cretaceous, with the latter being predominant. Additionally, there exists a Paleozoic hydrothermal Hg mineralization event at the northeastern edge of the region.

Supplementary Materials: The following supporting information can be downloaded at https://www.mdpi.com/article/10.3390/min14070669/s1: Table S1: Uranium and lead contents and isotopic results of carbonate minerals; Figure S1: Tera–Wasserburg U-Pb plots for WC-1.

Author Contributions: Methodology, J.L., Y.Z. and X.H.; sample collection, J.L. and Y.Z.; software, Y.G. and X.L.; data analysis, Y.Z., Y.G. and X.L.; writing—original draft preparation, J.L. and Y.Z.; writing—review and editing, J.L. and X.H.; funding acquisition, Y.Z. and X.H. All authors have read and agreed to the published version of the manuscript.

Funding: This research was jointly funded by the Guizhou Provincial Basic Research Program (Natural Science) (ZK [2024] general 348), the Hubei Provincial Natural Science Foundation of China (No. 2023AFD210), and the National Natural Science Foundation of China (No. 42072091).

Data Availability Statement: The data that support the findings of this study are available from the authors upon reasonable request.

Acknowledgments: We express our gratitude to Dai Zhihui and Tang Yanwen for their invaluable assistance in conducting LA-ICP-MS analysis. Additionally, we extend our sincere appreciation to Gao Wei and Huang Yong for their insightful guidance throughout the experiment and manuscript preparation.

Conflicts of Interest: The authors declare no conflicts of interest.

References

1. Tu, G.C. Two unique mineralization areas in southwest China. *Bull. Mineral. Petrol. Geochem.* **2002**, *21*, 1–2. (In Chinese with English Abstract)
2. Hu, R.Z.; Fu, S.L.; Huang, Y.; Zhou, M.F.; Fu, S.H.; Zhao, C.H.; Wang, Y.J.; Bi, X.W.; Xiao, J.F. The giant South China Mesozoic low-temperature metallogenic domain: Reviews and a new geodynamic model. *J. Asian Earth Sci.* **2016**, *137*, 9–34. [CrossRef]
3. Hu, X.L.; Gong, Y.J.; Zeng, G.P.; Zhang, Z.J.; Wang, J.; Yao, S.Z. Multistage pyrite in the Getang sediment-hosted disseminated gold deposit, southwestern Guizhou Province, China: Insights from textures and in situ chemical and sulfur isotopic analyses. *Ore Geol. Rev.* **2018**, *89*, 1–16. [CrossRef]
4. Hu, X.L.; Zeng, G.P.; Zhang, Z.J.; Li, W.T.; Liu, W.H.; Gong, Y.J.; Yao, S.Z. Gold mineralization associated with Emeishan basaltic rocks: Mineralogical, geochemical, and isotopic evidences from the Lianhuashan ore field, southwestern Guizhou Province, China. *Ore Geol. Rev.* **2018**, *95*, 604–619. [CrossRef]
5. Su, W.C.; Hu, R.Z.; Xia, B.; Xia, Y.; Liu, Y.P. Calcite Sm-Nd isochron age of the Shuiyindong Carlin-type gold deposit, Guizhou, China. *Chem. Geol.* **2009**, *258*, 269–274. [CrossRef]
6. Chen, M.H.; Mao, J.W.; Li, C.; Zhang, Z.Q.; Duan, Y. Re-Os isochron ages for arsenopyrite from Carlin-like gold deposits in the Yunnan-Guizhou-Guangxi "golden triangle", southwestern China. *Ore Geol. Rev.* **2015**, *64*, 316–327. [CrossRef]
7. Pi, Q.H.; Hu, R.Z.; Xiong, B.; Li, Q.L.; Zhong, R.C. In situ SIMS U-Pb dating of hydrothermal rutile: Reliable age for the Zhesang Carlin-type gold deposit in the golden triangle region, SW China. *Miner. Deposita* **2017**, *52*, 1179–1190. [CrossRef]
8. Huang, Y.; Hu, R.Z.; Bi, X.W.; Fu, S.L.; Peng, K.Q.; Gao, W.; Oyebamiji, A.; Zhaanbaeva, A. Low-temperature thermochronology of the Carlin-type gold deposits in southwestern Guizhou, China: Implications for mineralization age and geological thermal events. *Ore Geol. Rev.* **2019**, *115*, 103178. [CrossRef]
9. Gao, W.; Hu, R.Z.; Hofstra, A.H.; Li, Q.L.; Zhu, J.J.; Peng, K.Q.; Mu, L.; Huang, Y.; Ma, J.W.; Zhao, Q. U-Pb Dating on Hydrothermal Rutile and Monazite from the Badu Gold Deposit Supports an Early Cretaceous Age for Carlin-Type Gold Mineralization in the Youjiang Basin, Southwestern China. *Econ. Geol.* **2021**, *116*, 1355–1385. [CrossRef]
10. Jin, X.Y.; Zhao, J.X.; Feng, Y.X.; Hofstra, A.H.; Deng, X.D.; Zhao, X.F.; Li, J.W. Calcite U-Pb dating unravels the age and hydrothermal history of the giant Shuiyindong Carlin-type gold deposit in the golden triangle, South China. *Econ. Geol.* **2021**, *116*, 1253–1265. [CrossRef]
11. Wang, R.X.; Wang, Q.F.; Zhao, J.X.; Groves, D.I.; Kirkland, C.L.; Feng, X.Y.; Uysal, I.T.; Yang, L.; Deng, J. Carbonate U-Pb and illite Rb-Sr geochronology of sediment-hosted gold: A case study of Yata gold deposit. *Chem. Geol.* **2023**, *621*, 121352. [CrossRef]

12. Hu, R.Z.; Peng, J.T.; Ma, D.S.; Su, W.C.; Shi, C.H.; Bi, X.W.; Tu, G.C. Epoch of large-scale low-temperature mineralizations in southwestern Yangtze massif. *Miner. Depos.* **2007**, *26*, 583–596. (In Chinese with English Abstract)
13. Zhang, Z.Z.; Gong, Y.J.; Jin, S.C.; Qi, S.L.; Gan, J.M.; Zhang, L.G. Sm-Nd isochron age of the Dazhuyuan fluorite deposit in northeastern Guizhou and its geological significance. *Acta Petrol. Mineral.* 2018; 37, 949–958. (In Chinese with English Abstract)
14. Luo, K.; Zhou, J.; Cugerone, A.; Zhou, M.; Feng, Y.; Jin, Z.; Zhao, J.A. Paleozoic mercury mineralisation event in South China: In situ U-Pb dating and chemical compositions of calcite from the Jianyan Hg deposit. *Sci. China Earth Sci.* **2023**, *66*, 1877–1891. [CrossRef]
15. Li, Q.; Parrish, R.R.; Horstwood, M.S.A.; McArthur, J.M. U-Pb dating of cements in Mesozoic ammonites. *Chem. Geol.* **2014**, *376*, 76–83. [CrossRef]
16. Coogan, L.A.; Parrish, R.R.; Roberts, N.M.W. Early hydrothermal carbon uptake by the upper oceanic crust: Insight from in situ U-Pb dating. *Geology* **2016**, *44*, 147–150. [CrossRef]
17. Roberts, N.M.W.; Walker, R.J. U-Pb geochronology of calcite-mineralized faults: Absolute timing of rift-related fault events on the northeast Atlantic margin. *Geology* **2016**, *44*, 531–534. [CrossRef]
18. Nuriel, P.; Weinberger, R.; Kylanderclark, A.R.C.; Hacker, B.R.; Craddock, J.P. The onset of the Dead Sea transform based on calcite age-strain analyses. *Geology* **2017**, *45*, 587–590. [CrossRef]
19. Methner, K.; Mulch, A.; Fiebig, J.; Wacker, U.; Gerdes, A.; Graham, S.A.; Chamberlain, C.P. Rapid Middle Eocene temperature change in western North America. *Earth Planet. Sci. Lett.* **2016**, *450*, 132–139. [CrossRef]
20. Hu, R.Z.; Zhou, M.F. Multiple Mesozoic mineralization events in South China-an introduction to the thematic issue. *Mineral. Depos.* **2012**, *47*, 579–588. [CrossRef]
21. Yan, J.; Hu, R.Z.; Liu, S.; Lin, Y.T.; Zhang, J.C.; Fu, S.L. NanoSIMS element mapping and sulfur isotope analysis of Au-bearing pyrite from Lannigou Carlin-type Au deposit in SW China: New insights into the origin and evolution of Au-bearing fluids. *Ore Geol. Rev.* **2018**, *92*, 29–41. [CrossRef]
22. Zhang, X.J.; Xiao, J.F. Zircon U-Pb geochronology, Hf isotope and geochemistry study of the Late Permian diabases in the northwest Guangxi autonomous region. *Bull. Mineral. Petrol. Geochem.* **2014**, *33*, 163–176. (In Chinese with English Abstract)
23. Zhu, J.J.; Hu, R.Z.; Richards, J.P.; Bi, X.W.; Stern, R.; Lu, G. No genetic link between Late Cretaceous felsic dikes and Carlin-type Au deposits in the Youjiang basin, Southwest China. *Ore Geol. Rev.* **2017**, *84*, 328–337. [CrossRef]
24. Liu, S.; Su, W.; Hu, R.; Feng, C.; Gao, S.; Coulson, I.M.; Wang, T.; Feng, G.; Tao, Y.; Xia, Y. Geochronological and geochemical constraints on the petrogenesis of alkaline ultramafic dikes from southwest Guizhou Province, SW China. *Lithos* **2010**, *114*, 253–264. [CrossRef]
25. Zhou, Y.F. The application of regional gravity to the deep geology and mineralization prognosis in Guangxi. *Geol. Guangxi* **1993**, *6*, 15–24. (In Chinese with English Abstract)
26. Chen, M.H.; Mao, J.W.; Frank, P.B.; Norman, T.; Uttley, P. Structural features and metallogenesis of the Carlin-type Jinfeng (Lannigou) gold deposit, Guizhou Province, China. *Ore Geol. Rev.* **2011**, *43*, 217–234.
27. Zhuo, Y.Z.; Hu, R.Z.; Xiao, J.F.; Zhao, C.H.; Huang, Y.; Yan, J.; Li, J.W.; Gao, W.; Li, J.X. Trace elements and C-O isotopes of calcite from Carlin-type gold deposits in the Youjiang Basin, SW China: Constraints on ore-forming fluid compositions and sources. *Ore Geol. Rev.* **2019**, *113*, 103067. [CrossRef]
28. Yan, J.; Mavrogenes, J.A.; Liu, S.; Coulson, I.M. Fluid properties and origins of the Lannigou Carlin-type gold deposit, SW China: Evidence from SHRIMP oxygen isotopes and LA-ICP-MS trace element compositions of hydrothermal quartz. *J. Geochem. Explor.* **2020**, *215*, 106546. [CrossRef]
29. Su, W.C.; Dong, W.D.; Zhang, X.C. Carlin-type gold deposits in the Dian-Qian-Gui "golden Triangle" of southwest China. *Rev. Econ. Geol.* **2018**, *20*, 157–185.
30. Chen, D.Y. Discovery of rich-thallium ore body in paragenesis ore deposit of mercury and thallium and research of its mineralization mechanism. *J. Guizhou Univ. Technol.* **1989**, *18*, 1–20. (In Chinese with English Abstract)
31. Huang, G.S.; Du, Y.Y. The features and genesis of micrograined and disseminated gold deposit in Shan Dan Hg-ore zone. *Geol. Guizhou*, 1993; 10, 1–9. (In Chinese with English Abstract)
32. Zhao, F.Q.; Gu, S.Y.; Mao, Q.; Deng, Z.X.; Guo, Z.D. LA-ICP-MS trace element characteristics of pyrites from the Lanmuchang Hg-Tl deposit in Xingren County, Guizhou Province and the Tl mineralization of the deposit. *Acta Miner. Sin.* **2022**, *42*, 692–704. (In Chinese with English Abstract)
33. Zhuo, Y.Z. A Comparative Study of the Genetic Mechanism of the Carlin-Type Gold Deposit and Mercury Deposit in Southwest China. Ph.D. Thesis, Institute of Geochemistry, Chinese Academy of Sciences, Guiyang, China, 2019. (In Chinese with English Abstract).
34. Chen, D.Y.; Zou, Z.X. Studying situation about Lanmuchang type thallium (mercury) ore deposits in Southwestern Guizhou. *Geol. Guizhou* **2000**, *17*, 271–277. (In Chinese with English Abstract)
35. Zhuo, Y.Z.; Huang, Y.; Li, J.W.; Gao, W.; Li, J.X. Fluid properties and sources of Sixiangchang carbonate-associated mercury deposit, southwest China. *Acta Geochim.* **2019**, *38*, 670–682. [CrossRef]
36. Liu, J.Z.; Wang, Z.P.; Yang, C.F. Discriminant index and significance of structural body alteration of Carlin-type gold deposits in Yunnan-Guizhou-Guangxi and its surrounding areas, China. *Gold Sci. Technol.* **2022**, *30*, 532–539. (In Chinese with English Abstract)

37. Roberts, N.M.W.; Rasbury, E.T.; Parrish, R.R.; Smith, C.J.; Horstwood, M.S.A.; Condon, D.J. A Calcite reference material for LA-ICP-MS U-Pb geochronology. *Geochem. Geophys. Geosyst.* **2017**, *18*, 2807–2814. [CrossRef]
38. Liu, Y.S.; Hu, Z.C.; Gao, S.; Guinther, D.; Xu, J.; Gao, C.G.; Chen, H.H. In situ analysis of major and trace elements of anhydrous minerals by LA-ICP-MS without applying an internal standard. *Chem. Geol.* **2008**, *257*, 34–43. [CrossRef]
39. Ludwig, K.R. *ISOPLOT 3.75, a Geochronological Toolkit for Microsoft Excel*; Berkeley Geochronological Center Special Publication: Berkeley, CA, USA, 2012.
40. Gao, W.; Hu, R.Z.; Huang, Y.; Zhu, J.J.; Li, Q.L.; Mei, L.; Bi, X.W.; Liu, J.Z. Hydrothermal apatite as a robust U-Th-Pb chronometer for the Carlin-type gold deposits in the Youjiang basin, SW China. *Miner. Depos.* **2024**, *59*, 109–131. [CrossRef]
41. Li, J.X.; Hu, R.Z.; Zhao, C.H. Sulfur isotope and trace element compositions of pyrite determined by NanoSIMS and LA-ICP-MS: New constraints on the genesis of the Shuiyindong Carlin-like gold deposit in SW China. *Miner. Depos.* **2020**, *55*, 1279–1298. [CrossRef]
42. Xie, Z.J.; Xia, Y.; Cline, J.S. Magmatic origin for sediment-hosted Au deposits, Guizhou Province, China: In situ chemistry and sulfur isotope composition of pyrites, Shuiyindong and Jinfeng Deposits. *Econ. Geol.* **2018**, *113*, 1627–1652. [CrossRef]
43. Xie, Z.J.; Xia, Y.; Cline, J.S. Are there Carlin-type gold deposits in China? A comparison of the Guizhou, China, deposits with Nevada, USA, deposits. *Rev. Econ. Geol.* **2018**, *20*, 187–233.
44. Palinkaš, S.S.; Hofstra, A.H.; Percival, T.J. Comparison of the Allchar Au-As-Sb-Tl Deposit, Republic of Macedonia, with Carlin-type gold deposits. *Rev. Econ. Geol.* **2018**, *20*, 335–363.
45. Guillong, M.; Wotzlaw, J.F.; Looser, N.; Laurent, O. New analytical and data evaluation protocols to improve the reliability of U-Pb LA-ICP-MS carbonate dating. *Geochronology* **2020**, *2*, 155–167. [CrossRef]
46. Gao, W.; Hu, R.Z.; Li, Q.L.; Liu, J.Z.; Li, X.H. Research advances on the geochronology of Carlin-type gold deposits in the Youjiang Basin, southwestern China. *Earth Sci. Front.* **2024**, *31*, 267–283. (In Chinese with English Abstract)
47. Zhu, L.M.; Hu, R.Z. The coexistence and separation of gold and antimony in micro-disseminated gold deposits in southwest Guizhou and thermodynamic analysis. *Sci. China Ser. D*, 1999; 29, 481–488. (In Chinese)
48. Peng, J.T.; Hu, R.Z.; Jiang, G.H. Samarium-Neodymium isotope system of fluorites from the Qinglong antimony deposit, Guizhou Province: Constraints on the mineralizing age and ore-forming materials' sources. *Acta Petrol. Sin.* **2003**, *19*, 785–791. (In Chinese with English Abstract)
49. Wang, D.H.; Qin, Y.; Wang, C.H.; Chen, Y.C.; Gao, L. Mineralization pedigree for epithermal Hg, Sb, Au deposits in Guizhou Province-Taking the Dachang Sb deposit, the Zimudang Au deposit and the Luanyantang Hg deposit for examples. *Geotecton. Metallog.* **2012**, *36*, 330–336. (In Chinese with English Abstract)
50. Mao, J.W.; Cheng, Y.B.; Chen, M.H. Major types and time-space distribution of Mesozoic ore deposits in South China and their geodynamic settings. *Miner. Depos.* **2013**, *48*, 267–294.
51. Hu, R.Z.; Gao, W.; Fu, S.L.; Su, W.C.; Peng, J.T.; Bi, X.W. Mesozoic intraplate metallogenesis in South China. *Earth Sci. Front.* **2024**, *31*, 226–238. (In Chinese with English Abstract)
52. Wang, Y.J.; Fan, W.M.; Zhang, G.W. Phanerozoic tectonics of the South China Block: Key observations and controversies. *Gondwana Res.* **2013**, *23*, 1273–1305. [CrossRef]
53. Li, J.H.; Zhang, Y.Q.; Dong, S.W. Cretaceous tectonic evolution of South China: A preliminary synthesis. *Earth Sci. Rev.* **2014**, *134*, 98–136. [CrossRef]
54. Zhang, H.J.; Lv, Q.T.; Wang, X.L. Seismically imaged lithospheric delamination and its controls on the Mesozoic magmatic province in South China. *Nat. Commun.* **2023**, *14*, 2718. [CrossRef] [PubMed]
55. Chu, Y.; Lin, W.; Faure, M.; Wang, Q.C.; Ji, W.B. Phanerozoic tectonothermal events of the Xuefengshan Belt, central South China: Implications from U-Pb age and Lu-Hf determinations of granites. *Lithos* **2012**, *150*, 243–255. [CrossRef]
56. Shu, L.S. Predevonian tectonic evolution of South China: From Cathaysian block to Caledonian period folded orogenic belt. *Geol. J. China Univ.* **2006**, *12*, 418–431. (In Chinese with English Abstract)
57. Faure, M.; Shu, L.S.; Wang, B.; Charvet, J.; Choulet, F.; Monie, P. Intracontinental subduction: A possible mechanism for the Early Palaeozoic Orogen of SE China. *Terra Nova* **2009**, *21*, 360–368. [CrossRef]
58. Charvet, J.; Shu, L.; Faure, M.; Choulet, F.; Wang, B.; Lu, H.; Breton, N.L. Structural development of the Lower Paleozoic belt of South China: Genesis of an intracontinental orogen. *J Asian Earth Sci.* **2010**, *39*, 309–330. [CrossRef]

minerals

MDPI

Article

Petrogenesis of the Dalaku'an Mafic–Ultramafic Intrusion in the East Kunlun, Xinjiang: Constraints from the Mineralogy of Amphiboles

Yazhou Fan [1,*], Yali Deng [1], Zhaode Xia [2,3], Minghao Ren [1,*] and Jianhan Huang [1]

1 College of Geosciences and Engineering, North China University of Water Resources and Electric Power, Zhengzhou 450046, China; dengyali0929@163.com (Y.D.); huangjianhan@ncwu.edu.cn (J.H.)

2 College of Earth Science and Resources, Chang'an University, Xi'an 710054, China; karlde@163.com

3 Key Laboratory of Western China's Mineral Resources and Geological Engineering, Ministry of Education, Xi'an 710054, China

* Correspondence: fanyazhou@ncwu.edu.cn (Y.F.); renmh@ncwu.edu.cn (M.R.); Tel.:+86-0371-69127572 (Y.F.)

Abstract: The Dalaku'an mafic–ultramafic intrusion, located in the western segment of the East Kunlun, presents conducive conditions for the magmatic Cu-Ni sulfide deposits. According to the detailed petrographic observation, the amphiboles within distinct rock types were analyzed by EPMA analysis. The crystallization conditions, such as temperature, pressure, oxygen fugacity, and water content of the magma, were calculated to explore the genesis of the intrusion. The amphiboles were divided into three types: Amp-I, characterized by low silicon content but enrichment of aluminum, titanium, and alkali, predominantly comprising Tschermakitic hornblende and Magnesio-hornblende with mantle-derived traits; Amp-II, exhibiting elevated silicon content but diminished levels of aluminum, titanium, and alkali, primarily constituted of Magnesio-hornblende; whereas Amp-III manifests as Actinolitic hornblende, indicative of crustal origins. The calculated temperatures of amphiboles ranged between Amp-I (955–880) °C, Amp-II (852–774) °C, and Amp-III (761–760) °C; the pressures ranged between Amp-I (454–274) MPa, Amp-II (194–93) MPa, and Amp-III (101–84) MPa; the oxygen fugacities (△NNO) ranged between Amp-I (0.93–2.17), Amp-II (1.55–2.52), and Amp-III (1.89); and the water contents (H_2O_{melt}) ranged from (6.69–8.67) to (5.90–7.32). The magma experienced multiple stages of crystallization and underwent complex magma evolution at different depths. The high oxygen fugacity and water content could be attributed to the subduction of the oceanic crust. The magma source of the Dalaku'an intrusion was metasomatized by fluids from subducting plates, thereby originating within a post-collision extension.

Keywords: amphibole; crystallization condition; genetic mineralogy; Dalaku'an intrusion; East Kunlun; Xinjiang

1. Introduction

Mafic–ultramafic intrusions in convergent margin settings have been extensively studied, including the Central Asian Orogenic Belt, East Kunlun Orogenic Belt, Appalachian Orogen, and Variscan Orogenic Belt [1]. Despite the widespread distribution of mafic–ultramafic rocks, only a few of them host sulfide mineralization. The East Kunlun Orogenic Belt (EKOB) is an important polymetallic mineralization belt in China [2], which extends for approximately 1500 km long in the northern part of the Qinghai-Tibet Plateau, bounded by the NE–SW trending Altyn Tagh fault to the west and the Qinling Orogenic Belt to the northeast [3]. The Qaidam Block and Bayan Har-Qiangtang Block occur in the north and south, respectively [2,3] (Figure 1a). The mafic–ultramafic rocks widespread in the EKOB are classified into two groups that are distinguished according to their emplacement in different tectonic settings: (1) ophiolite-type mafic–ultramafic blocks [4], similar to magnesian ultramafic rocks without the metallogenic relation of Cu-Ni sulfide deposits, such as Heishan-Qingshuiquan-Wenquan ophiolites and Muzitag-Buqingshan-A'nyemaqen-De'erni ophiolites that represent remnant sections of the Early Paleozoic Proto-Tethys and the Late Paleozoic-Early Mesozoic Paleo-Tethys, respectively [4–10] (Figure 1a); (2) (layered) mafic–ultramafic intrusions related to extension [11,12] (Figure 1a), such as Xiarihamu [11,13–15], Shitoukengde [16–20], Binggounan [21,22], Akechukesai [23] and Dalaku'an [24]. Each provides important information on the chemical history of the mafic–ultramafic parental mantle source and its tectonic setting. Among these mafic–ultramafic intrusions, Xiarihamu intrusion contains super-large nickel deposits [13], and Shitoukengde and Binggounan also have nickel mineralization [14–16,22,25–27], making the East Kunlun region a favorable area for magmatic copper–nickel sulfide deposits [12].

The Dalaku'an mafic–ultramafic intrusion yields the highest nickel and copper grades, 0.65 wt% and 1.4 wt% [24], respectively, showing a characteristic of ferruginous mafic-ultramafic intrusions [24,28,29]. The formation age of the Dalaku'an mafic–ultramafic intrusion is 244 Ma ± 2 Ma (Zircon U-Pb LA-ICP-MS) [24], indicating a Middle Triassic age, which coincided with the transition period of the Paleo-Tethys Ocean subduction closure to intracontinental collision orogeny [12,30]. The mantle source of this intrusion may be the enriched lithospheric mantle transformed by plate fluid replacement according to the geochemical and isotopic evidence [24]. The petrography and olivine composition indicate that the parent melt was saturated in sulfur during the olivine crystallization process, and then sulfide segregation occurred [24,28,29], which implied favorable conditions for the formation of magmatic Cu-Ni sulfide deposits. The Cu-Ni sulfide mineralization of the mafic–ultramafic intrusion may be restricted by magmatic evolution conditions such as temperature, pressure, oxygen fugacity, and water content. The crystallization of amphibole is controlled by these physical and chemical conditions [11,12]. Therefore, the chemical composition of amphibole can reflect the magmatic process [31–40] and the tectonic environment [33,41,42]. Various amphiboles, as common minerals, occur in distinct rock types of the Dalaku'an intrusion [24,28]; however, their types, compositions, and genetic significance are still neither adequate nor systematic.

This study focuses on various amphiboles within the Dalaku'an intrusion, leveraging detailed petrographic characteristics and mineral chemistry to (1) identify the types of amphiboles, (2) elucidate the temperature, pressure, oxygen fugacity, and (3) constrain the petrogenesis of the Dalaku'an intrusion.

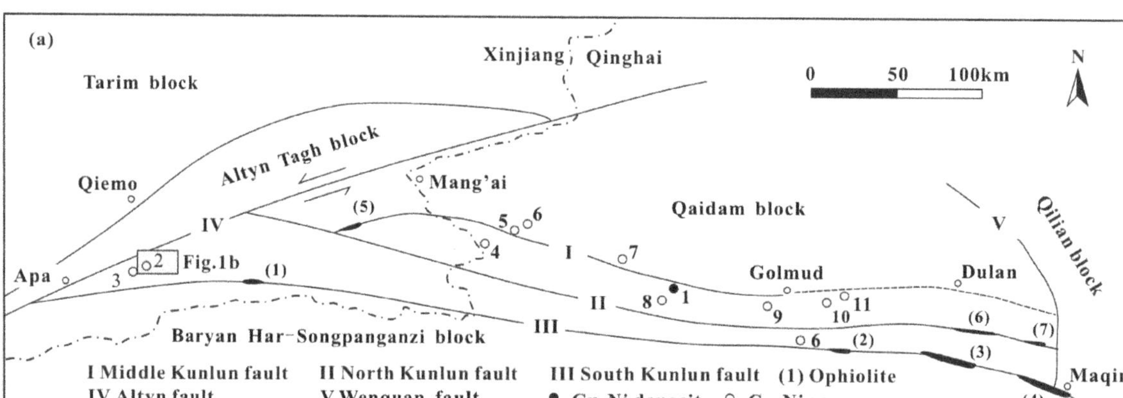

Figure 1. *Cont.*

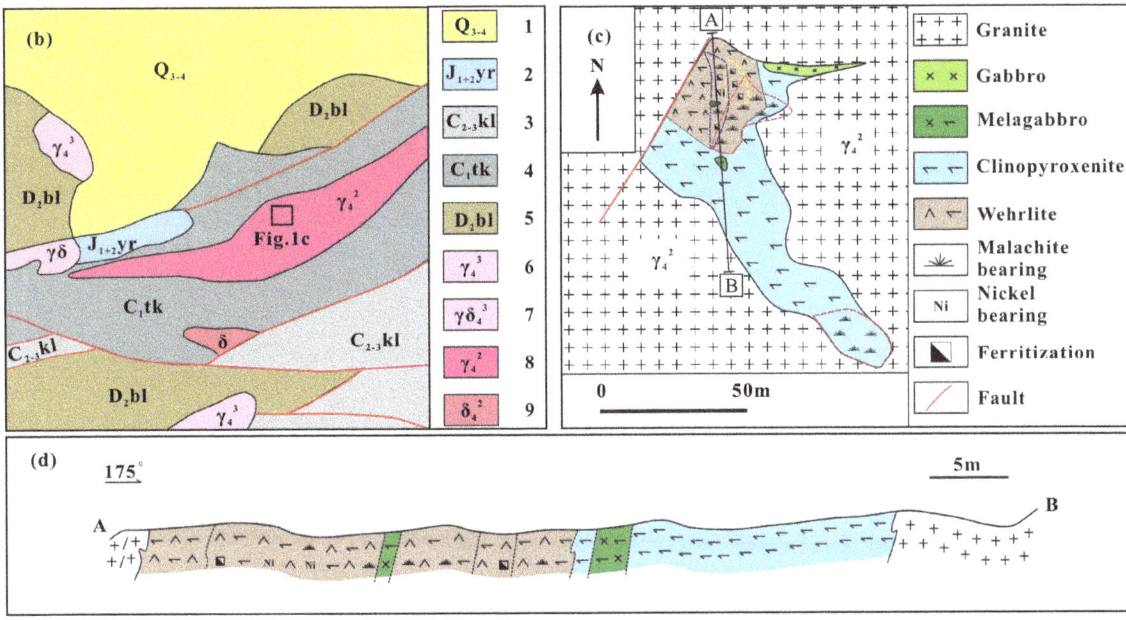

Figure 1. The tectonic and geological maps of the Dalaku'an intrusion. (**a**) Tectonic units of the East Kunlun (modified after [2]). Cu-Ni deposits: 1—Xiarihamu; Cu-Ni occurrences: 2—Dalaku'an, 3—Jikelikuole, 4—Lalinggaolihegounao, 5—Maxingdawannan, 6—Binggounan, 7—Kaimuqihe, 8—Akechukesai, 9—Detangou, 10—Bairiqiligou, 11—Shitoukengde. Ophiolites: (1) Muztag, (2) Buqingshan, (3) A'nyemaqen, (4) De'erni, (5) Heishan, (6) Qingshuiquan, (7) Wenquan. (**b**) The geological sketch map of the study area. Legend: 1—Upper Pleistocene, 2—Yerqiang Group, 3—Kalamilan Group, 4—Tuokuzidaban Group, 5—Bulakebashi Group, 6—the Late Variscan granite, 7—the Late Varican granodiorite, 8—the Middle Varican granite, 9—the Middle Varican diorite. (**c**) Lithologic map of the Dalaku'an intrusion. (**d**) Geological section (A-B); legend as in (**c**).

2. Geological Background

The Dalaku'an mafic–ultramafic intrusion is located about 30 km southwest of Qiemo County, belonging to a part of the Late Paleozoic Kalamilan trench-arc system in the southern belt of the East Kunlun orogeny ([28,43–45] (Figure 1a). The main strata in the region include the Middle Devonian Brakbashi Group, the Lower Carboniferous Tuokuzidaban Group, the Middle–Upper Carboniferous Kalamilanhe Group, the Lower Middle Jurassic Yerqiang Group, and the Upper Pleistocene (Figure 1b). The magmatic intrusions are mainly Variscan granite–diorite and Indosinian mafic–ultramafic pluton ([24,46]) (Figure 1b,c). The Dalaku'an mafic–ultramafic intrusion is composed of two sections from east to west. The eastern segment spans about 96 m in length and 9–12 m in width, showing intrusive contact and local fault contact with Variscan granite [24] (Figure 1c,d). The Dalaku'an intrusion is mainly composed of ultramafic rocks and a small amount of mafic rocks (Figure 1c,d). Ultramafic rocks include lherzolite, wehrlite, olivine websterite, and clinopyroxenite, while mafic rocks include melagabbro and gabbro [24,28]. Wehrlite, clinopyroxenite, and (mela-)gabbro are exposed on the surface (Figure 2a). The significant accumulations of malachite, azurite, and limonite on the surface indicate a positive potential for mineralization, which was produced by the oxidation of sulfides such as pentlandite, chalcopyrite, pyrite, and pyrrhotite [24] (Figure 2b).

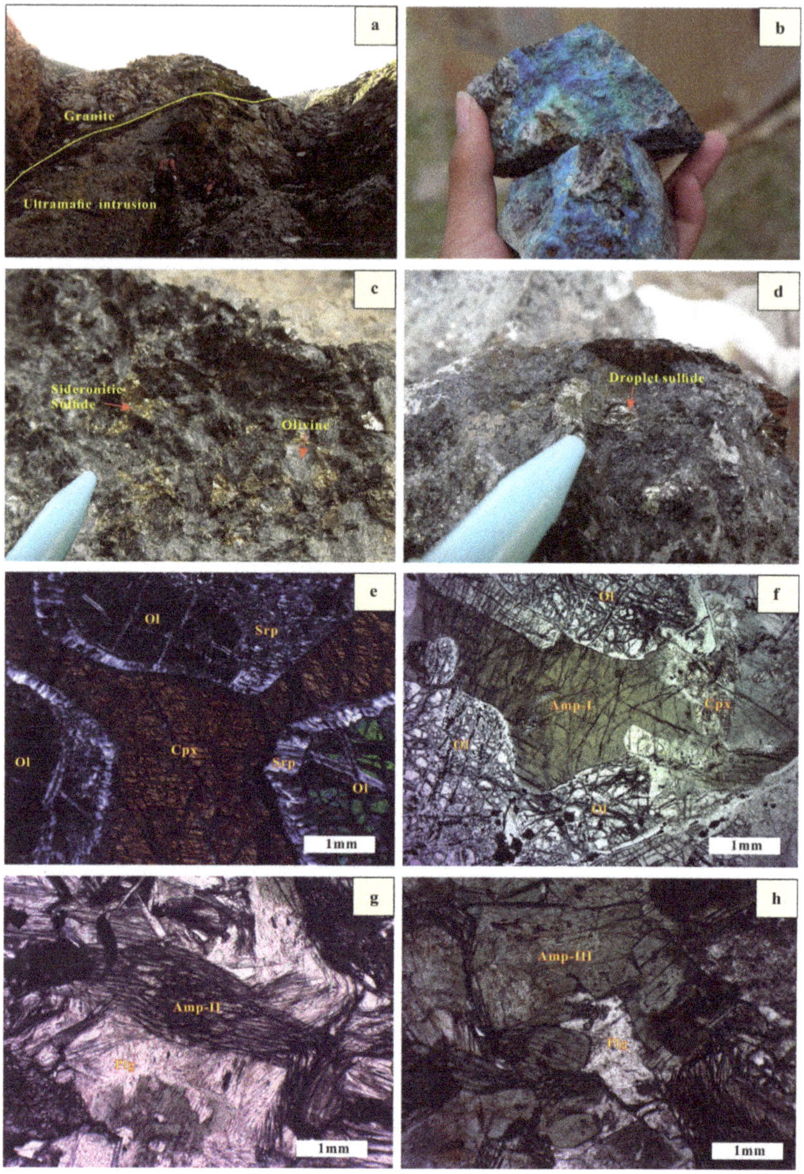

Figure 2. Field photographs and microphotographs of the Dalaku'an intrusion: (**a**) field occurrence of the ultramafic rocks intruding the surrounding granite; (**b**) sulfide mineralization with malachite and azurite in olivine websterite [24]; (**c**) sideronitic sulfide assemblages in lherzolite [24]; (**d**) sulfide droplets in olivine websterite; (**e**) interstitial clinopyroxene between the cumulus olivine in wehrlite; (**f**) brown amphiboles (Amp-I) with olivine poikryst as irregular interstitial minerals between cumulate olivines in the peridotites and pyroxenites; (**g**) idiomorphic and semi-idiomorphic amphiboles (Amp-II) distributed in gabbros and melagabbros; (**h**) dark grey-green amphiboles (Amp-III) contain opaque minerals. The mineral abbreviations: Plg—plagioclase; Cpx—clinopyroxene; Amp—amphibole; Ol—olivine; Srp—serpentine.

3. Petrography and Analytical Methods

3.1. Petrography

In this study, we collected 21 rock samples, including lherzolite, wehrlite, olivine websterite, clinopyroxenite, and (mela-)gabbro. Among peridotites and pyroxenites, the sulfides such as pentlandite, pyrrhotite, chalcopyrite, and pyrite occur in the textures of star-dot form, spot shape, disseminated, and sideronitic forms [24,29] (Figure 2c,d).

Peridotites are mainly composed of olivine (45%–60%), clinopyroxene (30%–40%), orthopyroxene (5%–10%), and a small amount of amphibole (0%–5%). Olivines (chrysolite, Fo = 85–89 [24]) as cumulate minerals are rounded and granular, with a particle size of 2 mm–5 mm, and underwent serpentinized along the cracks and edges to precipitate dust-like magnetites (Figure 2e,f). Clinopyroxene (augite and diopside [28]) as interstitial minerals are medium–coarse-grained, semi-automorphic–heteromorphic (Figure 2e), and orthopyroxene (bronzite, En = 81–82 [24]) as a corona may be the result of reaction between olivine and the residual melt. Amphiboles are also interstitial minerals between cumulus olivines, with some containing both olivine and clinopyroxene, thus manifesting a poikilitic texture characterized by a brown and irregular appearance (Figure 2f), classified as Type-I amphiboles (Amp-I).

Pyroxenites are composed of clinopyroxene (65%–80%), olivine (10%–30%), orthopyroxene (10%–20%), and a few amphiboles. Clinopyroxene and orthopyroxene exhibit a short columnar, granuloblastic, and mosaic texture with particle sizes of 1 mm–2 mm; the dihedral angle between the three particles is similar, about 120°. Some pyroxenes have been amphibolized. Olivines are mostly rounded, with a particle size of 0.5 mm–1.0 mm, and mostly wrapped by pyroxenes to form the poikilitic texture.

Gabbros are mostly distributed in the ultramafic rock in a lens shape or discontinuously distributed at the edge of the intrusion and have been classified into melagabbro and gabbro according to the modal abundance of clinopyroxene and plagioclase. Melagabbro mainly consists of clinopyroxene (65%–75%) and plagioclase (20%–30%), while gabbro consists of clinopyroxene (35%–50%) and plagioclase (40%–60%). A small amount of primary amphiboles (5%–10%) occur in both, with medium to fine-grained texture, exhibiting a grain with a greenish to yellow-green hue, and displaying idiomorphic and semi-idiomorphic features (Figure 2g), which are classified as Type-II amphiboles (Amp-II).

Type-III amphiboles (Amp-III) are mainly distributed in pyroxenites, melagabbros, and gabbros, showing dark grey-green coloration and idiomorphic and semi-idiomorphic features. Additionally, these amphiboles locally containing opaque minerals could be products of late alteration in clinopyroxenes (Figure 2h).

3.2. Analytical Methods

The major element analysis of amphiboles was performed on the JEOL JXA-8100 electron probe (EMPA) at the Key Laboratory of Western China's Mineral Resources and Geological Engineering of the Ministry of Education, Chang'an University. The beam spot diameter was 1 μm, the acceleration voltage was 15 kV, and the test current was 20 nA. The peak and background counting times are as follows: Si, Fe, and Mg, 20 s and 10 s, respectively; Ca, 40 s and 20 s, respectively; and Ni, 60 s and 30 s, respectively. The standard samples used are as follows: Si and Mg are olivine, Fe is magnetite, Ca is diopside, and Ni is nickel metal. The relative accuracy of Si, Fe, and Mg is 2%, the relative accuracy of Ca and Ni is 5%, and the analysis results are corrected by the ZAF method, which takes into account the atomic number (Z), absorption (A) and fluorescence excitation (F).

4. Results

The composition data of amphiboles and the related calculated results are listed in Supplementary Table S1. The molecular formula for amphiboles is commonly written as $A_{0-1}B_2C_5[T_8O_{22}](OH)_2$. In this formula, the T position accommodates Si and Al; the C position may contain Al, Ti, Cr, Fe^{3+}, Mg, Fe^{2+}, and Mn; and the B position typically hosts Ca, Na, and the A position can accommodate Na and K. The cations are based on 23 oxygen atoms. According to the nomenclature principles recommended by the International Mineralogical Association [47–50], the ^{B}Ca (Bivalent cation ratio) values of amphiboles range from 1.68 to 1.87, exceeding the threshold of 1.5, thereby classifying them as calcic amphiboles. Calcic amphiboles are further classified by $^{A}(Na + K)$ values, which range from 0.03 to 0.43, falling below 0.5. These classifications are depicted in the TSi-Mg/(Mg + Fe^{2+}) diagram (Figure 3). The first type of amphiboles (Amp-I) comprises tschermakitic and magnesio hornblende. The second type (Amp-II) consists of magnesio hornblende, while the third (Amp-III) is featured by an actinolitic hornblende (Figure 3).

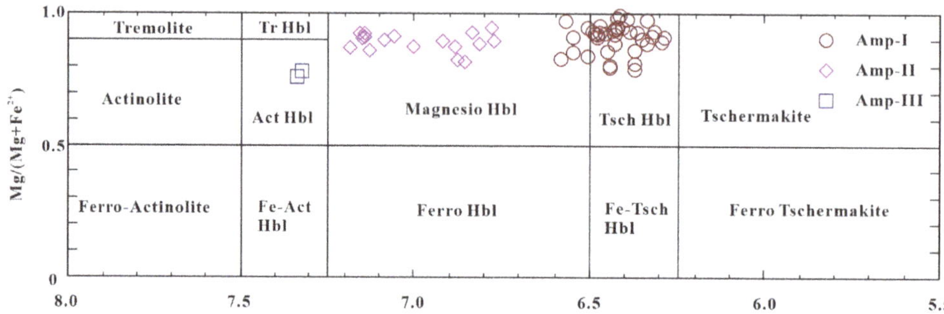

Figure 3. Classification and nomenclature of amphiboles. The classification areas are based on [47].

The chemical compositions of different types of amphiboles are shown in the Harker diagrams (Figure 4a–f). Amp-I is characterized by low silicon content yet abundant in aluminum, titanium, and alkali. Its SiO_2 content is 43.30–46.11 wt%, with an average of 44.64 wt%. TiO_2 content was 0.18–1.50 wt%, with an average of 0.84 wt%. Al_2O_3 content was 10.90–12.97 wt%, with an average of 11.79 wt%. The content of ALK (Na_2O + K_2O) was 1.96–2.43 wt%, with an average of 2.20 wt%. Amp-II exhibits intermediate features, with SiO_2 content ranging from 47.02 to 50.73 wt%, with an average value of 49.04 wt%. TiO_2 content was 0.20–0.93 wt%, with an average value of 0.37 wt%. Al_2O_3 content was 6.48–9.64 wt%, and the average value was 8.10 wt%. The ALK content was 1.04–1.81 wt%, with an average of 1.35 wt%. Amp-III, on the other hand, is enriched in silicon while exhibiting lower levels of titanium, aluminum, and alkali. Its SiO_2 content is 51.68–51.84 wt%, with an average value of 51.76%. TiO_2 content is 0.49 wt%, Al_2O_3 content is 6.12–6.92 wt%, and ALK content is 0.76–1.05 wt%.

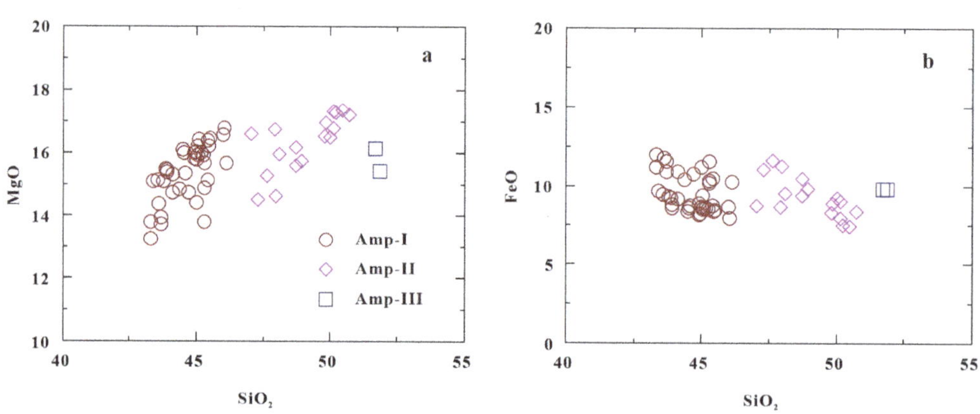

Figure 4. *Cont.*

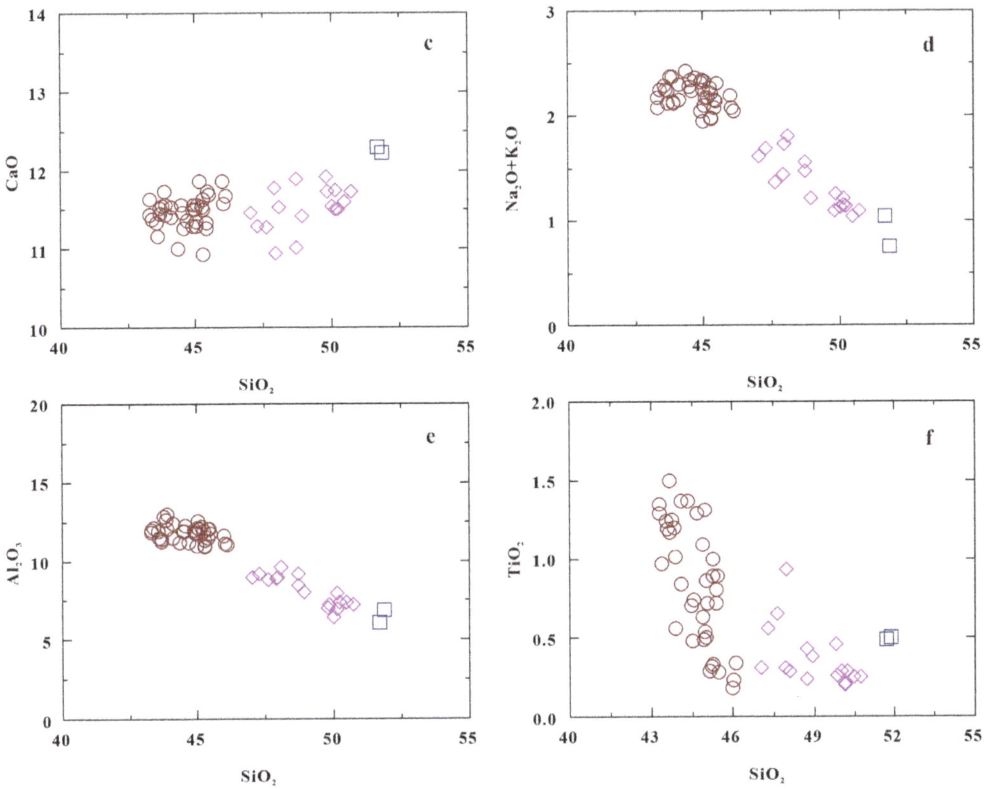

Figure 4. The Harker diagrams of the different types of amphiboles: (**a**) SiO_2 vs. MgO diagram; (**b**) SiO_2 vs. FeO diagram; (**c**) SiO_2 vs. CaO diagram; (**d**) SiO_2 vs. N_2O+K_2O diagram; (**e**) SiO_2 vs. Al_2O_3 diagram; (**f**) SiO_2 vs. TiO_2 diagram.

5. Discussion

5.1. Genesis of Amphiboles

Amphiboles may originate from various geological processes such as magmatic, hydrothermal, and metamorphic processes. Petrographic characteristics reveal that the first and second types of poikilitic and interstitial amphibole (Amp-I and Amp-II) exhibit features indicative of magmatic origin, while the third type (Amp-III) appears to have formed by late-stage hydrothermal alteration. The differentiation is evident in the (Ca + Na + K)-TSi diagram, where Amp-I and Amp-II are situated within the magmatic area, whereas Amp-III occupies the post-magmatic area [51] (Figure 5a). For magmatic amphiboles, the chemical compositions serve as an effective tool for distinguishing between mantle-derived, crust-derived, and mixed-source amphiboles ([47,51,52]. Crust-derived amphiboles typically exhibit an Al_2O_3 content below 10%, a Si/ (Si + Ti + Al) ratio exceeding 0.775, and a Mg value Mg# = [Mg/(Mg + Fe^{2+})] less than 0.5. In contrast, mantle-derived amphiboles feature an Al_2O_3 content exceeding 10%, a Si/(Si + Ti + Al) ratio below 0.756, and a Mg value greater than 0.7 [41]. Within the Dalaku'an intrusion, the Al_2O_3 contents and Mg values of Amp-I are 10.90–12.97 and 0.79–0.99, respectively. Meanwhile, the Al_2O_3 contents of Amp-II are 6.48–9.64, and the Mg values are 0.03–0.23. In the diagrams of Al-TSi and TiO_2-Al_2O_3 (Figure 5b,c), Amp-I is situated in the mantle source region, exhibiting Al_2O_3 contents consistent with the layered magmatic series. In contrast, Amp-II and Amp-III are situated in the crust source zone, which indicates that the Dalaku'an intrusion underwent a complex magmatic history probably related to the potential crustal contamination.

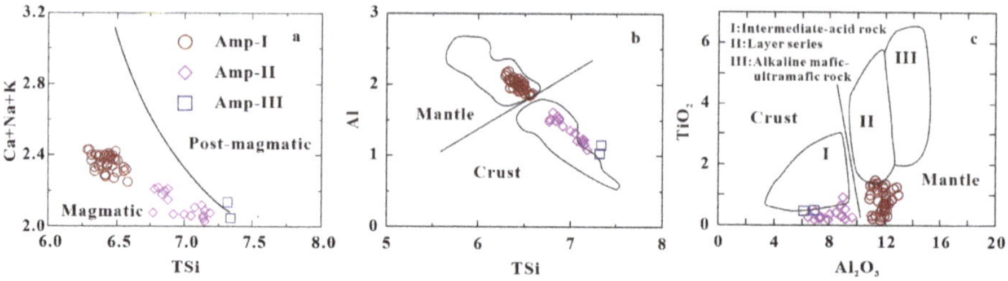

Figure 5. Genetic identification of amphiboles (the classification areas in (**a**) are based on [51]; the referred areas in (**b**,**c**) are based on [41]).

5.2. Temperature and Pressure

The chemical compositions of amphiboles serve as valuable indicators of the physical and chemical conditions, such as temperature and pressure during magma crystallization [32–40,52–63]. The Al^{IV}-Al diagram (Figure 6a) shows that the amphiboles of Amp-I are formed under high-pressure conditions, while those of Amp-II and Amp-III are formed in low-pressure conditions. According to the following thermobarometers [60–62]:

$$T(\pm 30\ ^\circ C) = -151.487 \times Si^* + 2041,\ R^2 = 0.84$$

In which $Si^* = Si + Al^{IV}/15 - 2Ti^{IV} - Al^{VI}/2 - Ti^{VI}/1.8 + Fe^{3+}/9 + Fe^{2+}/3.3 + Mg/26 + {}^{B}Ca/5 + Na/1.3 - {}^{A}Na/15 + {}^{A}[\ ]/2.3$.

$$P\ (\pm 44\ MPa) = 19.209e^{(1.438AlT)},\ R^2 = 0.99$$

R^2 is the correlation parameter, and ${}^{A}[\ ]$ is the vacancy of A-sit.

The crystallization temperatures of amphiboles within the Dalaku'an intrusion exhibit variations: Amp-I amphibole crystallizes within the range of 880–955 °C, with an average of 924 °C, while Amp-II amphibole crystallizes within the range of 774–852 °C, averaging at 813 °C. Amphibole of Amp-III crystallizes within a narrower range of 760–761 °C. Distinct types of amphiboles within the intrusion also demonstrate varying crystallization pressures: Amp-I amphibole ranges from 274 to 454 MPa, with an average of 344 MPa, corresponding to a continental depth of approximately 13 km. Amp-II amphibole ranges from 93 to 194 MPa, averaging 140 MPa, indicative of a continental depth of about 5.3 km. Amphibole of Amp-III, however, exhibits pressures between 84 and 101 MPa, falling below the lower limit of the formula's calculation. Hence, the calculated results for Amp-III serve as reference values only. The calculation results suggest that the Dalaku'an intrusion is a medium–shallow intrusion. However, it has undergone magma chamber evolution at various depths across multiple stages.

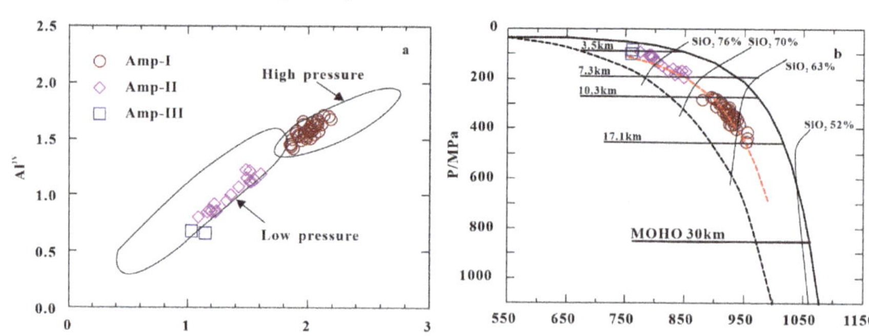

Figure 6. Temperature and pressure conditions of amphibole (the referred areas in (**a**) are based on [64]; the calculation model in (**b**) is based on [61]).

5.3. Oxygen Fugacity and Water Content of Magma

In addition to temperature and pressure, oxygen fugacity and water content of magma are crucial factors influencing the processes of magma evolution [60,61,64,65]. Experimental studies indicate that the Fe/(Fe + Mg) ratio in amphibole serves as a sensitive indicator for oxygen fugacity, with values less than 0.6 indicative of high oxygen fugacity systems. Values greater than 0.8 indicate a low oxygen fugacity system, and values between 0.6 and 0.8 represent medium oxygen fugacity [57]. The Fe/(Fe + Mg) ratios of amphiboles within the Dalaku'an intrusion range from 0.19 to 0.34, indicating a magmatic system with high oxygen fugacity. According to the oxygen fugacity formula of amphiboles [52,60,66]:

$$\triangle NNO = 1.644\ Mg^* - 4.01,\ R^2 = 0.89$$

In which $Mg^* = Mg + Si/47 - Al^{VI}/9 - 1.3Ti^{VI} + Fe^{3+}/3.7 + Fe^{2+}/5.2 - {}^{B}Ca/20 - {}^{A}Na/2.8 + {}^{A}[\ ]/9.5$

The error of the formula is ±0.22%.

During the crystallization of amphiboles within the Dalaku'an intrusion, the obtained oxygen fugacity values range from △NNO = 0.93 to 2.17 for Amp-I, averaging at 1.61, corresponding to $logfO_2$ of about −9.81. For Amp-II, the range is △NNO = 1.55 to 2.52, with an average of 2.17, corresponding to a $logfO_2$ of about −11.42. The average △NNO of Amp-III is 1.89, corresponding to a logfO2 of about −12.84. In the $logfO_2$-T diagram, the amphibole data all fall within the region between HM and NNO, with a variation range from ΔNNO + 0.58 to ΔNNO + 2.0 (with an average of ΔNNO + 1.79) (Figure 7a). These characteristics indicate that amphiboles in the Dalaku'an intrusion may have formed in a high oxygen fugacity environment. The relative oxygen fugacity varied limitedly during the evolution stages of magma chambers at different depths, suggesting weak processes such as assimilative mixing or magma mixing that have not significantly altered the magma composition [24]. Furthermore, there exists a strong correlation between the Al^{IV} of amphibole and its water content in the melt [61]. The formula for calculating water content is based on the molecular formula of amphibole [60,61]:

$$H_2O_{melt} = 5.215Al^* + 12.28,\ R^2 = 0.83$$

In which $Al^* = Al + Al^{IV}/13.9 - (Si + Ti^{VI})/5 - Fe^{2+}/3 - Mg/1.7 + ({}^{B}Ca + {}^{A}[])/1.2 + {}^{A}Na/2.7 - 1.56{}^{A}K - Fe^{\#}/1.6;\ Fe^{\#} = Fe^{3+}/(Fe_{total} + Mg + Mn)$

The error of this formula is ±0.41%. According to the formula, the water contents of the magma during the crystallization of amphibole within the Dalaku'an intrusion are as follows: Amp-I H_2O_{melt} =6.69–8.67, with an average of 7.67, and Amp-II is about 5.90–7.32, with a mean of 6.69 (Figure 7b).

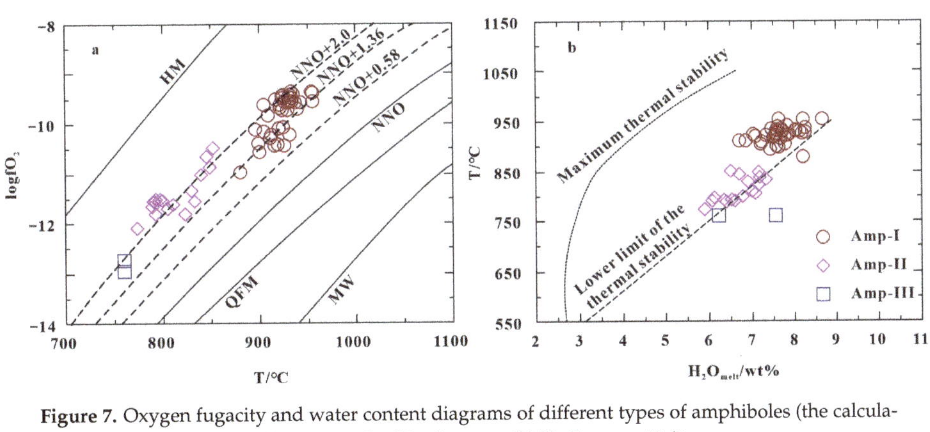

Figure 7. Oxygen fugacity and water content diagrams of different types of amphiboles (the calculation model is based on [61]). (**a**) T vs. $logfO_2$ diagram; (**b**) H_2O_{melt} vs. T diagram.

The water contents calculated from amphiboles within the Dalaku'an intrusion during crystallization exhibit relatively high values and a broad range of variation (Figure 7b). This observation may be attributed to the influence of oceanic subduction in the mantle source, suggesting that the original magma is relatively rich in water. In addition, discrepancies in water content among amphiboles at various evolutionary stages may stem from the abundance of amphibole crystals during the early crystallization period [61].

5.4. Magmatic Properties and Tectonic Environment

The composition characteristics of amphibole serve as indicators reflecting the properties and tectonic environment of magma to a certain extent [32–40,42,61]. In the K_2O-TiO_2 and ^{A}K-TAl^{IV} diagrams, the amphibole components all fall within the subalkaline and calc-alkaline regions (Figure 8a,b), indicating that the Dalaku'an intrusion belongs to the calc-alkaline series. The SiO_2-Na_2O diagram depicts amphibole characteristics typical of a subduction zone (Figure 8c).

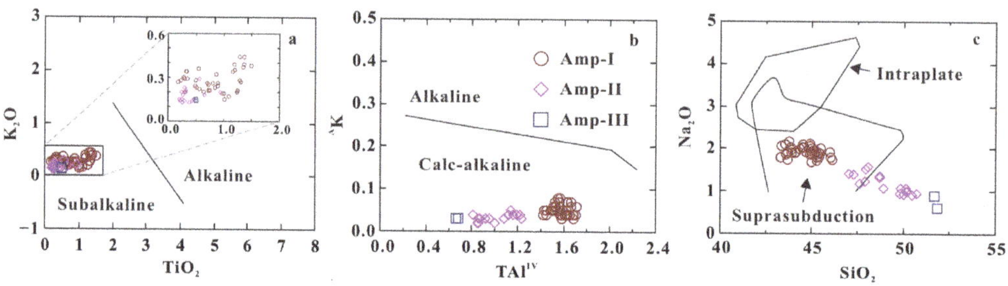

Figure 8. Magmatic properties and identification of tectonic environment of the Dalaku'an intrusion (the referred areas in (**a**, **b**, and **c**) are from [67], [61], and [42], respectively).

The East Kunlun Orogenic Belt (EKOB) was a continental composite orogenic belt with major orogenic periods such as the late Early Paleozoic and Triassic [2,6,68–73]. From Late Permian to Early Triassic, the Paleo-Tethys Ocean subducted northward under the East Kunlun terrae [6], which induced the formation of a large area of arc-related acidic and basic intrusions spreading nearly east–west ([2,6,69,70,74–81]. Since the Middle Triassic, the Paleo-Tethys Ocean has closed, and the collision between the Baryan Har-Songpanganzi block in the south and the East Kunlun terrane in the north has produced collisional granites and mafic–ultramafic intrusion [2,82–85]. The EKOB has entered the stage of collision orogeny and post-collision tectonic evolution [2,24,68,86–90]. The formation age of the Dalaku'an mafic–ultramafic intrusion is 244 Ma $\pm$ 2 Ma [24], indicating a Middle Triassic age, which coincided with the transition period of the Paleo-Tethys Ocean subduction closure to intracontinental collision orogeny [30]. The mantle source of the Dalaku'an intrusion may be the enriched lithospheric mantle transformed by plate fluid replacement according to the geochemical and isotopic evidence [24]. Generally, mafic intrusions formed under post-collisional extensional environments exhibit characteristics of island arc affinity. These intrusions are considered to result from the transformation of post-subduction plates and magmatic sources metasomatized by fluids [67]. This interpretation aligns with the formation of the Dalaku'an intrusion in a high-oxygen fugacity environment.

6. Conclusions

(1) Amphiboles within the Dalaku'an intrusion exhibit calcic amphibole characteristics and can be divided into three types: primarily tschermakitic hornblende and magnesio-hornblende for the first type, magnesio-hornblende for the second, and actinolitic hornblende for the third.

(2) The presence of amphiboles with both mantle and crustal origins suggests that the Dalaku'an intrusion has undergone complex magmatic processes.

(3) Crystallization temperatures for amphiboles range from (955–880) °C to (852–774) °C and then to (761–760) °C, while crystallization pressures span from 454 to 274 MPa to (194–93) MPa and then to (101–84) MPa. These findings imply that the Dalaku'an intrusion experienced multi-stage magma chamber evolution at various depths.

(4) Oxygen fugacity △NNO during crystallization ranges from (0.93–2.17) to (1.55–2.52) and to (1.89), falling between HM and NNO, indicative of a high oxygen fugacity environment. The H_2O_{melt} content in the magma, calculated from amphiboles, ranges from (6.69–8.67) to (5.90–7.32), suggesting relatively high water content and significant variation during crystallization. This variation may be linked to the influence of oceanic subduction in the mantle source, indicating that the original magma was relatively rich in water.

(5) The amphiboles exhibit characteristics typical of a subduction zone. Combined with regional data, it is inferred that the magmatic source area of the Dalaku'an intrusion was metasomatized by subducted plate fluids, forming within a post-collisional extensional background.

Supplementary Materials: The following supporting information can be downloaded at https://www.mdpi.com/article/10.3390/min14070651/s1, Table S1: the compositions of amphiboles in the Dalaku'an mafic–ultramafic intrusion from the East Kunlun, Xinjiang (%).

Author Contributions: Methodology, Y.F., Y.D. and Z.X.; validation, Y.D., J.H. and Y.F.; formal analysis, M.R.; investigation, Z.X., J.H. and M.R.; data curation, J.H.; writing—original draft preparation, Y.F. and M.R.; writing—review and editing, Y.F., Y.D., M.R. and J.H.; supervision, M.R. and J.H.; funding acquisition, Y.F. and J.H. All authors have read and agreed to the published version of the manuscript.

Funding: This study was jointly financially supported by the Natural Science Foundation of China (Nos. 42102039 and 42003033) and the high-level talent research project of North China University of Water Resources and Electric Power (No. 201940775).

Data Availability Statement: The data presented in this study are available in the supplement materials of this article.

Acknowledgments: We would like to express our gratitude to the Key Laboratory of Western Mineral Resources and Geological Engineering of the Ministry of Education for their support with the microprobe analysis. We are grateful to the anonymous reviewers who helped improve the paper and to the editors for handling, editing, and advising.

Conflicts of Interest: The authors declare no conflicts of interest.

References

1. Gao, J.F.; Wang, H.H. Permian mafic-ultramafic magmatism and sulfide mineralization in the Central Asian Orogenic Belt: A review. *J. Asian Earth Sci.* **2024**, *264*, 106071. [CrossRef]
2. Jiang, C.F.; Yang, J.S.; Feng, B.G.; Zhu, Z.Z.; Zhao, M.; Chai, Y.C.; Shi, X.D.; Wang, R.D.; Zhao, J.Q. *Opening-Closing Structure of the Kunlun Mountains*; Geological Publishing House: Beijing, China, 1992; pp. 1–217. (In Chinese with English abstract)
3. Mo, X.X.; Luo, Z.H.; Deng, J.F.; Yu, X.H.; Liu, C.D.; Kan, H.W.; Yuan, W.M.; Liu, Y.H. Growth of granite and crust in East Kunlun orogenic belt. *Geol. J. Univ.* **2007**, *13*, 403–414. (In Chinese with English abstract)
4. Wen, T.; Dong, J.L.; Wang, C.; Song, S.G. Two ophiolite belts in the east kunlun orogenic belt record evolution from the proto-tethys to paleo-tethys oceans. *Int. Geol. Rev.* **2022**, *65*, 1957–1976. (In Chinese with English abstract) [CrossRef]
5. Gao, Y.L.; Wu, X.N.; Zuo, G.C. The characters and tectonic significance of ophiolite first discoved in the East Kunlun area. *Bull. Xi'an Inst. Geol. Min. Res. Chin. Acad. Geol. Sci.* **1988**, *21*, 17–28. (In Chinese with English abstract)
6. Yang, J.S.; Robinson, P.T.; Jiang, C.Y.; Xu, Z.Q. Ophiolites of the Kunlun Mountains, China and their tectonic implications. *Tectonophysics* **1996**, *258*, 215–231. [CrossRef]
7. Zhu, Y.H.; Pan, Y.M.; Zhang, K.X.; Chen, N.S.; Wang, G.C.; Hou, G.J. Mineralogical characteristics and petrogenesis of ophiolites in East Kunlun orogenic belt, Qinghai Province. *Acta Mineral. Sin.* **2000**, *2*, 128–142. (In Chinese with English abstract)
8. Bian, Q.T.; Li, D.H.; Pospelov, I.; Yin, L.M.; Li, H.S.; Zhao, D.S.; Chang, C.F.; Luo, X.Q.; Gao, S.L.; Astrakhantsev, O.; et al. Age, geochemistry and tectonic setting of buqingshan ophiolites, north qinghai-tibet plateau, china. *J. Asian Earth Sci.* **2004**, *23*, 577–596. [CrossRef]
9. Lan, Z.L.; Li, J.L.; He, S.L. Ocean-continent subduction within the Paleotethyan archiopeligic ocean from Muztrag ophiolite. *Earth Sci. J. China Univ. Geosci.* **2007**, *3*, 322–328. (In Chinese with English abstract)

10. Meng, F.C.; Cui, M.H.; Wu, X.K.; Ren, Y.F. Heishan mafic–ultramafic rocks in the Qimantag area of Eastern Kunlun, NW China: Remnants of an early Paleozoic incipient island arc. *Gondwana Res.* **2015**, *2*, 745–759. [CrossRef]
11. Jiang, C.Y.; Ling, J.L.; Zhou, W.; Du, W.; Wang, Z.X.; Fan, Y.Z.; Song, Y.F.; Song, Z.B. Petrogenesis of the Xiarihamu Ni-bearing layered mafic-ultramafic intrusion,East Kunlun: Implications for its extensional island arc environment. *Acta Petrol. Sin.* **2015**, *4*, 1117–1136. (In Chinese with English abstract)
12. Liu, Y.G.; Zhang, J.W.; Feng, Z.X.; Yang, S.L.; Wang, Y.Z.; Li, J.Q.; Zhao, Z.Y.; Wang, Z.A.; Li, S.L.; Chen, Z.G.; et al. Exploration and research progress of magmatic copper-nickel-cobalt sulfide deposits in the north-eastern margin of the Qinghai−Tibetan Plateau. *Geol. China* **2023**, 1–46. (In Chinese with English abstract)
13. Li, S.J.; Sun, F.Y.; Gao, Y.W.; Zhao, J.W.; Li, L.S.; Yang, Q.A. Theoretical guidance and practice of small rock mass forming large ore—Enlightenment and significance of prospecting breakthrough of Summer Hamu Copper-nickel deposit in East Kunlun, Qinghai Province. *Northwest Geol.* **2012**, *45*, 185–191. (In Chinese with English abstract)
14. Li, C.S.; Zhang, Z.W.; Li, W.Y.; Wang, Y.L.; Sun, T.; Ripley, E.M. Geochronology, petrology and Hf–S isotope geochemistry of the newly-discovered Xiarihamu magmatic Ni–Cu sulfide deposit in the Qinghai–Tibet plateau, western China. *Lithos* **2015**, *216–217*, 224–240. [CrossRef]
15. Song, X.Y.; Yi, J.N.; Chen, L.M.; She, Y.W.; Liu, C.Z.; Dang, X.Y.; Yang, Q.A.; Wu, S.K. The Giant Xiarihamu Ni-Co Sulfide Deposit in the East Kunlun Orogenic Belt, Northern Tibet Plateau, China. *Econ. Geol.* **2016**, *111*, 29–55. [CrossRef]
16. Zhou, W.; Wang, B.Y.; Xia, M.Z.; Xia, Z.D.; Jiang, C.Y.; Dong, J.; Xie, E.S. Mineralogical characteristics and metallogenic potential analysis of the Mafic ultramafic rock mass in Shikengde, East Kunlun. *J. Petrol. Mineral.* **2016**, *35*, 81–96. (In Chinese with English abstract)
17. Zhang, Z.W.; Wang, Y.L.; Qian, B.; Liu, Y.G.; Zhang, D.Y.; Lv, P.R.; Dong, J. Metallogeny and tectonomagmatic setting of Ni-Cu magmatic sulfide mineralization, Number I Shitoukengde mafic-ultramafic complex, East Kunlun Orogenic Belt, NW China. *Ore Geol. Rev.* **2018**, *96*, 236–246. [CrossRef]
18. Wang, C.Y.; Zhang, Z.W.; Zhang, C.J.; Chen, C.H.; Li, Y.; Qian, B. Constraints on sulfide saturation by crustal contamination in the Shitoukengde Cu-Ni deposit, East Kunlun orogenic belt, northern QinghaiTibet Plateau, China. *Geosci. J.* **2020**, *25*, 401–415. [CrossRef]
19. Jia, L.H.; Mao, J.W.; Li, B.L.; Zhang, D.Y.; Sun, T.T. Geochronology and petrogenesis of the Late Silurian Shitoukengde mafic–ultramafic intrusion, NW China: Implications for the tectonic setting and magmatic Ni-Cu mineralization in the East Kunlun Orogenic Belt. *Int. Geol. Rev.* **2021**, *63*, 549–570. [CrossRef]
20. Mao, Y.J.; Zheng, X.Y.; Pan, T.; Yuan, F.L.; Huang, H.L.; Qin, K.Z. Redox-controlled olivine-sulfide equilibration of the Shitoukengde Ni–Cu deposit in Qinghai-Tibet Plateau and its implications. *Miner. Depos.* **2024**, *59*, 577–599. [CrossRef]
21. He, S.Y.; Sun, F.F.; Li, Y.P.; Li, D.S.; Yu, M.; Qian, Y.; Liu, Y.L.; Bai, G.L.; Zhao, M.Q.; Zhang, P.; et al. Geochemical Geochemical and Geochronological Significance of the Binggounan Garbbro in the Qiman Tage Region. *Bull. Mineral. Petrol. Geochem.* **2017**, *4*, 582–592. (In Chinese with English abstract)
22. Zhang, Z.W.; Wang, Y.L.; Qian, B.; Li, W.Y. Zircon SHRIMP U-Pb age of the Binggounan Magmatic Ni-Cu deposit in East Kunlun Mountains and its tectonic implications. *Acta Geol. Sin.* **2017**, *4*, 724–735. (In Chinese with English abstract)
23. Yan, J.M.; Sun, F.Y.; Li, B.L.; Li, L.; Zhang, W.H.; Yan, Z.P.; Zhang, Y.S. Geochronological, geochemical, and mineralogical characteristics of the Akechukesai-I mafic–ultramafic complex in the eastern Kunlun area of the northern Tibet Plateau, West China: Insights into ore potential. *Ore Geol. Rev.* **2020**, *121*, 103468. [CrossRef]
24. Xia, M.Z.; Fan, Y.Z.; Xia, Z.D.; Rui, H.C.; Jiang, C.Y. The chronology, geochemistry and metallogenic conditions of the Mafic ultramafic rock mass in Dalaku'an, East Kunlun, Xinjiang. *Acta Petrol. Sin.* **2018**, *34*, 2380–2392. (In Chinese with English abstract)
25. Yan, J.M. A slab break-off model for mafic–ultramafic igneous complexes in the East Kunlun Orogenic Belt, northern Tibet: Insights from early Palaeozoic accretion related to post-collisional magmatism. *Int. Geol. Rev.* **2018**, *61*, 1171–1188. [CrossRef]
26. Liu, Y.G.; Li, W.Y.; Jia, Q.Z.; Zhang, Z.W.; Wang, Z.A.; Zhang, Z.B.; Zhang, J.W.; Qian, B. The Dynamic Sulfide Saturation Process and a Possible Slab Break-off Model for the Giant Xiarihamu Magmatic Nickel Ore Deposit in the East Kunlun Orogenic Belt, Northern Qinghai-Tibet Plateau, China. *Econ. Geol.* **2018**, *113*, 1383–1417. [CrossRef]
27. Li, L.; Sun, F.Y.; Wang, L.; Zhang, Y.J.; Wang, C.; Qian, Y.; Zhao, T.F.; Yan, J.M.; Tan, S.C. Tectonomagmatic setting and magmatic Cu-PGE metallogeny during the late Neoproterozoic in the Eastern Kunlun Orogen, NW China: Insights from the Hatu Cu-Pt-Pd-bearing mafic-ultramafic intrusion. *Precambrian Res.* **2023**, *395*, 107147. [CrossRef]
28. Fan, Y.Z.; Xia, Z.D.; Xia, M.Z.; Jiang, C.Y.; Meng, F.C. Mineral chemistry and properties of mafic ultramafite in Dalakuan, East Kunlun, Xinjiang. *J. Petrol. Mineral.* **2018**, *37*, 454–466. (In Chinese with English abstract)
29. Wang, M.; Xia, Z.D.; Rui, H.C.; Zhao, Y.M.; Xia, M.Z.; Jing, D.L. Mineral characteristics and metallogenic significance of Dalaku Bank Cu-Ni deposit in southern margin of Tarim. *Geol. Explor.* **2016**, *52*, 429–437. (In Chinese with English abstract)
30. Liu, C.D.; Mo, X.X.; Luo, Z.H.; Yu, X.H.; Chen, H.W.; Li, S.W.; Zhao, X. Mixing events between the crust- and mantle-derived magmas in Eastern Kunlun: Evidence from zircon SHRIMP II chronology. *Chin. Sci. Bull.* **2004**, *49*, 828–834. [CrossRef]
31. Duan, X.P.; Meng, F.C.; Fan, Y.Z. The constraints of kaersutite and pargasite on metallogeny in Xiarihamu mafic-ultramafic intrusion, East Kunlun. *Acta Petrol. Sin.* **2019**, *6*, 1819–1832. (In Chinese with English abstract)
32. Perchuk, L.L. *Equilibria of Rock-Forming Minerals*; Nauka: Moscow, Russia, 1970; 320p.
33. Ulrych, J. Clinopyroxenes in the Cenozoic volcanic of the Ceske Stredohori MTS: A review. *Acta Univ. Carol.* **1986**, *4*, 367–376.
34. Brown, E. The crossite content of Ca-amphibole as a guide to pressure of metamorphism. *J. Petrol.* **1977**, *18*, 53–72. [CrossRef]

35. Hynes, A. A comparison of amphiboles from medium and low pressure metabasites. *Contrib. Mineral. Petrol.* **1982**, *81*, 119–125. [CrossRef]
36. Helmy, H.M.; El Mahallawi, M.M. Gabbro Akarem mafic–ultramafic complex, Eastern Desert, Egypt: A Late Precambrian analogue of Alaskan-type complexes. *Mineral. Petrol.* **2003**, *77*, 85–108. [CrossRef]
37. Helmy, H.M.; Yoshikawa, M.; Shibata, T.; Arai, S.; Tamura, A. Corona structure from arc mafic-ultramafic cumulates: The role and chemical characteristics of latemagmatic hydrous liquids. *J. Mineral. Petrol. Sci.* **2008**, *103*, 333–344. [CrossRef]
38. Abdel Halim, A.H.; Helmy, H.M.; Abd El-Rahman, Y.M.; Shibata, T.; El Mahallawi, M.M.; Yoshikawa, M.; Arai, S. Petrology of the Motaghairat mafic-ultramafic complex, Eastern Desert, Egypt: A high-Mg post-collisional extension-related layered intrusion. *J. Asian Earth Sci.* **2016**, *116*, 164–180. [CrossRef]
39. Khedr, M.Z.; El-Awady, A.; Arai, S.; Hauzenberger, C.; Tamura, A.; Stern, R.J.; Morishita, T. Petrogenesis of the ~740 Korab Kansi mafic-ultramafic intrusion, South Eastern desert of Egypt: Evidence of Ti-rich ferropicritic magmatism. *Gondwana Res.* **2020**, *82*, 48–72. [CrossRef]
40. Khedr, M.Z.; Kamh, S.; Al Desouky, A.A.; Takazawa, E.; Hauzenberger, C.; Whattam, S.A.; El-Awady, A. Remote sensing and geochemical investigations of sulfide-bearing metavolcanic and gabbroic rocks (Egypt): Constraints on host-rock petrogenesis and sulfide genesis. *Gondwana Res.* **2023**, *119*, 282–312. [CrossRef]
41. Jiang, C.Y.; An, S.Y. Chemical composition and petrological significance of calcareous hornblende in igneous rocks. *Mineral. Petrol.* **1984**, *3*, 1–9. (In Chinese with English abstract)
42. Coltorti, M.; Bonadiman, C.; Faccini, B.; Grégoire, M.; O'Reilly, S.Y.; Powell, W.G. Amphiboles from suprasubduction and intraplate lithospheric mantle. *Lithos* **2007**, *99*, 68–84. [CrossRef]
43. Gong, X.P.; Ma, H.D.; Yang, X.K.; Li, G.T.; Wang, Q.M. Characteristics, evolution and significance of the Muztag-Whale Lake fault zone. *Geotecton. Metallog.* **2004**, *28*, 418–427. (In Chinese with English abstract)
44. Yin, F.G.; Pan, G.T.; Li, X.Z.; Meng, D.B.; Bai, D.Y.; Liu, W. Geological and geochemical characteristics of ophiolitic melanomics in the middle part of Kunlun Orogenic belt. *Geotecton. Metallog.* **2004**, *28*, 194–200. (In Chinese with English abstract)
45. Han, H.W.; Wei, M.Y.; Mu, L.X.; Yu, J.H.; Zhang, Y.S.; Zhou, Z.Y.; Hu, J.M.; Feng, J.X.; Zhang, C.J. Genesis of East Kunlun Katlisi copper zinc deposit. *Geotecton. Metallog.* **2007**, *31*, 77–82. (In Chinese with English abstract)
46. Bureau of Geology and Mineral Resources of Xinjiang. *Regional Geology of Xinjiang Uygur Autonomous Region (BGMRX)*; Geological Publishing House: Beijing, China, 1993; pp. 657–717. (In Chinese with English abstract)
47. Leake, B.E. Nomenclature of Amphiboles. *Mineral. Mag.* **1978**, *42*, 533–563. [CrossRef]
48. Leake, B.E.; Woolley, A.R.; Arps, C.E.S.; Birch, W.D.; Gilbert, M.C.; Grice, J.D.; Hawthorne, F.C.; Kato, A.; Kisch, H.J.; Krivovichev, V.G.; et al. Nomenclature of amphiboles: Report of the subcommittee on amphiboles of the international mineralogical association, commission on new minerals and mineral names. *Can. Mineral.* **1997**, *35*, 219–246.
49. Leake, B.E.; Woolley, A.R.; Birch, W.D.; Burke, E.A.J.; Ferraris, G.; Grice, J.D.; Hawthorne, F.C.; Kisch, H.J.; Krivovichev, V.G.; Schumacher, J.C.; et al. Nomenclature of amphiboles: Additions and revisions to the International Mineralogical Associations amphibole nomenclature. *Mineral. Mag.* **2004**, *68*, 209–215. [CrossRef]
50. Hawthorne, F.C.; Oberti, R. On the Classification of Amphiboles. *Can. Mineral.* **2007**, *44*, 1–21. [CrossRef]
51. Leake, B.E. On aluminous and edenitic hornblendes. *Mineral. Mag.* **1971**, *38*, 389–407. [CrossRef]
52. Hollister, L.S.; Grissom, G.C.; Peters, E.K.; Stowell, H.H.; Sisson, V.B. Confirmation of the empirical correlation of Al in hornblende with pressure of solidification of calc-alkaline plutons. *Am. Mineral.* **1987**, *72*, 231–239.
53. Hammarstrom, J.M.; Zen, E. Aluminum in hornblende; an empirical igneous geobarometer. *Am. Mineral.* **1986**, *71*, 1297–1313.
54. Holloway, J.R.; Burnham, C.W. Melting Relations of Basalt with Equilibrium Water Pressure Less than Total Pressure. *J. Petrol.* **1972**, *13*, 1–29. [CrossRef]
55. Schmidt, M.W. Amphibole composition in tonalite as a function of pressure: An experimental calibration of the Al-in-hornblende barometer. *Contrib. Mineral. Petrol.* **1992**, *110*, 304–310. [CrossRef]
56. Adam, J.; Green, T.H. The effects of pressure and temperature on the partitioning of Ti, Sr and REE between amphibole, clinopyroxene and basanitic melts. *Chem. Geol.* **1994**, *117*, 219–233. [CrossRef]
57. Anderson, J.L.; Smith, D.R. The effects of temperature and fO_2 on the Al-in hornblende barometer. *Am. Mineral.* **1995**, *80*, 549–559. [CrossRef]
58. Ernst, W.G.; Liu, J. Experimental phase-equilibrium study of Al- and Ti-contents of calcic amphibole in MORB-A semiquantitative thermobarometer. *Am. Mineral.* **1998**, *83*, 952–969. [CrossRef]
59. King, P.L.; Hervig, R.L.; Holloway, J.R.; Delaney, J.S.; Dyar, M.D. Partitioning of Fe3+/Fe total between amphibole and basanitic melt as a function of oxygen fugacity. *Earth Planet. Sci. Lett.* **2000**, *178*, 97–112. [CrossRef]
60. Ridolfi, F.; Puerini, M.; Renzulli, A.; Menna, M.; Toulkeridis, T. The magmatic feeding system of El Reventador volcano (Sub-Andean zone, Ecuador) constrained by texture, mineralogy and thermobarometry of the erupted products. *J. Volcanol. Geotherm. Res.* **2008**, *176*, 94–106. [CrossRef]
61. Ridolfi, F.; Renzulli, A.; Puerini, M. Stability and chemical equilibrium of amphibole in calc-alkaline magmas: An overview, new thermobarometric formulations and application to subduction-related volcanoes. *Contrib. Mineral. Petrol.* **2010**, *160*, 45–66. [CrossRef]
62. Ridolfi, F.; Renzulli, A. Calcic amphiboles in calc-alkaline and alkaline magmas: Thermobarometric and chemometric empirical equations valid up to 1130 °C and 2.2 GPa. *Contrib. Mineral. Petrol.* **2012**, *163*, 877–895. [CrossRef]

63. Krawczynski, M.J.; Grove, T.L.; Behrens, H. Amphibole stability in primitive arc magmas: Effects of temperature, h2o content, and oxygen fugacity. *Contrib. Mineral. Petrol.* **2012**, *164*, 317–339. [CrossRef]
64. Candela, P.A.; Holland, H.D. The partitioning of copper and molybdenum between silicate melts and aqueous fluids. *Geochim. Et Cosmochim. Acta* **1984**, *48*, 373–380. [CrossRef]
65. Richards, J.P. Tectono-Magmatic Precursors for Porphyry Cu-(Mo-Au) Deposit Formation. *Econ. Geol.* **2003**, *98*, 1515–1533. [CrossRef]
66. Erdmann, S.; Martel, C.; Pichavant, M.; Kushnir, A.R.L. Amphibole as an archivist of magmatic crystallization conditions: Problems, potential, and implications for inferring magma storage prior to the paroxysmal 2010 eruption of Mount Merapi, Indonesia. *Contrib. Mineral. Petrol.* **2014**, *167*, 1016. [CrossRef]
67. Molina, J.F.; Scarrow, J.H.; Montero, P.; Bea, F. High-Ti amphibole as a petrogenetic indicator of magma chemistry: Evidence for mildly alkalic-hybrid melts during evolution of Variscan basic–ultrabasic magmatism of Central Iberia. *Contrib. Mineral. Petrol.* **2009**, *158*, 69–98. [CrossRef]
68. Li, R.B.; Pei, X.Z.; Li, Z.C.; Pei, L.; Chen, G.C.; Chen, Y.X.; Liu, C.J.; Wang, S.M. Paleo-Tethys Ocean subduction in eastern section of East Kunlun Orogen: Evidence from the geochronology and geochemistry of the Wutuo pluton. *Acta Petrol. Sin.* **2018**, *11*, 3399–3421. (In Chinese with English abstract)
69. Li, R.B.; Pei, X.Z.; Li, Z.C.; Pei, L.; Chen, G.C.; Liu, Z.Q.; Chen, Y.X.; Liu, C.J.; Wang, M.; Zhang, M. Paleo-Tethyan Ocean Evolution and Indosinian Orogenesis in the East Kunlun Orogen, Northern Tibetan Plateau. *Minerals* **2022**, *12*, 1590. [CrossRef]
70. Chen, G.C.; Pei, X.Z.; Li, R.B.; Li, Z.C.; Chen, Y.X.; Liu, C.J.; Pei, L. Multiple Sources of Indosinian Granites and Constraints on the Tectonic Evolution of the Paleo-Tethys Ocean in East Kunlun Orogen. *Minerals* **2022**, *12*, 1604. [CrossRef]
71. Pan, Z.C.; Sun, F.Y.; Cong, Z.C.; Tian, N.; Xin, W.; Wang, L.; Zhang, Y.J.; Wu, D.Q. Petrogenesis and Tectonic Implications of the Triassic Granitoids in the Ela Mountain Area of the East Kunlun Orogenic Belt. *Minerals* **2022**, *12*, 880. [CrossRef]
72. Dong, Y.P.; Hui, B.; Sun, S.S.; Yang, Z.; Zhang, F.F.; He, D.F.; Sun, J.P.; Shi, X.H. Multiple orogeny and geodynamicsfrom Proto-Tethysto Paleo-Tethys ofthe Central China Orogenic Belt. *Acta Geol. Sin.* **2022**, *10*, 3426–3448. (In Chinese with English abstract)
73. Yang, J.S.; Shi, R.D.; Wu, C.L.; Wang, X.B.; Robinson, P.T. Dur'ngoi ophiolite in East Kunlun, northeast Tibetan plateau: Evidence for Paleo-Tethyan suture in Northwest China. *J. Earth Sci.* **2009**, *20*, 303–331. [CrossRef]
74. Xiong, F.H.; Ma, C.Q.; Zhang, J.Y.; Liu, B. The origin of mafic microgranular enclaves and their host granodiorites from East Kunlun, Northern Qinghai-Tibet Plateau: Implications for magma mixing during subduction of Paleo-Tethyan lithosphere. *Min. Pet.* **2012**, *104*, 211–224. [CrossRef]
75. Li, R.B.; Pei, X.Z.; Li, Z.C.; Liu, Z.Q.; Chen, G.C.; Chen, Y.X.; Wei, F.H.; Gao, J.M.; Liu, C.J.; Pei, L. Geological charcteristics of Late Palaeozoic-Mesozoic unconformities and their response to some significant tectonic events in eastern part of Eastern Kunlun. *Earth Sci. Front.* **2012**, *5*, 244–254. (In Chinese with English abstract)
76. Ma, C.Q.; Xiong, F.H.; Yin, S.; Wang, L.X.; Gao, K. Intensity and cyclicity of orogenic magmatism: An example from a Paleo-Tethyan granitoid batholith, Eastern Kunlun. northern Qinghai-Tibetan Plateau. *Acta Petrol. Sin.* **2015**, *12*, 3555–3568. (In Chinese with English abstract)
77. Shi, L.C.; Chang, G.H.; Qi, S.S.; Chen, G.T.; Zhao, M.F.; Xu, B. The discovery of Dazaohuogou-Wanbaogou Late Permian epicontinental arc volcanic rocks in Eastern Kunlun Mountains and its significance. *Geol. Bull. China* **2016**, *7*, 1115–1122. (In Chinese with English abstract)
78. Li, Z.C.; Pei, X.Z.; Pei, L.; Liu, C.J.; Xu, L.L.; Li, R.B.; Lin, H.; Wang, M.; Ji, S.; Qin, L.; et al. Petrogenesis and Geochronology of A1-Type Rhyolites in the Late Late Triassic of the East Kunlun Orogenic Belt: Constraints on the End of the Paleo-Tethys Orogenic Event. *Minerals* **2023**, *13*, 290. [CrossRef]
79. Yan, D.D.; Chu, Z.Q.; Liu, Z.Y.; Wang, W.; Xiong, F.H. Petrogenesis of Early Triassic Felsic Volcanic Rocks in the East Kunlun Orogen, Northern Tibet: Implications for the Paleo-Tethyan Tectonic and Crustal Evolution. *Minerals* **2023**, *13*, 607. [CrossRef]
80. Chen, J.J.; Fu, L.B.; Leng, C.B.; Zhao, X.; Ma, J.; Gao, H.Z.; Xia, Y. Origin of Quartz Diorite and Mafic Enclaves in the Delong Gold-Copper Deposit and Evaluation of the Gold-Copper Mineralization Potential. *Minerals* **2023**, *13*, 1202. [CrossRef]
81. Norbu, N.; Tang, L.; Li, J.C.; Kong, H.L.; Li, Y.Z.; Jia, Q.Z.; Xu, Y. Petrogenesis of Middle Triassic Adakite-like Intrusions in the Asiha Orogenic Gold Deposit, East Kunlun Orogenic Belt, China. *Minerals* **2023**, *13*, 74. [CrossRef]
82. Luo, Z.H.; Ke, S.; Cao, Y.Q.; Deng, J.; Chen, H.W. Late Indosinian mantle-derived magmatism in the East Kunlun. *Geol. Bull. China* **2002**, *6*, 292–297. (In Chinese with English abstract)
83. Ding, S.; Huang, H.; Niu, Y.L.; Zhao, Z.D.; Yu, X.H.; Mo, X.X. Geochemistry, geochronology and petrogenesis of East Kunlun high Nb-Ta rhyolites. *Acta Petrol. Sin.* **2011**, *12*, 3603–3614. (In Chinese with English abstract)
84. Chen, G.C.; Pei, X.Z.; Li, R.B.; Li, Z.C.; Liu, C.J.; Chen, Y.X.; Pei, L.; Li, X.B. Age and lithogenesis of Keri syenogranite from eastern part of East Kunlun Orogenic Belt: Constraint on the Middle Triassic tectonic evolution of East Kunlun. *Acta Petrol. Sin.* **2018**, *3*, 567–585. (In Chinese with English abstract)
85. Chen, G.C.; Pei, X.Z.; Li, R.B.; Li, Z.C.; Pei, L.; Liu, C.J.; Chen, Y.X. Origin of granitoids and coeval mafic microgranular enclaves in the East Kunlun Orogen, northern Tibetan Plateau: Insights from petrography, U-Pb geochronology, geochemical and isotope analyses. *J. Asian Earth Sci.* **2024**, *269*, 106117. [CrossRef]

86. Ao, C.; Sun, F.Y.; Li, B.L.; Wang, G.; Li, L.; Li, S.J.; Zhao, J.W. Geochemical characteristics, U-Pb chronology and tectonic significance of Xiaojianshan gabbro in Qimantage area, East Kunlun. *Geotecton. Metallog.* **2015**, *39*, 1176–1184. (In Chinese with English abstract)
87. Huang, H.; Niu, Y.L.; Nowell, G.; Zhao, Z.D.; Yu, X.H.; Zhu, D.C.; Mo, X.X.; Ding, S. Geochemical constraints on the petrogenesis of granitoids in the East Kunlun Orogenic belt, northern Tibetan Plateau: Implications for continental crust growth through syn-collisional felsic magmatism. *Chem. Geol.* **2014**, *370*, 1–18. [CrossRef]
88. Ding, Q.F.; Jiang, S.Y.; Sun, F.Y. Zircon U-Pb geochronology, geochemical and Sr-Nd-Hf isotopic compositions of the Triassic granite and diorite dikes from the Wulonggou mining area in the Eastern Kunlun Orogen, NW China: Petrogenesis and tectonic implications. *Lithos* **2014**, *205*, 266–283. [CrossRef]
89. Wang, Y.L.; Zhang, Z.W.; Zhang, J.W.; Qian, B.; Liu, Y.G.; You, M.X. Early Mesozoic mantle-derived magmatic events and their geological significance in the East Kunlun orogenic belt. *Geol. Explor.* **2017**, *53*, 855–866. (In Chinese with English abstract)
90. Tindle, A.G.; Webb, P.C. Probe-AMPH—A spreadsheet program to classify microprobe-derived amphibole analyses. *Comput. Geosci.* **1994**, *20*, 1201–1228. [CrossRef]

minerals

MDPI

Article

Geology and Geochronology of Magmatic–Hydrothermal Breccia Pipes in the Yixingzhai Gold Deposit: Implications for Ore Genesis and Regional Exploration

Li-Zhong Zhang [1,2,3], Wen-Sheng Gao [1,2] and Xiao-Dong Deng [1,2,*]

1 State Key Laboratory of Geological Processes and Mineral Resources, China University of Geosciences, Wuhan 430074, China
2 School of Earth Resources, China University of Geosciences, Wuhan 430074, China
3 Mineral Exploration Institute, Zijin Mining Group Co., Xiamen 361006, China
* Correspondence: dengxiaodong@cug.edu.cn

Abstract: Magmatic–hydrothermal breccia pipes are widespread in numerous major porphyry and epithermal gold deposits globally, representing significant repositories of metal resources and serving as potential indicators for exploration targeting. More than ten breccia pipes occur in the Central Taihangshan District (CTD) of the North China Craton. Some of these breccia pipes host gold mineralization and are proposed to be related to the adjacent lode gold mineralization. However, the lack of detailed geological constraints make this hypothesis ambiguous. To address this, the present study conducted comprehensive field observations, drill core logging, an in situ sulfur isotope analysis of pyrite, and the $^{40}Ar/^{39}Ar$ dating of adularia along a 1400 m section of the Tietangdong breccia pipe at Yixingzhai. Three distinct breccia facies were identified at Tietangdong, exhibiting variable proportions across the entire section, including a massive skarn breccia; polymictic, skarn matrix-supported breccia; and polymictic, intrusive rock cement chaotic breccia. Furthermore, adularia $^{40}Ar/^{39}Ar$ dating indicates a syn-/post-gold mineralization age of 136 ± 1.5 Ma, coinciding with the age of post-breccia felsite dike. The deepest sampled pyrite displays $\delta^{34}S$ values of ~2.7‰, strongly indicating a magmatic–hydrothermal signature. These results, when combined with the geological, geochronological, and isotopic studies on the adjacent lode gold mineralization, further suggest a close genetic relationship between the breccia pipes and the lode Au mineralization, paving the way for their utilization as effective indicators for gold targeting within the CTD.

Keywords: North China Craton; Central Taihangshan District; breccia pipes; gold mineralizations; sulfur isotope; adularia dating

1. Introduction

Porphyry and epithermal gold deposits represent a crucial and economically significant class within the spectrum of gold deposits [1,2]. Magmatic–hydrothermal breccia pipes have been reported in some giant porphyry and epithermal gold deposits [3–7], such as the Cripple Creek gold deposit, USA [8,9], Mt. Polley Cu-Ag-Au porphyry deposit, Canada [10], Kellan gold deposit, Indonesia [11], and Ladolam epithermal gold deposit, Papua New Guinea [12]. As such, the magmatic–hydrothermal breccia pipe is a potential indicator for porphyry and epithermal gold deposit exploration [13]. The magmatic–hydrothermal breccia pipe is characterized by the mechanical disruption of wall rocks resulting from the release of overpressured magmatic–hydrothermal fluids or the subsurface emplacement of intrusions [3]. In contrast, other breccia pipes, such as phreatic breccia and pebble dikes, share some similar textures with magmatic–hydrothermal breccias, but lack significant mineralization [3]. Therefore, determining the genesis of breccia pipes is crucial in evaluating the relationships between breccia pipes and mineralization.

Numerous breccia pipes have been identified in the Central Taihangshan District (CTD) within the central North China Craton (NCC) (Figure 1b, Refs. [14,15]). This district hosts numerous early Cretaceous hydrothermal ore deposits, encompassing porphyry Cu-Mo [16], skarn Fe-Au, lode Au [17,18], and volcanic-hosted Pb-Zn deposits [19]. More than ten breccia pipes have been found adjacent to these deposits, such as the Yixingzhai, Gaofan, Tanshang, Zhijiadi, and Chakou deposits (Figure 1b) [14]. It is uncertain whether these breccia pipes are associated with their adjacent hydrothermal mineralization due to poor geological constraints.

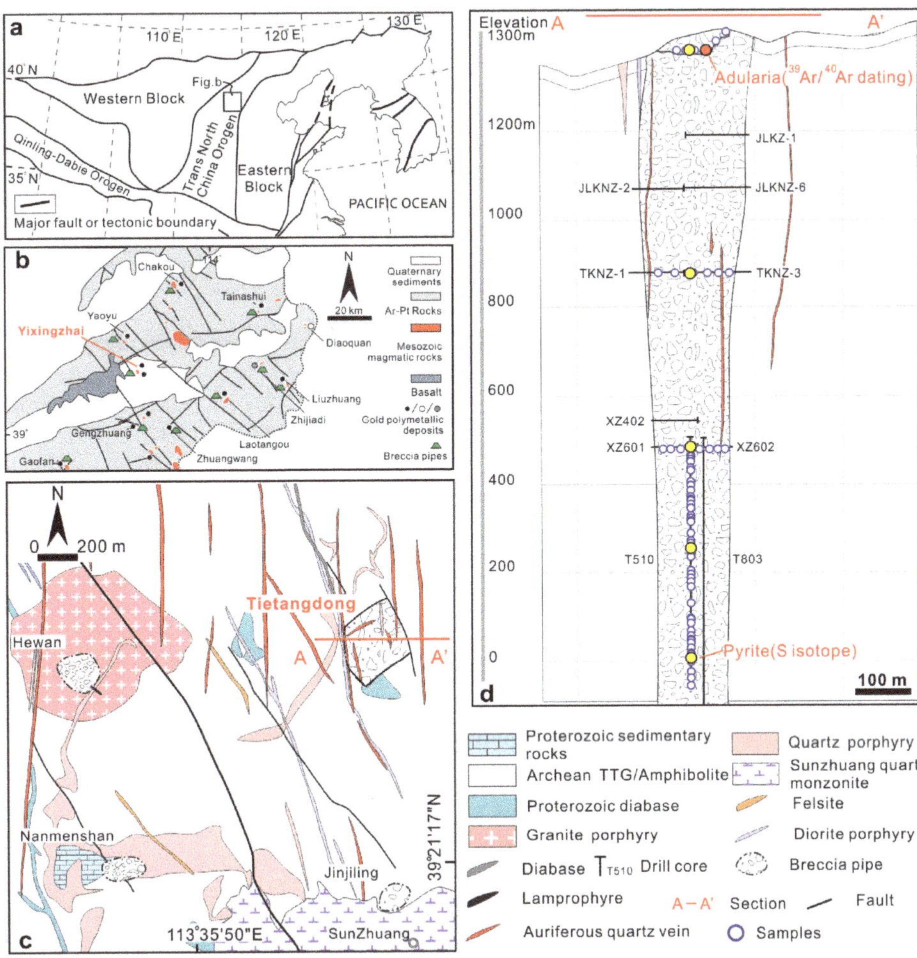

Figure 1. (**a**) Geological map showing the tectonic division of the NCC and the location of the CTD. (**b**) Regional geological map illustrating the distribution of the early Cretaceous magma, breccia pipes, and gold polymetallic deposits in the CTD. Geological map (**c**) and cross-section along the A-A' transect (**d**) of the Tietangdong breccia pipe (modified after [14]).

The Yixingzhai gold deposit (Lat. 39°21′51″ N, Long. 113°35′46″ E) is the largest gold deposit in the CTD (Figure 1c, Ref. [20]). It has sixteen lode Au ore bodies and four breccia pipes that also host gold mineralization. The Tietangdong breccia pipe is the most gold-rich breccia pipe (~20 t Au). Previous works [21–23] have suggested a skarn crypto-explosive origin for Tietangdong based on samples collected from the shallow part of the breccia pipe.

Moreover, the authors of [22] analyzed zoning textures within garnet and epidote grains, inferring several magmatic–hydrothermal pulses and multiple crypto-explosion events in Tietangdong. However, our understanding of breccia facies and their spatial distribution is still poor due to the inability to obtain deep samples, impacting insights into fully comprehending the genesis of gold mineralization within the Tietangdong breccia pipe. Furthermore, the absences of syn-gold mineralized geochronological and isotopic fluid source tracing studies limit our understanding of its relationship with the adjacent lode Au mineralization, thereby diminishing the efficacy of utilizing breccia pipes as a targeting indicator. The Zijin Mining Company recently initiated a series of drilling campaigns at the Tietangdong breccia pipe, with some drill holes reaching a depth of ~1400 m below the surface. These drilling campaigns provide an opportunity to address the issues mentioned above. The present study conducted comprehensive field observations and drill core logging across a deep cross-section (1400 m in depth), the $^{40}Ar/^{39}Ar$ dating of syn-ore adularia, and an in situ sulfur isotope analysis of pyrite grains associated with the native gold. The results collectively provide valuable insights into the ore's genesis and regional gold exploration.

2. Geological Background

The CTD is located in the northern part of the ca. 1.85 Ga Trans-North China Orogen (Figure 1a, Ref. [24]). Within the CTD, the basement rocks comprise Archean to Paleoproterozoic tonalite–trondhjemite–granodiorite gneiss (TTG), amphibolite, and banded iron formation [24,25]. The NCC remained relatively stable from the late Paleoproterozoic to the Paleozoic, accumulating a substantial sequence of carbonate and clastic sediments [26]. The deep NE-trending faults in the basement, together with the NW-trending faults, constitute the primary structural framework of the CTD during the Precambrian. Subsequently, these faults experienced reactivation during the Mesozoic [14,26]. Along these faults, numerous intermediate to felsic plutons, stocks, and dikes intruded the ancient basement of the CTD (Figure 1b [27]). These felsic intrusions, with zircon U-Pb ages at 142–130 Ma [18,28,29], are interpreted as products of crust–mantle interactions resulting from the lithospheric destruction of the NCC [30]. Many porphyry Cu-Mo deposits, skarn Fe-Au deposits, and lode Au deposits formed within or around these early Cretaceous intrusions (Figure 1b). Previous molybdenite Re-Os, garnet U-Pb, titanite U-Pb, and phlogopite $^{40}Ar/^{39}Ar$ geochronological studies revealed that the mineralization formed in the period of 140–130 Ma, concurrent with the early Cretaceous intrusions [17,18,21,30–32]. Numerous breccia pipes have been identified in the CTD [15]. These pipes are structurally controlled by the NW- and NE-trending faults and have spatial relationships with these early Cretaceous intrusions and mineralization (Figure 1b).

The Yixingzhai deposit is the largest gold deposit in the central CTD (Figure 1b). It is characterized by four breccia pipes and sixteen NW-trend auriferous quartz veins hosted in the Archean to Paleoproterozoic TTG basement. Additionally, amphibolite, metamorphic diabase, and carbonate wall rocks have been identified in the ore field (Figure 1c, Ref. [14]). Yixingzhai features multiple generations of 135–140 Ma intermediate to felsic intrusive dikes, stocks, and plutons (Figure 1c), including the Hewan granite porphyry, Nanmenshan quartz porphyry, and Sunzhuang quartz monzonite (Figures 1c and 2a–d) [17,18,20,21,30]. Four breccia pipes in the Yixingzhai gold deposit are identified along these intrusions: Tietangdong, Hewan, Nanmenshan, and Jinjiling (Figure 1c).

Sixteen lode Au ore bodies are hosted in Paleo–Proterozoic TTG gneiss and/or granite porphyry, ranging in thickness from 0.05 to 4.30 m (Figure 1c). These ore bodies extend for lengths of 150–2300 m and downdip for 500–1100 m. They exhibit a diverse mineralogical composition, including quartz, hematite, pyrite, chalcopyrite, sphalerite, and galena, with the dominant alteration characterized by sericite and chlorite [17,30].

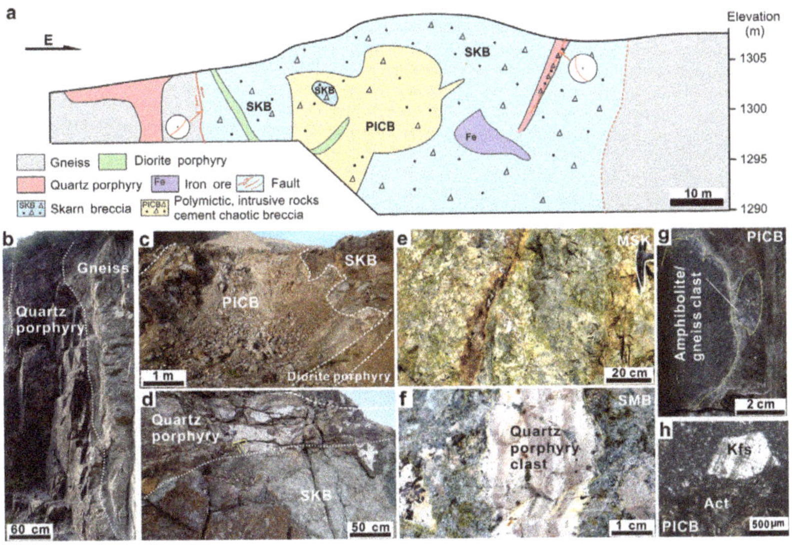

Figure 2. (**a**) Schematic cross-section showing the spatial distribution of different breccia facies, faults, and dikes in Tietangdong. Stereographic maps of faults along the margin of the Tietangdong breccia pipe are presented. (**b**) Quartz porphyry cutting the gneiss at the western part of the Tietangdong breccia pipe. (**c**) Outcrop of polymictic, intrusive rock cement chaotic breccia (PICB) and skarn breccia (SKB) in the middle of the Tietangdong breccia pipe. (**d**) Quartz porphyry cutting the SKB at the middle of the Tietangdong breccia pipe. (**e**) Massive skarn breccia (MSK) at the Tietangdong breccia pipe, displaying aggregations of massive prograde and retrograde skarn minerals with fine to medium grain sizes. Definitive contact boundaries between clasts and matrix/cement are absent. (**f**) Polymictic, skarn matrix-supported breccia (SMB) in the Tietangdong breccia pipe, exhibiting a matrix-supported texture. Polymictic, rounded to sub-rounded clasts include quartz porphyry, felsite, gneiss, and amphibolite. (**g**) PICB at the Tietangdong breccia pipe, where gray felsite (**h**) cements rounded gneiss and amphibolite clasts. Abbreviations: Kfs—K feldspar, Act—actinolite.

3. Sample and Methods

More than 300 samples, ranging from the open pit to the deep drill cores, were collected for petrographical observation and breccia facies identification (Figure 1d). The classification of breccia pipe facies follows the methodologies established in [11,12]. Specifically, the matrix comprises tiny grains of rock powder or skarn minerals that are derived from the carbonate clasts undergoing metasomatism. The cement component denotes the intrusive rocks that cement the clasts within the breccia pipe. Over 100 double-polished thin sections were examined under an optical microscope to investigate the mineralogy and textural features based on their optical properties, such as the refractive index, birefringence, interference colors, and extinction characteristics. In addition to optical microscopy, carbon-coated sections were further studied using backscattered electron (BSE) imaging with a Quanta 450 FEG scanning electron microscope (SEM) at the State Key Laboratory of Geological Processes and Mineral Resources (GPMR), China University of Geosciences (CUG). The accelerating voltage was set to 15 kV, with a beam current of 20 nA. The SEM-BSE technique provided valuable information about the elemental contrast and micro-textural relationships, which are crucial for identifying minerals with similar optical properties. In cases where further confirmation was necessary, energy-dispersive X-ray spectroscopy (EDS) was employed to analyze the chemical composition of the minerals, ensuring accurate identification. Detailed sample descriptions and logging results are listed in Supplementary Material S1 (Table S1) and illustrated in Figure 3.

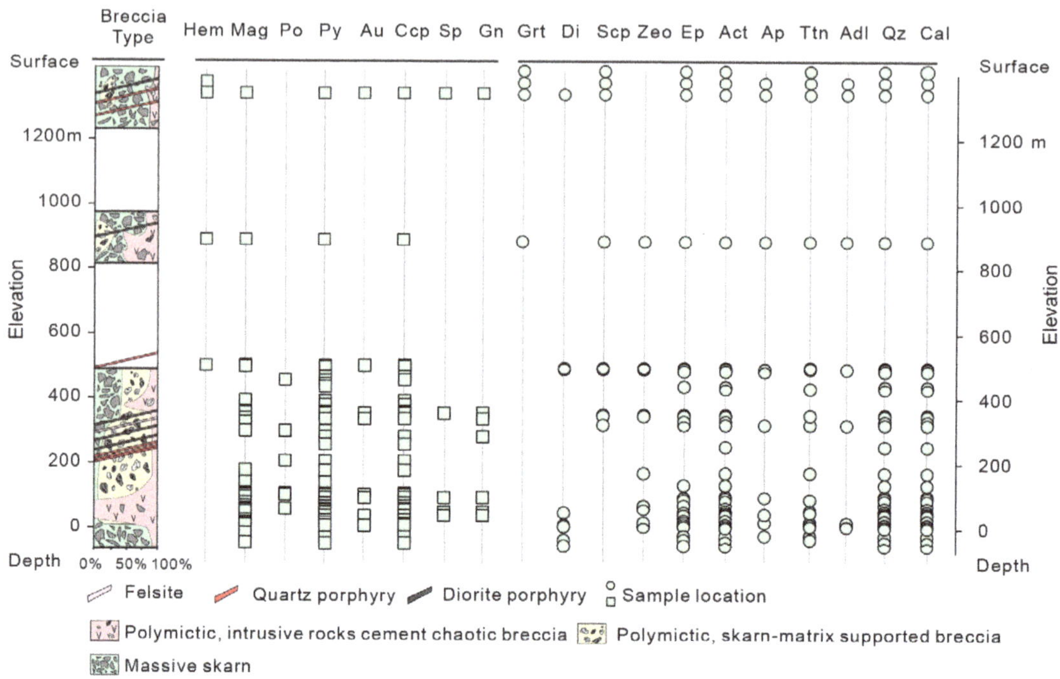

Figure 3. Cross-sectional illustration depicting breccia facies, intrusive rocks, and mineral associations in the Tietangdong breccia pipe. Abbreviations: Hem—hematite, Mag—magnetite, Po—pyrrhotite, Py—pyrite, Au—native gold, Ccp—chalcopyrite, Sp—sphalerite, Gn—galena, Grt—garnet, Di—diopside, Scp—scapolite, Zeo—zeolite, Ep—epidote, Act—actinolite, Ap—apatite, Ttn—titanite, Adl—adularia, Qz—quartz, Cal—calcite.

3.1. Adularia $^{40}Ar/^{39}Ar$ Dating

The sample TTD-10 was collected from the open pit for adularia $^{40}Ar/^{39}Ar$ dating, with adularia and calcite cementing the retrograde epidote and actinolite (Figure 4c). After careful petrographic observation, adularia was extracted through a series of steps: crushing, sieving, washing in distilled water using an ultrasonic bath for 10 min, and air-drying for two days. Finally, 0.5 to 2.0 mm adularia grains were handpicked under a binocular microscope and validated via SEM-EDS before $^{40}Ar/^{39}Ar$ dating.

The mineral separates were irradiated with ZBH-2506 biotite standards (132.7 ± 0.2 Ma [33]) in the China Mianyang Research Reactor (CMRR) at the China Academy of Engineering for 48 h. After cooling for three months, the argon isotope ratios were analyzed using the incremental heating method on a multi-collector Argus VI noble gas mass spectrometer connected to a Coherent 50W CO_2-IR laser device at the $^{40}Ar/^{39}Ar$ laboratory at CUG. The experimental details of the method were provided in [34]. The $^{40}Ar/^{39}Ar$ results were calculated and plotted using the ArArCALC (version 2.52b [35]) software. All dates were reported using 5.543×10^{-10} a^{-1} as the total decay constant for ^{40}K [36], with reactor correction factors set at 8.984×10^{-4} for $(^{39}Ar/^{37}Ar)_{Ca}$, 2.673×10^{-4} for $(^{36}Ar/^{37}Ar)_{Ca}$, and 5.97×10^{-3} for $(^{40}Ar/^{39}Ar)_K$.

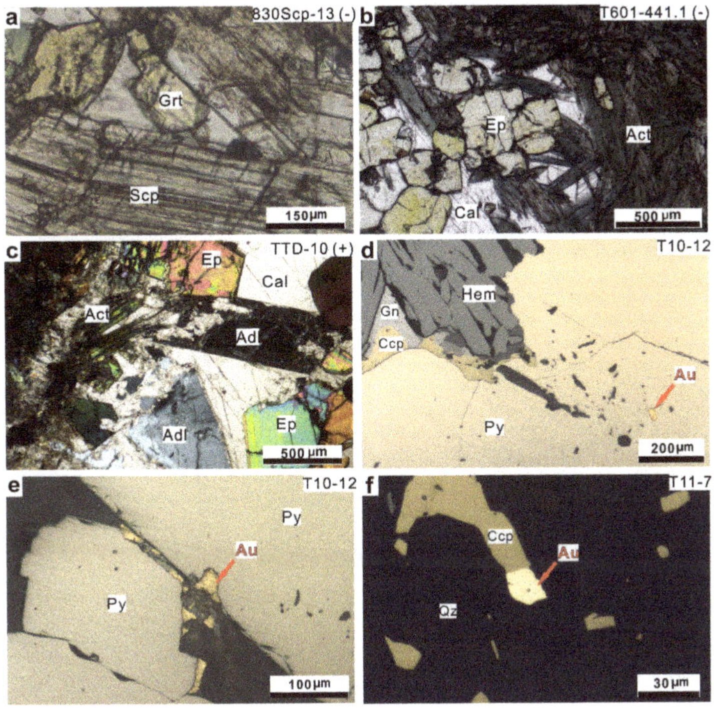

Figure 4. Microphotographs illustrating mineral associations and the occurrences of gold grains at the Tietangdong breccia pipe. (**a**) Garnet coexists with scapolite during the prograde skarn stage. (**b**) Epidote coexists with actinolite during the retrograde skarn stage, cemented by late-stage calcite. (**c**) Calcite and adularia at the quartz–sulfide–calcite stage cement the early stage epidote and actinolite. Microphotographs show anhedral native gold either as inclusions (**d**) or filling the interstitial of pyrite and chalcopyrite (**e**,**f**) at the Tietangdong breccia pipe. Abbreviations: Au—native gold, Hem—hematite, Py—pyrite, Ccp—chalcopyrite, Sp—sphalerite, Gn—galena, Grt—garnet, Scp—scapolite, Ep—epidote, Act—actinolite, Qz—quartz, Cal—calcite, Adl—adularia, '+' —under cross-polarized light, '-' —under plane-polarized light.

3.2. *In Situ Sulfur Isotope Analysis*

Five samples from the Tietangdong breccia pipe were selected for in situ sulfur isotope analysis. In situ sulfur isotope analyses of pyrite that related to gold inclusions were carried out using an NWR Femto^{UC} femtosecond system (New Wave Research, Fremont, CA, USA) coupled with a Neptune Plus multi-collector inductively coupled plasma mass spectrometer (MC-ICPMS, Thermo Fisher Scientific, Bremen, Germany) at the GPMR, CUG. In the laser ablation system, helium was used as the carrier gas for the ablation cell and was mixed with argon (make-up gas) after the ablation cell. Single-spot ablation mode was employed. The laser fluence was kept constant at ~2.5 J/cm^2. To mitigate the downhole fractionation effect, a large spot size (40 μm) and slow pulse frequency (4 Hz) were utilized [37]. The Neptune Plus, equipped with nine Faraday cups fitted with 10^{11} Ω resistors, collected isotopes ^{32}S, ^{33}S, and ^{34}S in static mode. The X skimmer cone and Jet sample cone in Neptune Plus were used to improve the signal intensity. Nitrogen (4 mL/min) was introduced into the central gas flow to diminish the polyatomic interferences. Medium resolution was applied in all measurements, with the revolving power kept at greater than 5000.

The standard-sample bracketing method (SSB) was employed to correct instrumental mass fractionation. The external standard of pyrite PPP-1 (δ^{34}S = 5.40‰ ± 0.16‰) was used to correct the mass fractionation of the sulfur isotope in various sulfide samples. In addition, the in-house references of pyrrhotite YP136 (δ^{34}S = 1.50‰ ± 0.30‰ [38]) were repeatedly analyzed to verify the calibration accuracy. The analyzed δ^{34}S results of YP136 were 1.44‰ ± 0.18‰ (n = 10, 2σ). All data reductions for the LA-MC-ICPMS analysis were conducted using the ISO-Compass software [39].

4. Results

4.1. Breccia Facies and Intrusive Sequences

Based on the clasts composition, matrix characteristics, and cement types [11,12], the breccia facies of the Tietangdong breccia pipe can be classified into two main facies: the skarn breccia (SKB) and the polymictic intrusive rock cement-supported chaotic breccia (PICB) (Table 1).

Table 1. Classification of breccia facies in the Tietangdong breccia pipe.

Breccia Facies		Interpretation	Component				Internal Organization and Mineralization	Spatial Distribution
			Clast	Matrix	Cement	Open Space		
Skarn breccia (SKB)	Massive skarn (MSK)	Magmatic–hydrothermal breccia	Aggregates of massive prograde and retrograde skarn minerals (Grt, Scp, Di, Act, Ep, Qz, Chl, and Cal) with fine to medium grain sizes (2–64 mm), commonly lacking definitive boundaries between the clasts and the cement; ore minerals include Mag, Hem, Py, Ccp, Sp, Gn, and Au			<5 vol.% vugs	Massive skarn ores, mainly bearing low to high grades of gold and iron	Mainly distributed at the top of the Tietangdong breccia pipe
	Polymictic, skarn matrix-supported breccia (SMB)	Magmatic–hydrothermal breccia	5–70 vol.%; generally rounded to sub-rounded; polymictic clasts, including felsite, gneiss, amphibolite, and a few skarn carbonate clasts; the clasts measure 5–30 mm, with blocks measuring up to 20 cm occasionally	30–95 vol.%; skarn minerals (Scp, Act, Ep, Py, Ccp) that are predominantly 1–2 mm	-	<5 vol.% vugs	Chaotic; mainly matrix supported; no high grade of gold	
Polymictic, intrusive rock cement chaotic breccia (PICB)		Intrusive breccia	10–75 vol.%; rounded to sub-rounded; polymictic clasts, including partly skarn carbonate clasts, gneiss, amphibolite, and some felsite clasts; the clasts measure 5–60 mm	-	15–90 vol.%; igneous rocks (felsite, granite porphyry, and diorite porphyry)	-	Chaotic; cement supported to clast supported; rarely mineralized	Mainly distributed at the bottom of the Tietangdong breccia pipe

The SKB can be categorized as a massive skarn breccia (MSK) and a polymictic skarn matrix-supported breccia (SMB). The MSK is characterized by aggregates of massive prograde and retrograde skarn minerals, including garnet, scapolite, diopside, actinolite, epidote, quartz, chlorite, and calcite (Table 1), with a grain size of 2–64 mm (Figure 4a,b). It has indistinct boundaries between the clasts and the cement (Figure 2e). Ore minerals in the MSK consist of magnetite, hematite, and pyrite (Figure 3). The MSK is widely distributed at the top of the Tietangdong breccia pipe (Figures 2a and 3). The SMB comprises polymictic clasts and a skarn matrix (Figure 2f, Table 1). Polymictic clasts comprise felsite, gneiss, amphibolite, and partly skarn carbonate clasts. Individual clasts are generally rounded to sub-rounded, measuring 5–30 mm, with blocks up to 20 cm occasionally. The matrix is mainly composed of skarn minerals, including scapolite, actinolite, epidote, and some sulfide minerals (pyrite and chalcopyrite), predominately in the size of 1–2 mm (Figure 2f). The SMB is primarily distributed at the top of the Tietangdong breccia pipe (Figures 2a and 3).

The PICB (Figure 2c,g, Table 1) exhibits chaotic characteristics and transitions from cement-supported to clast-supported. Polymictic clasts encompass partly skarn carbonate, gneiss, amphibolite, and some felsite clasts (Figure 2g). These individual clasts generally display rounded to sub-rounded shapes, with sizes ranging from 5 to 60 mm (Figure 2g).

The cement mainly consists of fine-grained intrusive rocks (Figure 2g,h). The PICB is mainly distributed at the base of the Tietangdong breccia pipe (Figure 3).

Three generations of intrusive dikes have been identified through field relationships (Figure 3). The first pre-breccia stage is represented by intrusive rock fragments of the clasts, such as the felsite clasts in the SMB (Figure 2f). The second syn-breccia stage is characterized by the felsic intrusive cement in the PICB (Figure 2g). The third post-breccia stage is represented by intrusive dikes that cut the Tietangdong breccia pipe (Figures 2d and 3).

4.2. Mineralization of and Alteration in Tietangdong Breccia Pipe

The Tietangdong breccia pipe exhibits three hydrothermal alteration and mineralization stages: the prograde skarn stage, retrograde skarn stage, and quartz–sulfide–calcite stage. The prograde skarn stage consists of garnet, diopside, scapolite, and minor magnetite (Figures 3 and 4a). It is primarily concentrated in the shallow part of the breccia pipe (Figure 3, 800–1300 m above sea level). The retrograde skarn stage is characterized by scapolite, epidote, actinolite, magnetite, and minor sulfide minerals, with minor native gold (Figures 3 and 4b). It is distributed throughout the breccia pipe, often replacing or cementing the prograde skarn minerals. The quartz–sulfide–calcite stage is represented by pervasive quartz, sulfides, calcite, and adularia (Figures 3 and 4c–f). This stage is identified as the dominant gold mineralization stage. Native gold is mainly present as inclusions (5–35 μm) or filling the intergranular space of the texturally homogeneous pyrite and chalcopyrite of this stage (Figure 4d–f), with tiny irregular native Au grains (1–5 μm) sporadically adjacent along the margin of hematite, quartz, calcite, and chlorite grains [22]. The quartz–sulfide–calcite stage is distributed throughout the Tietangdong breccia pipe (Figure 3). The mineral paragenesis of the Tietangdong breccia pipe is summarized in Figure 5. In comparison to previous works [22,23], scapolite and adularia were first identified in this study. Although molybdenite was reported in an earlier work [22], we did not observe this mineral in this study (Figure 5).

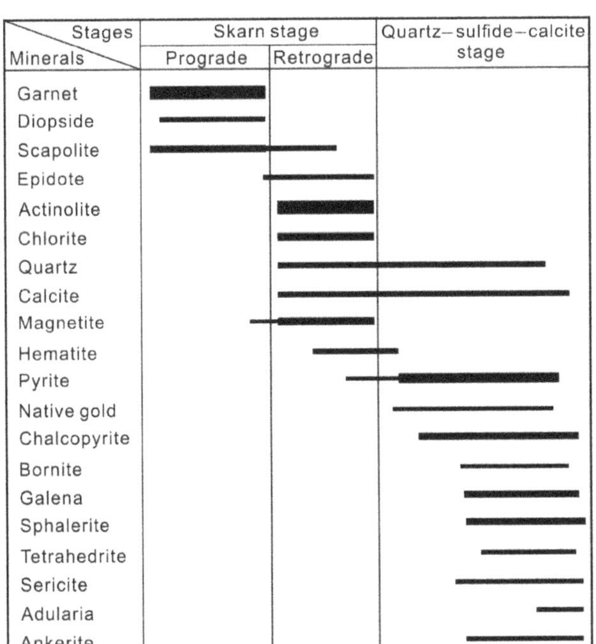

Figure 5. Mineral paragenesis in the Tietangdong breccia pipe.

The breccia facies, intrusive sequences, and mineral associations along the section in the Tietangdong breccia pipe are illustrated in Figure 3. The results show that the proportions of MSK and SMB decrease with depth. However, the PICB increases with the depth in the Tietangdong breccia pipe (Figure 3). In addition, Figure 3 reveals a vertical zonation of sulfide and oxide mineral associations along the 1400 m section. This zonation encompasses two segments: (1) hematite and magnetite-bearing breccia above approximately 510 m above sea level and (2) magnetite, pyrrhotite, and pyrite-bearing breccia below 510 m above sea level (Figure 3).

4.3. $^{40}Ar/^{39}Ar$ Dating of Adularia

The $^{40}Ar/^{39}Ar$ dating results obtained for the adularia grains in sample TTD10 are illustrated as the $^{40}Ar/^{39}Ar$ spectrum and inverse isochron plots in Figure 6, with detailed data in Supplementary Material S2 (Table S2). The adularia $^{40}Ar/^{39}Ar$ spectra show anomalously old apparent ages at low-temperature steps, followed by several continuous plateau ages (135.9 $\pm$ 1.5 Ma, 2σ level, full external uncertainty considering the decay constant, $\Sigma^{39}Ar = 70\%$) at mid–high temperature steps (Figure 6a). The old apparent ages at low-temperature steps are interpreted as ^{39}Ar loss in intercrystallite and/or damaged sites of adularia through recoil during irradiation [40]. Then, argon in the tight crystallographic reservoir of adularia releases at mid–high temperature steps and forms the plateau age (Figure 6a), which can be interpreted as the timing of adularia formation. In addition, the adularia grains also yield an isochron age of 135.2 $\pm$ 1.7 Ma (Figure 6b), which is consistent with its plateau age. This consistency supports the adularia formed at ca. 136 Ma in the Tietangdong breccia pipe.

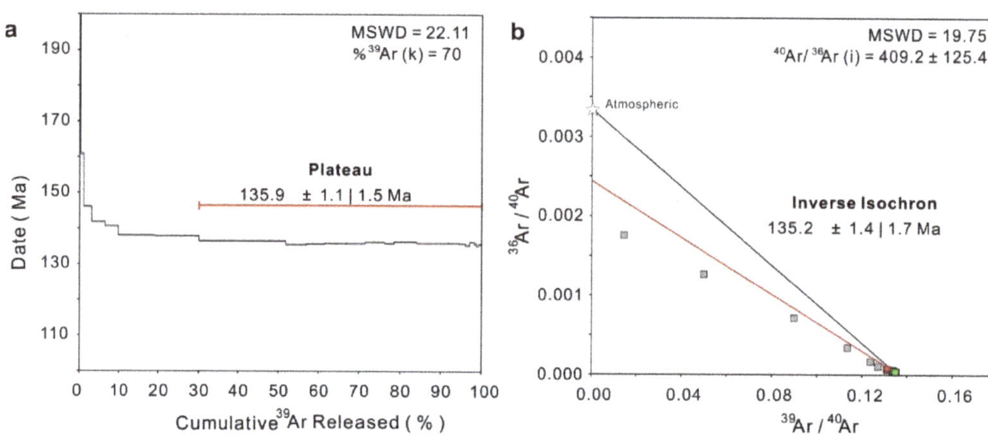

Figure 6. Laser-heated $^{40}Ar/^{39}Ar$ date spectra (**a**) and inverse isochron (**b**) of adularia from the Tietangdong breccia pipe.

4.4. In Situ Sulfur Isotopes of Pyrite

Sixteen spot analyses of pyrite from the Tietangdong breccia pipe yielded $\delta^{34}S$ values ranging from 2.03‰ to 9.01‰, with a mean of 4.08‰ (Figure 7a; Table 2). These results strongly illustrate decreased pyrite $\delta^{34}S$ values with increasing depth within the Tietangdong breccia pipe (Figure 7b). Specifically, at elevations of 15 m, 258 m, 510 m, 830 m, and 1380 m, the $\delta^{34}S$ values exhibit ranges of 2.48‰–2.73‰, 2.03‰–2.63‰, 2.10‰–4.22‰, 3.98‰–9.01‰, and 5.29‰–6.59‰, respectively.

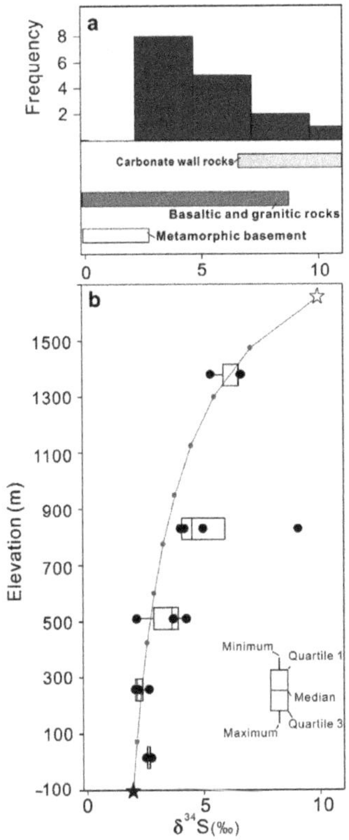

Figure 7. (**a**) Comparison of sulfur isotope results from the Tietangdong breccia pipe. The boxes represent the reported isotopic ranges of sulfur isotope values from different reservoirs [26,41,42]. (**b**) Depth profile of pyrite sulfur isotope from the Tietangdong breccia pipe. The linear regression curve illustrates a two-end member mixing trend between the magmatic–hydrothermal fluids (assuming 2‰ of $\delta^{34}S$ and 1000 ppm of sulfur contents at an elevation of −100 m) and the ore field carbonate wall rocks (assuming 10‰ of $\delta^{34}S$ and 200 ppm of sulfur contents at an elevation of 1450 m).

Table 2. The sulfur isotope analysis results in the Tietangdong breccia pipe.

Sample no.	Elevation (m)	Minerals	$\delta^{34}S$ (‰)	2SD (‰)	Secondary Standard	$\delta^{34}S$ (‰)	2SD (‰)
T601-494.7-01	15	Pyrite	2.48	0.10	YP136	1.33	0.12
T601-494.7-02	15	Pyrite	2.69	0.10	YP136	1.56	0.16
T601-494.7-03	15	Pyrite	2.73	0.11	YP136	1.44	0.16
T601-251.6-01	258	Pyrite	2.63	0.10	YP136	1.49	0.09
T601-251.6-02	258	Pyrite	2.03	0.10	YP136	1.51	0.11
T601-251.6-03	258	Pyrite	2.16	0.08	YP136	1.39	0.18
T11-16-2-01	1380	Pyrite	6.59	0.11	YP136	1.27	0.12
T11-16-2-02	1380	Pyrite	6.56	0.11	YP136	1.49	0.15
T11-16-2-03	1380	Pyrite	5.29	0.11	YP136	1.41	0.14
830-9-01	830	Pyrite	3.98	0.13	YP136	1.55	0.12
830-9-02	830	Pyrite	4.13	0.12			
830-9-03	830	Pyrite	9.01	0.14			
830-9-04	830	Pyrite	4.94	0.16			
T510-3-2-01	510	Pyrite	3.67	0.10			
T510-3-2-02	510	Pyrite	4.22	0.11			
T510-3-2-03	510	Pyrite	2.10	0.10			

5. Discussion

5.1. Discrimination and Genesis of Breccia Facies

Detailed field observation and drill core logging identified three breccia facies at Tietangdong, including massive skarn breccia (MSK), polymictic, skarn matrix-supported breccia (SMB), and polymictic, intrusive rock cement chaotic breccia (PICB) (Figure 3). The presence of high-salinity fluid inclusions in skarn minerals [31] in the MSK and SMB suggests a magmatic–hydrothermal origin of the Tietangdong breccia pipe [43]. The identification of PICB further substantiates this conclusion. The PICB exhibits felsite as cement for both wall rock and skarn clasts, providing evidence of the direct involvement of intrusive rocks in the formation of the Tietangdong breccia pipe. Moreover, the increasing prevalence of PICB in deeper segments of the breccia pipe (Figure 3) indicates heightened intrusive activities. The widespread occurrence of felsite and diorite clasts in PICB and SMB, along with post-breccia felsite and diorite dikes (Figures 2–4), suggests significant intrusive processes both preceding and succeeding the breccia pipe formation. In summary, our comprehensive geological observations across the 1400 m section confirm the magmatic–hydrothermal origin of the Tietangdong breccia pipe.

5.2. The Age of Breccia-Hosted Au Mineralization at Tietangdong

Adularia is prevalent at the quartz–sulfide–calcite stage (Figure 4c). Given the lower temperature of adularia crystallization [44], in line with the low closure temperature of the $^{40}Ar/^{39}Ar$ system [45], adularia is a promising candidate for constraining the time of syn-/post-gold mineralization at Tietangdong. Consequently, the obtained date of 135.9 $\pm$ 1.5 Ma (Figure 6) may represent the age of gold mineralization in the Tietangdong breccia pipe. Although this age coincides well with the pre-gold garnet U-Pb dating result of 139 $\pm$ 4 Ma [21], it exhibits a slight inconsistency with the zircon U-Pb age of the post-breccia quartz porphyry (141 $\pm$ 1 Ma, ref. [21]). This discrepancy might be derived from the application of different radiometric dating systems and analytical methods. After accounting for their systematic and the decay constant uncertainties of $^{40}Ar/^{39}Ar$ (~2%, total relative uncertainty [45–47]) and U-Pb (~3%, total relative uncertainty [45–47]), the overlapping results around ca. 136 Ma are revealed. In summary, the spatial and temporal evidence strongly support a close genetic relationship between gold mineralization and the intrusive rocks at the Tietangdong breccia pipe. Furthermore, the age of Tietangdong gold mineralization is consistent with the zircon U-Pb age of the syn-/post-lode Au Sunzhuang pluton (*ca.* 135 Ma, refs. [18,30]), indicating a potential genetic relationship between the breccia-hosted gold mineralization and lode Au mineralization. The sulfur isotope results below provide additional support for this hypothesis.

5.3. The Source of Breccia-Hosted Au Mineralization at Tietangdong

Pyrite is one of the most predominant sulfide minerals, enclosing gold grains (Figure 4) within the Tietangdong breccia. Furthermore, no sulfate has been identified in the whole section. Consequently, the sulfur isotopes of pyrite could be a reliable proxy for the sulfur isotope of hydrothermal fluids at Tietangdong [48]. The sulfur isotopes of the pyrite from the Tietangdong breccia pipe exhibit a variation of 2.03‰–9.01‰ (Figure 7a). Despite aligning with the sulfur range of the regional metamorphic basement ($\delta^{34}S$ values of -10.0‰ to 2.0‰, ref. [26]), the absence of an early Cretaceous metamorphic event in the CTD precludes the involvement of metamorphic fluids [49]. In addition, the identification of daughter crystal-bearing fluid inclusions in garnet grains [21] indicates high-salinity ore-forming fluids, distinguishing them from the typical low-salinity and CO_2-rich fluid inclusions associated with metamorphic hydrothermal fluids [50].

Most sulfur isotopes fall within the global magmatic–hydrothermal fluids range ($\delta^{34}S$ values of -1‰ to 8‰, ref. [41], Figure 7), implying a magmatic–hydrothermal origin of the ore-forming fluid. Furthermore, the $\delta^{34}S$ values exhibit a positive elevation-related fractionation trend within the Tietangdong breccia pipe (Figure 7). This trend could be explained using a scenario involving gradually diminishing water–rock interactions between exsolved magmatic–hydrothermal fluids ($\delta^{34}S$ values ~2‰) and the carbonate wall rocks ($\delta^{34}S$ values ~10‰) [42] with the increasing depth of the breccia pipe. As Figure 3 illustrates, the decreasing proportion of skarn breccia implies a reduced involvement of carbonate wall rocks at greater depths. This shift would supply a less heavy sulfur isotope for the deep hydrothermal system than the shallow part. Therefore, lighter sulfur isotope values are revealed at depth, while heavier sulfur isotope values are retained in shallow areas. Under this assumption, a simple two-end member mixed model was employed, as depicted in Figure 7b. The simulated model aligns well with our sulfur isotope results. This alignment suggests a more pronounced magmatic–hydrothermal signature in the deeper part of the Tietangdong breccia pipe. It further implies promising potential for gold mineralization in the deep section. This view has been supported by the subsequent drill campaigns conducted by Zijin Mining Company.

An alternative interpretation of the sulfur isotope fractionation observed in this study could be linked to the redox state of hydrothermal fluids. In hydrothermal systems, an oxidized state promotes the formation of SO_4^{2-} ligands, which preferentially incorporate the heavy sulfur isotope, resulting in the depletion of the heavy sulfur isotope in pyrite [48,51,52]. However, our geological observations present a contradiction to this theoretical expectation. The presence of hematite in the shallow part and pyrrhotite in the deeper sections (Figure 3) suggests a theoretically lighter sulfur isotope composition in the shallow region and a heavier one at the base. This contradiction excludes the redox hypothesis.

The sulfur isotope composition of the deep section of the Tietangdong breccia pipe (ca. 2.7‰, Table 2) closely resembles the sulfur isotope composition of the lode Au (0.7‰–3.2‰, Ref. [30]). The gold mineralization age at Tietangdong breccia pipe shares a similar mineralization age with the lode Au in the Yixingzhai deposit. Moreover, both the breccia-hosted Au and lode Au orebodies are controlled by the ore-field NW-/NE-trending faults [14]. Therefore, we propose that the Tietangdong breccia pipe and the lode Au are part of the same magmatic–hydrothermal mineralization system.

5.4. Genetic Model and Exploration Implications

Based on field investigations, geochronology, and sulfur isotope analysis, we propose a genetic evolution model for the Tietangdong breccia pipe and the lode Au mineralization in Yixingzhai. During the Mesozoic, NW-/NE-trending faults in Yixingzhai became active, establishing connections with deep magma reservoirs. These activities facilitated the intrusion of felsite and diorite dikes and the formation of hydrothermal conduits at ca. 135–140 Ma (Figure 8a). Along these conduits, the magmatic–hydrothermal fluids, exsolved from the deep magma reservoirs through second boiling [4], were trapped in the shallow part due to the low permeability of superficial TTG and carbonate wall rocks. Subsequently, the conjunction between the faults and the dikes, identified as structural weak points in the ore field, lost stability due to the higher pressure of the accumulated volatiles compared to the resistance of the wall rocks. This process led to a series of fault displacements, wall rock ruptures, skarn alterations, and gold mineralization. Concurrently, the continued intrusion of felsic dikes contributed to the formation of intrusive breccia in the deeper sections of the breccia pipe (Figure 8a–b). These breccia and intrusive processes could have happened several times, as revealed through the geochemistry of the distinct zonation of garnet and epidote grains [22]. Simultaneously, lode Au mineralization occurred at the Yixingzhai TTG adjacent to the Tietangdong breccia, which hosted Au mineralization. Ultimately, the Tietangdong breccia pipe was intersected by late-stage quartz and diorite dikes, followed by subsequent erosion (Figure 8c). These late-stage quartz and diorite dikes are relatively fresh and lack gold mineralization (Figures 2d and 3), contrasting with previous observations [22].

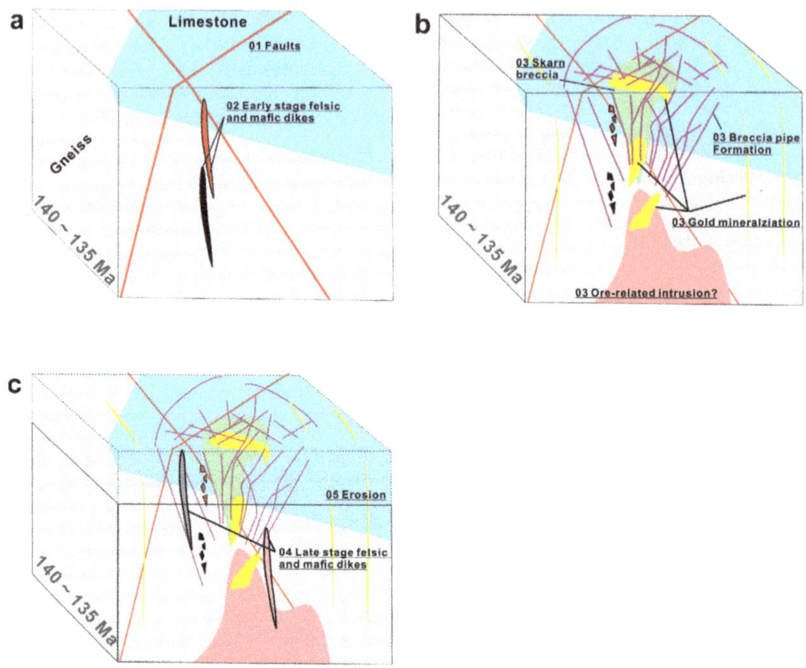

Figure 8. Schematic representations of the formation of Tietangdong breccia pipes (modified after [22]). (**a**) Pre-mineralization intrusion events. (**b**) Breccia pipe formation and gold mineralization processes. (**c**) Post-mineralization intrusion and erosion events.

This study provides valuable insights for exploration within the CTD. Firstly, this study reveals the dominance of magmatic–hydrothermal fluids in the formation of the breccia pipe, as evidenced by the abundance of intrusive rock clasts, pre-/syn-/post-breccia intrusions (Figures 2 and 3), adularia age, and sulfur isotope results (Figure 7). Additionally, a strong genetic relationship between the Tietangdong breccia pipe and gold mineralization was demonstrated in this study. This correlation suggests that breccia pipes could be a good indicator for gold targeting within the CTD. The intersections of regional NW- and NE-trending faults (Figure 1c) favor the formation of breccia pipes, which, in turn, may contribute to targeting new gold deposits, highlighting substantial exploration potential for gold in this region.

6. Conclusions

The present study synthesizes detailed field observations, the adularia $^{40}Ar/^{39}Ar$ dating of 135.9 ± 1.5 Ma, and a pyrite sulfur isotope analysis of 2.03‰–9.01‰, conclusively establishing a comprehensive geological formation model of the Tietangdong breccia pipe. Detailed field observations and drill core logging results categorize the breccia into massive skarn breccia (MSK); polymictic, skarn matrix-supported breccia (SMB); and polymictic, intrusive rock cement chaotic breccia (PICB). Furthermore, the sources and age of the Tietangdong breccia pipe align closely with those of the lode Au mineralization at the Yixingzhai deposit, indicating their potential genetic links. Numerous breccia pipes have been identified in the CTD, exhibiting geological characteristics akin to those of the Tietangdong breccia pipe and representing high potential for future gold targeting. The proposed integrated approach underscores the significance of geological, geochronological, and isotopic evidence in understanding ore genesis and informing regional gold exploration strategies.

Supplementary Materials: The following supporting information can be downloaded at https://www.mdpi.com/article/10.3390/min14050496/s1: Table S1: Breccia facies logging results of the Tietangdong breccia pipe; Table S2: Ar-Ar geochronology of adularia from the Tietangdong breccia pipe.

Author Contributions: Conceptualization, L.-Z.Z. and W.-S.G.; methodology, L.-Z.Z. and W.-S.G.; writing—original draft preparation, L.-Z.Z.; writing—review and editing, X.-D.D. and W.-S.G.; supervision, X.-D.D.; funding acquisition, X.-D.D. All authors have read and agreed to the published version of the manuscript.

Funding: This work was supported by the MOST Special Fund from the GPMR (Grant No. MSFGPMR2022-5), China University of Geosciences (Wuhan) and Zijin Mining Co. (Grant No. 2021026037). Wensheng Gao is financially supported by the China Scholarship Council (Grant No. 202106410102).

Data Availability Statement: All data, models, or codes that support the findings of this study are available from the corresponding author upon reasonable request.

Acknowledgments: We thank the Academic Editor for handling the manuscript and providing insightful comments, as well as four anonymous reviewers for their constructive feedback, which greatly improved the manuscript. Xiao-Rui Deng, Yu-Xiang Li, and Yang Yang are thanked for their help during the fieldwork and laboratory analysis. Jian-Pei Lu is thanked for the excellent sample preparation.

Conflicts of Interest: Zhang Li-Zhong is employed by Zijin Mining Co. Deng Xiao-Dong has received research grants from Zijin Mining Co. However, the funding sponsors played no role in the design of the study; in the collection, analysis, or interpretation of data; in the writing of the manuscript; or in the decision to publish the results.

References

1. Frimmel, H.E. Earth's Continental Crustal Gold Endowment. *Earth Planet. Sci. Lett.* **2008**, *267*, 45–55. [CrossRef]
2. Ulrich, T.; Günther, D.; Heinrich, C.A. Gold Concentrations of Magmatic Brines and the Metal Budget of Porphyry Copper Deposits. *Nature* **1999**, *399*, 676–679. [CrossRef]
3. Sillitoe, R.H. Porphyry Copper Systems. *Econ. Geol.* **2010**, *105*, 3–41. [CrossRef]
4. Sillitoe, R.H. Ore-Related Breccias in Volcanoplutonic Arcs. *Econ. Geol.* **1985**, *80*, 1467–1514. [CrossRef]
5. Chen, J.; Cooke, D.R.; Piquer, J.; Selley, D.; Zhang, L.; White, N.C. Hydrothermal Alteration, Mineralization, and Structural Geology of the Zijinshan High-Sulfidation Au-Cu Deposit, Fujian Province, Southeast China. *Econ. Geol.* **2019**, *114*, 639–666. [CrossRef]
6. Wang, B.; Zhou, T.-F.; Fan, Y.; Chen, J.; Liu, Y.-N.; Chen, Y. Breccia, Hydrothermal Alteration and Structural Geology of the Huangtun Porphyry Au-Cu Deposit in the Middle-Lower Yangtze Metallogenic Belt, Eastern China. *Ore Geol. Rev.* **2020**, *120*, 103414. [CrossRef]
7. Sun, Y.; Wang, X.; Zhang, Y.; Li, D.; Shan, W.; Geng, K.; Wei, P.; Liu, Q.; Xie, W.; Chi, N. Cu and S Isotopes of the Main Sulfides and Indicative Significance in the Qibaoshan Cu-Au Polymetallic Ore District, Wulian County, Shandong Province, North China Craton. *Minerals* **2023**, *13*, 723. [CrossRef]
8. Jensen, E.P. Magmatic and Hydrothermal Evolution of the Cripple Creek Gold Deposit, Colorado, and Comparisons with Regional and Global Magmatic-Hydrothermal Systems Associated with Alkaline Magmatism. Ph.D. Thesis, University of Arizona, Tucson, AZ, USA, 2003.
9. Thompson, T.B.; Trippel, A.D.; Dwelley, P.C. Mineralized Veins and Breccias of the Cripple Creek District, Colorado. *Econ. Geol.* **1985**, *80*, 1669–1688. [CrossRef]
10. Pass, H.E. Breccia-Hosted Chemical and Mineralogical Zonation Patterns of the Northeast Zone, Mt. Polley Cu-Ag-Au Alkalic Porphyry Deposit, British Columbia, Canada. Ph.D. Thesis, University of Tasmania, Hobart, Australia, 2010.
11. Davies, A.G.S. Geology and Genesis of the Kelian Gold Deposit, East Kalimantan, Indonesia. Ph.D. Thesis, University of Tasmania, Hobart, Australia, 2002.
12. Blackwell, J.L. Characteristics and Origins of Breccias in a Volcanic Hosted Alkalic Epithermal Gold Deposit, Ladolam, Lihir Island, Papua New Guinea. Ph.D. Thesis, University of Tasmania, Hobart, Australia, 2010.
13. Sillitoe, R.H. Intrusion-Related Gold Deposits. In *Gold Metallogeny and Exploration*; Springer: Dordrecht, The Netherlands, 1991; pp. 165–209.
14. Jing, S.-H. *Study of the Geology and Mineralization Rules of the Yixingzhai Gold Deposit, Fanshi County, Shanxi Province*; Unpublished Geological Report; 211 Geological Team of Shanxi Bureau of Geology and Mineral Resources: Taiyuan, China, 1985.
15. Li, S.; Ma, X. Crypto Explosive Rock and Its Ore-Forming Sense to Gold-Silver Deposit in the Northeastern Shanxi Province, China. *Contrib. Geol. Miner. Resour. Res.* **1999**, *14*, 8–14.

16. Zhang, J.-Q.; Li, S.-R.; Santosh, M.; Li, Q.; Niu, S.-D.; Lu, J.; Yan, L.-N. The Diaoquan Ag-Cu Polymetallic Skarn Mineralization in Central North China Craton: Timing, Source and Genetic Model. *Ore Geol. Rev.* **2019**, *104*, 745–764. [CrossRef]
17. Zhang, J.-Q.; Li, S.-R.; Santosh, M.; Niu, S.-D.; Li, Q.; Lu, J. The Magmatic–Hydrothermal Mineralization Systems of the Yixingzhai and Xinzhuang Gold Deposits in the Central North China Craton. *Ore Geol. Rev.* **2017**, *88*, 416–435. [CrossRef]
18. Zhang, J.-Q.; Li, S.-R.; Santosh, M.; Li, Q.; Niu, S.-D.; Li, Z.-D.; Zhang, X.-G.; Jia, L.-B. Timing and Origin of Mesozoic Magmatism and Metallogeny in the Wutai-Hengshan Region: Implications for Destruction of the North China Craton. *J. Asian Earth Sci.* **2015**, *113*, 677–694. [CrossRef]
19. Liu, T.; Xiong, S.-F.; Jiang, S.-Y.; Li, H.-L.; Chen, Q.-Z.; Jiang, H. Genesis of the Zhijiadi Ag-Pb-Zn Deposit, Central North China Craton: Constraints from Fluid Inclusions and Stable Isotope Data. *Geofluids* **2017**, *2017*, 4153618. [CrossRef]
20. Zhang, L.-Z. Element Geochemistry and Age of the Altered-Porphyry Gold Mineralization in Yixingzhai Gold Deposit, Fanshi City, Shanxi Province. *Geol. Miner. Resour. South China* **2018**, *34*, 134–141. [CrossRef]
21. Zhang, L.-Z.; Chen, L.; Wang, G.; Deng, X.-D.; Li, J.-W. Garnet U-Pb Dating Constraints on the Timing of Breccia Pipes Formation and Genesis of Gold Mineralization in Yixingzhai Gold Deposit, Shanxi Province. *Eart. Scie.* **2020**, *45*, 108. [CrossRef]
22. Zhang, J.-Q.; Li, S.-R.; Santosh, M.; Luo, J.-Y.; Li, C.-L.; Song, J.-Y.; Lu, J.; Liang, X. The Genesis and Gold Mineralization of the Crypto-Explosive Breccia Pipe in the Yixingzhai Gold Region, Central North China Craton. *Geol. J.* **2020**, *55*, 5664–5680. [CrossRef]
23. Shao, Y.; Zhang, Y.; Zhang, J.; Ding, Z.; Chen, X.; Liu, Z. Formation Mechanism of Breccia Pipe Type in Yixingzhai Gold Deposit. *J. Cent. South Univ. Technol.* **2008**, *15*, 89–94. [CrossRef]
24. Zhao, G.; Sun, M.; Wilde, S.A.; Sanzhong, L. Late Archean to Paleoproterozoic Evolution of the North China Craton: Key Issues Revisited. *Precambrian Res.* **2005**, *136*, 177–202. [CrossRef]
25. Tang, L.; Santosh, M. Neoarchean-Paleoproterozoic Terrane Assembly and Wilson Cycle in the North China Craton: An Overview from the Central Segment of the Trans-North China Orogen. *Earth-Sci. Rev.* **2018**, *182*, 1–27. [CrossRef]
26. Tian, Y.Q.; Yu, K.R.; Wang, A.J.; Xu, W.L. Geodynamic of Veined Gold Deposit Mineralization of Wutaishan-Hengshan Area. *J. Geol. Miner. Resour. North China* **1998**, *13*, 301–456.
27. Chang, J.; Audétat, A.; Li, J.-W. Tectono-Magmatic Controls on Decratonic Gold Deposits. *Contrib. Mineral. Petrol.* **2021**, *176*, 69. [CrossRef]
28. Deng, X.-D.; Li, J.-W.; Wen, G. U-Pb Geochronology of Hydrothermal Zircons from the Early Cretaceous Iron Skarn Deposits in the Handan-Xingtai District, North China Craton. *Econ. Geol.* **2015**, *110*, 2159–2180. [CrossRef]
29. Chang, Z.-G.; Dong, G.-C.; Somarin, A.K. U-Pb Dating and Trace Element Composition of Zircons from the Gujiao Ore-Bearing Intrusion, Shanxi, China: Implications for Timing and Mineralization of the Guojialiang Iron Skarn Deposit. *Minerals* **2020**, *10*, 316. [CrossRef]
30. Li, S.-R.; Santosh, M.; Zhang, H.-F.; Luo, J.-Y.; Zhang, J.-Q.; Li, C.-L.; Song, J.-Y.; Zhang, X.-B. Metallogeny in Response to Lithospheric Thinning and Craton Destruction: Geochemistry and U–Pb Zircon Chronology of the Yixingzhai Gold Deposit, Central North China Craton. *Ore Geol. Rev.* **2014**, *56*, 457–471. [CrossRef]
31. Deng, X.-R.; Deng, X.-D.; Zhang, L.-Z.; Gao, W.-S.; Li, Y.-X.; Yang, H.-P.; Li, Z.-B. U-Pb Geochronology and Trace Element Analysis of Titanite from the Diaoquan Porphyry-Skarn Polymetallic Deposit, North China: Implication for Cu-Ag-Mo Mineralization. *Earth Sci.* **2023**, *48*, 3327–3341. [CrossRef]
32. Li, J.; Li, N.; Wang, M.; Song, Y.; Tang, Z.; Zhang, P.; Wang, G.; Zhang, L. Formation of the Miaoan Au-Polymetallic Deposit in the Northern Taihang Mountain, North China Craton: Ore Geology, Geochronological and Geochemical Perspectives. *Minerals* **2022**, *12*, 1144. [CrossRef]
33. Wang, S. Age Determinations of ^{40}Ar-^{40}K, ^{40}Ar-^{39}Ar and Radiogenic ^{40}Ar Released Characteristics on K-Ar Geostandards of China. *Chin. J. Geol.* **1983**, *18*, 315–323.
34. Bai, X.-J.; Qiu, H.-N.; Liu, W.-G.; Mei, L.-F. Automatic ^{40}Ar/^{39}Ar Dating Techniques Using Multicollector ARGUS VI Noble Gas Mass Spectrometer with Self-Made Peripheral Apparatus. *J. Earth Sci.* **2018**, *29*, 408–415. [CrossRef]
35. Koppers, A.A.P. ArArCALC—Software for ^{40}Ar/^{39}Ar Age Calculations. *Comput. Geosci.* **2002**, *28*, 605–619. [CrossRef]
36. Steiger, R.H.; Jäger, E. Subcommission on Geochronology: Convention on the Use of Decay Constants in Geo- and Cosmochronology. *Earth Planet. Sci. Lett.* **1977**, *36*, 359–362. [CrossRef]
37. Fu, J.; Hu, Z.; Li, J.; Yang, L.; Zhang, W.; Liu, Y.; Li, Q.; Zong, K.; Hu, S. Accurate Determination of Sulfur Isotopes (δ^{33}S and δ^{34}S) in Sulfides and Elemental Sulfur by Femtosecond Laser Ablation MC-ICP-MS with Non-Matrix Matched Calibration. *J. Anal. At. Spectrom.* **2017**, *32*, 2341–2351. [CrossRef]
38. Li, R.; Xia, X.; Yang, S.; Chen, H.; Yang, Q. Off-Mount Calibration and One New Potential Pyrrhotite Reference Material for Sulfur Isotope Measurement by Secondary Ion Mass Spectrometry. *Geostand. Geoanal. Res.* **2019**, *43*, 177–187. [CrossRef]
39. Zhang, W.; Hu, Z.; Liu, Y. Iso-Compass: New Freeware Software for Isotopic Data Reduction of LA-MC-ICP-MS. *J. Anal. At. Spectrom.* **2020**, *35*, 1087–1096. [CrossRef]
40. Turner, G.; Cadogan, P.H. Possible Effects of ^{39}Ar Recoil in ^{40}Ar-^{39}Ar Dating. In Proceedings of the Fifth Lunar Science Conference, Houston, TX, USA, 18–22 March 1974; Volume 2, pp. 1601–1615.
41. Hoefs, J. *Stable Isotope Geochemistry*; Springer Textbooks in Earth Sciences, Geography and Environment; Springer International Publishing: Cham, Switzerland, 2021; ISBN 978-3-030-77691-6.
42. Luo, G.; Ono, S.; Huang, J.; Algeo, T.J.; Li, C.; Zhou, L.; Robinson, A.; Lyons, T.W.; Xie, S. Decline in Oceanic Sulfate Levels during the Early Mesoproterozoic. *Precambrian Res.* **2015**, *258*, 36–47. [CrossRef]

43. Meinert, L.D.; Dipple, G.M.; Nicolescu, S. World Skarn Deposits. In *One Hundredth Anniversary Volume*; Society of Economic Geologists: Littleton, CO, USA, 2005; ISBN 978-1-887483-01-8.
44. Dong, G.; Morrison, G.W. Adularia in Epithermal Veins, Queensland: Morphology, Structural State and Origin. *Miner. Depos.* **1995**, *30*, 11–19. [CrossRef]
45. Chiaradia, M.; Schaltegger, U.; Spikings, R.; Wotzlaw, J.-F.; Ovtcharova, M. How Accurately Can We Date the Duration of Magmatic-Hydrothermal Events in Porphyry Systems?—An Invited Paper. *Econ. Geol.* **2013**, *108*, 565–584. [CrossRef]
46. Horstwood, M.S.A.; Košler, J.; Gehrels, G.; Jackson, S.E.; McLean, N.M.; Paton, C.; Pearson, N.J.; Sircombe, K.; Sylvester, P.; Vermeesch, P.; et al. Community-Derived Standards for LA-ICP-MS U-(Th-)Pb Geochronology—Uncertainty Propagation, Age Interpretation and Data Reporting. *Geostand. Geoanal. Res.* **2016**, *40*, 311–332. [CrossRef]
47. Chelle-Michou, C.; Schaltegger, U. U-Pb Dating of Mineral Deposits: From Age Constraints to Ore-Forming Processes. In *Isotopes in Economic Geology, Metallogenesis and Exploration*; Huston, D., Gutzmer, J., Eds.; Mineral Resource Reviews; Springer International Publishing: Cham, Switzerland, 2023; pp. 37–87. ISBN 978-3-031-27896-9.
48. Ohmoto, H. Systematics of Sulfur and Carbon Isotopes in Hydrothermal Ore Deposits. *Econ. Geol.* **1972**, *67*, 551–578. [CrossRef]
49. Zhu, R.; Fan, H.; Li, J.; Meng, Q.; Li, S.; Zeng, Q. Decratonic Gold Deposits. *Sci. China Earth Sci.* **2015**, *58*, 1523–1537. [CrossRef]
50. Phillips, G.N.; Evans, K.A. Role of CO_2 in the Formation of Gold Deposits. *Nature* **2004**, *429*, 860–863. [CrossRef]
51. Ohmoto, H.; Rye, R.O.; Barnes, H.L. *Geochemistry of Hydrothermal Ore Deposits*; John Willey &. Sons Inc.: Hoboken, NJ, USA, 1979; Volume 517, p. 612.
52. Hutchison, W.; Finch, A.A.; Boyce, A.J. The Sulfur Isotope Evolution of Magmatic-Hydrothermal Fluids: Insights into Ore-Forming Processes. *Geochim. Cosmochim. Acta* **2020**, *288*, 176–198. [CrossRef]

MDPI AG
Grosspeteranlage 5
4052 Basel
Switzerland
Tel.: +41 61 683 77 34

Minerals Editorial Office
E-mail: minerals@mdpi.com
www.mdpi.com/journal/minerals

www.ingramcontent.com/pod-product-compliance
Lightning Source LLC
Chambersburg PA
CBHW040834250926
51850CB00007BA/1315
* 9 7 8 3 7 2 5 8 6 6 2 8 1 *